NON-LINEAR STATIC AND CYCLIC ANALYSIS OF STEEL FRAMES WITH SEMI-RIGID CONNECTIONS

Elsevier Science Internet Homepage
>   http://www.elsevier.nl   (Europe)
>   http://www.elsevier.com   (America)
>   http://www.elsevier.co.jp (Asia)

Consult the Elsevier homepage for full catalogue information on all books, journals and electronic products and services.

Elsevier Titles of Related Interest

SRIVASTAVA
Structural Engineering World Wide 1998 (CD-ROM Proceedings with Printed Abstracts Volume, 702 papers)
ISBN: 008-042845-2

OWENS
Steel in Construction (CD-ROM Proceedings with Printed Abstracts Volume, 268 papers)
ISBN: 008-042997-1

GODOY
Thin-Walled Structures with Structural Imperfections: Analysis and Behavior
ISBN: 008-042266-7

FUKUMOTO
Structural Stability Design
ISBN: 008-042263-2

USAMI & ITOH
Stability and Ductility of Steel Structures
ISBN: 008-043320-0

GUEDES-SOARES
Advances in Safety and Reliability (3 Volume Set)
ISBN: 008-042835-5

BJORHOVDE, COLSON & ZANDONINI
Connections in Steel Structures III
ISBN: 008-042821-5

CHAN & TENG
ICASS '96, Advances in Steel Structures (2 Volume Set)
ISBN: 008-042830-4

FRANGOPOL, COROTIS & RACKWITZ
Reliability and Optimization of Structural Systems
ISBN: 008-042826-6

TANABE
Comparative Performance of Seismic Design Codes for Concrete Structures
ISBN: 008-043021-X

Related Journals

Free specimen copy gladly sent on request: Elsevier Science Ltd, The Boulevard, Langford Lane, Kidlington, Oxford, OX5 1GB, UK

Advances in Engineering Software
CAD
Composite Structures
Computer Methods in Applied Mechanics & Engineering
Computers and Structures
Construction and Building Materials
Engineering Failure Analysis
Engineering Fracture Mechanics
Engineering Structures
Finite Elements in Analysis and Design
International Journal of Solids and Structures
International Journal of Fatigue
Journal of Constructional Steel Research
NDT & E International
Structural Safety
Thin-Walled Structures

To Contact the Publisher

Elsevier Science welcomes enquiries concerning publishing proposals: books, journal special issues, conference proceedings, etc. All formats and media can be considered. Should you have a publishing proposal you wish to discuss, please contact, without obligation, the publisher responsible for Elsevier's civil and structural engineering publishing programme:

>   Dr James Milne
>   Publisher, Engineering and Technology
>   Elsevier Science Ltd
>   The Boulevard, Langford Lane   Phone: +44 1865 843891
>   Kidlington, Oxford   Fax: +44 1865 843920
>   OX5 1GB, UK   E.mail: j.milne@elsevier.co.uk

General enquiries, including placing orders, should be directed to Elsevier's Regional Sales Offices – please access the Elsevier homepage for full contact details (homepage details at top of this page).

NON-LINEAR STATIC AND CYCLIC ANALYSIS OF STEEL FRAMES WITH SEMI-RIGID CONNECTIONS

S.L. Chan
Department of Civil and Structural Engineering,
The Hong Kong Polytechnic University

and

P.P.T. Chui
Ove Arup and Partners (Hong Kong) Limited
(Formerly Department of Civil and Structural Engineering,
The Hong Kong Polytechnic University)

ELSEVIER
2000
Amsterdam – Lausanne – New York – Oxford – Shannon – Singapore – Tokyo

ELSEVIER SCIENCE Ltd
The Boulevard, Langford Lane
Kidlington, Oxford OX5 1GB, UK

© 2000 Elsevier Science Ltd.   All rights reserved.

This work is protected under copyright by Elsevier Science, and the following terms and conditions apply to its use:

Photocopying

Single photocopies of single chapters may be made for personal use as allowed by national copyright laws. Permission of the Publisher and payment of a fee is required for all other photocopying, including multiple or systematic copying, copying for advertising or promotional purposes, resale, and all forms of document delivery. Special rates are available for educational institutions that wish to make photocopies for non-profit educational classroom use.

Permissions may be sought directly from Elsevier Science Rights & Permissions Department, PO Box 800, Oxford OX5 1DX, UK; phone: (+44) 1865 843830, fax: (+44) 1865 853333, e-mail: permissions@elsevier.co.uk. You may also contact Rights & Permissions directly through Elsevier's home page (http://www.elsevier.nl), selecting first 'Customer Support', then 'General Information', then 'Permissions Query Form'.

In the USA, users may clear permissions and make payments through the Copyright Clearance Center, Inc., 222 Rosewood Drive, Danvers, MA 01923, USA; phone: (978) 7508400, fax: (978) 7504744, and in the UK through the Copyright Licensing Agency Rapid Clearance Service (CLARCS), 90 Tottenham Court Road, London W1P 0LP, UK; phone: (+44) 171 631 5555; fax: (+44) 171 631 5500. Other countries may have a local reprographic rights agency for payments.

Derivative Works

Tables of contents may be reproduced for internal circulation, but permission of Elsevier Science is required for external resale or distribution of such material.
Permission of the Publisher is required for all other derivative works, including compilations and translations.

Electronic Storage or Usage

Permission of the Publisher is required to store or use electronically any material contained in this work, including any chapter or part of a chapter.

Except as outlined above, no part of this work may be reproduced, stored in a retrieval system or transmitted in any form or by any means, electronic, mechanical, photocopying, recording or otherwise, without prior written permission of the Publisher.

Address permissions requests to: Elsevier Science Rights & Permissions Department, at the mail, fax and e-mail addresses noted above.

Notice

No responsibility is assumed by the Publisher for any injury and/or damage to persons or property as a matter of products liability, negligence or otherwise, or from any use or operation of any methods, products, instructions or ideas contained in the material herein. Because of rapid advances in the medical sciences, in particular, independent verification of diagnoses and drug dosages should be made.

First edition 2000
Library of Congress Cataloging-in-Publication Data

Non-linear static and cyclic analysis of steel frames with semi-rigid connections/edited by Siu-Lai Chan and Pui-Tak Chui.
    p.cm.
  ISBN 0-08-042998-X
    1. Structural frames - - Design. 2. Structural analysis (Engineering) 3. Steel, Structural.
I. Chan, Siu-Lai. II. Chui, Pui-Tak.
TA660.F7 N66 2000
  624.1'773 - - dc21

                        99-046371

British Library Cataloguing in Publication Data
A catalogue record from the British Library has been applied for.
ISBN: 0 08 042998 X

∞ The paper used in this publication meets the requirements of ANSI/NISO Z39.48-1992 (Permanence of Paper).

Printed in The Netherlands.

TO OUR PARENTS

FOREWORD

Since introduction of the limit state design philosophy focussing on the structural behaviour at limit states instead of controlling the allowable material stress used in the old permissible stress design concept, the true structural behaviour is ever sought by large-scale testing and numerical modelling. For the ultimate limit state, the collapse behaviour is of paramount importance under the action of static or dynamic loads. The safety of a structure can then be insured by inspecting the response of a structure under the action of design or expected loads during the life of a structure.

Based on second-order analysis allowing for the practical nature of external forces and structures such as load eccentricity and initial imperfections, the "Nonlinear Integrated Design and Analysis (NIDA)" concept has been applied by the authors in the design of a number of unconventional structures such as steel trusses, space frames, steel and bamboo scaffolding and pre-tensioned steel trusses. In NIDA, the linear structural model can be used since it allows modelling by a single element for each member and the uncertainty in assuming an effective length and for checking of structural instability is eliminated. However, this approach relies on the currently used first-plastic-hinge concept and collapse behaviour allowing for geometrical and material non-linearities of an indeterminate structure is not studied. As a result, the advantages of indeterminacy and reserve in strength after the first plastic hinge cannot be fully exploited.

The availability of low-cost personal computers in the past decade or so has provided a golden opportunity for the use of sophisticated and refined methods of analysis requiring extensive computational time and effort. This includes the collapse analysis of steel frames allowing for geometrical and material non-linearities and under time-dependent loads. On the other hand, reliable and simple kinematic formulations for geometrical change and material yielding are still essential along with the powerful computing machines of huge memory and high speed, not only because of saving in computer time, but they also will provide the practitioners with a physical insight into the structural behaviour. Engineers may like to have a physical grasp and to examine the significance of varying some parameters such as the effect of adopting more flexible joints on the overall response of a structure. This study is trivial by the present method but becomes too coarse and uncertain for the linear analysis with empirical formulae such as the Rankine–Merchant equation and too complicated by the finite shell element analysis. A general feel is that the mathematical computer model should be a true reflection of the overall structural characteristics so that the finite plate elements should be used to model a plate structure with two lateral dimensions much larger than the third dimension or thickness. Similarly a skeletal structure should be modelled by the one-dimensional beam-column element instead of a patch of shell elements. On the other hand, the shell elements should be more appropriate for study of local effects and stress concentration, but not for overall structural design and analysis in a design office.

This book is devoted to the discussion and studies of simple and efficient numerical procedures used in a personal computer for large deflection and elasto-plastic analysis of steel frames under the attack of static and dynamic loads. The text is designed for use by senior undergraduates and post-graduate students specialised in non-linear analysis of steel structures. This includes B.Eng., B.Sc. and M.Sc. students taking courses in related subject and M.Ph. and Ph.D. students researching advanced structural engineering and steel structures. This text can be used by practising engineers involved in advanced structural analysis and design of steel structures. The arrangement of this text is in the order of complexity. In Chapter 1, the basic fundamental behaviour and philosophy for design of structural steel is discussed. Emphasis is placed on different modes of buckling and the inter-relationship between different types of analysis. An introduction to the well-known P-δ and P-Δ effects is also given. Different levels of refinement for non-linear analysis are described and their limitations and advantages are summarised. In Chapter 2, the basic matrix method of analysis is presented. The linear stiffness and transformation matrices are derived from the first principle. Several examples of linear analysis of semi-rigid jointed frames are included. It is evident that one must have a good understanding of linear analysis before handling a second-order non-linear analysis. In Chapter 3, the linearized bifurcation and second-order large deflection are compared and the detailed procedure for a second-order analysis based on the Newton–Raphson scheme is described with the aid of a flow-chart. This load control iterative scheme can be used in the investigation of the structural response at design load level, but it is handicapped by divergence at limit load. In Chapter 4, various solution schemes for tracing of post-buckling equilibrium paths are introduced and the Minimum Residual Displacement control method with arc-length load step control is employed for the post-buckling analysis of two and three dimensional structures. Connections play an important role in the behaviour of moment-resistant steel frames. Chapter 5 is addressed to the non-linear behaviour and modelling of semi-rigid connections. Several numerical functions for description of moment versus rotation curves of typical connection types are introduced. The inclusion of semi-rigid connections and material yielding to the static analysis of steel frames is described in Chapter 6. The scope of the work in this Chapter is of general practical interest since the "Advanced Analysis" can be implemented by the described solution method with incorporation of design requirements stipulated in various design codes such as out-of-plumbness and member initial imperfections. In Chapter 7, the cyclic response of steel frames with semi-rigid joints but elastic material characteristics is studied. This assumption allows us to concentrate on the influence of semi-rigid connections over the global structural response and the effects of using various connection types on the structural behaviour. In the last Chapter 8, the combined effects of semi-rigid connections and plastic hinges on steel frames under time-dependent loads are studied using a simple springs-in-series model. For computational effectiveness and efficiency, the concentrated plastic hinge concept is used throughout these studies. The plastic hinge model adopts the section assemblage concept in which material in web takes axial force and in flange takes bending moments. This model is general, does not require pre-requisite plastic moment-axial force interactive function and can be applied to many types of sections.

This book is structured to study the non-linear behaviour of steel frames under static and time-dependent loads in a way of increasing difficulty with the first few Chapters covering simple material and the last few Chapters dealing with more complicated problems with various sources of non-linearity. The text is hoped to make a contribution to the implementation of the fascinating and elegant non-linear analysis and design concept to practical steel structures. With this computational skill and technique, better understanding of structural behaviour can be modelled and obtained, based on which more innovative structural schemes can be invented and constructed safely and economically.

ACKNOWLEDGEMENTS

We are thankful to the Hong Kong Polytechnic University for providing resources in the completion of this book. We would also like to thank our overseas colleagues, who include Professors W.F. Chen, S. Kitipornchai, D.A. Nethercot and N.S. Trahair, for their encouragement and discussion. Special thanks should go to the students of the first author, who include but are not limited to Dr. G.W.M. Ho, Dr. A.K.W. So, Mr. C.Y. Yau and Dr. Z.H. Zhou, for their valuable contributions to his research partly described in this book. The authors are further indebted to Ove Arup and Partners (Hong Kong) Limited, for their encouragement. Appreciation should further be honoured to Miss Louisa McAra for her patience in proof-reading the text and Miss Freda Chow for the typing of some of this text. The first author would further like to express his gratitude to his wife, Shui-Wai, and his two sons, Lok-Man and Lok-Yan, for their bearance in allowing him to use their leisure time in writing this book. Without their support, this book would never have been completed. The second author would like to thank Miss Freda Chow, for her support in the course of writing this book. The first author is thankful to the Research Grant Council of the Hong Kong Special Administrative Region on the projects "Stability and non-linear analysis of glased frames" (B-Q080/95), "Static and dynamic advanced analysis for design of steel structures" (B-Q193/97) and "Nonlinear analysis of steel–concrete composite frame structures under gravity and lateral cyclic loads" (B-Q035/94), of which part of the findings are contained in this book. Finally, we gratefully acknowledge our pleasant cooperation with the publisher, Elsevier Science Ltd. and its Publisher Dr. J. Milne and his assistant Mrs. R. Davies and Dr. R. Maliukevičius of VTEX Ltd.

S.L. Chan
P.P.T. Chui

CONTENTS

FOREWORD vii

ACKNOWLEDGEMENTS xi

Chapter 1. BEHAVIOUR, ANALYSIS AND DESIGN OF STEEL FRAMES   1

1.1	Introduction	1
1.2	Probability of failure	3
1.3	Loads	3
	1.3.1 Dead loads	3
	1.3.2 Live loads	3
	1.3.3 Wind and earthquake loads	4
	1.3.4 Internal forces	4
	1.3.5 Load factor and load combinations	4
	1.3.6 Collapse load factor	4
1.4	Material properties	6
1.5	Plastic failure	8
1.6	Instability of a structure	9
	1.6.1 Euler buckling	9
	1.6.2 Lateral buckling of beams	10
	1.6.3 Torsional buckling	11
	1.6.4 Local buckling	13
1.7	The P-δ effect	14
1.8	The P-Δ effect	16
1.9	Design of steel structures	17
1.10	Types of analysis	18
	1.10.1 Linear analysis	18
	1.10.2 Elastic bifurcation analysis	19
	1.10.3 First order elastic–perfectly plastic analysis of rigid-jointed frames	20
	1.10.4 The Merchant–Rankine's formula for collapse load of rigid-jointed frames	21
	1.10.5 Second-order elastic analysis	22
	1.10.6 Second-order inelastic analysis	24
	1.10.7 Dynamic and cyclic analysis	24
	1.10.8 Advanced analysis	25
	1.10.9 Objectives and structure of this book	26
	References	27

Chapter 2. FIRST-ORDER LINEAR ELASTIC ANALYSIS FOR RIGID AND SEMI-RIGID FRAMES ... 29

2.1 Introduction ... 29
2.2 Linear analysis ... 30
2.3 Coordinate systems ... 32
    2.3.1 The global coordinate system ... 32
    2.3.2 The local coordinate system ... 32
2.4 The principle of the total potential energy for element stiffness formulation ... 36
2.5 Element stiffness for semi-rigid jointed members ... 41
2.6 Assembling of the stiffness matrix ... 45
2.7 Member force determinations ... 47
2.8 Checking of equilibrium ... 48
2.9 Examples ... 48
    2.9.1 The bending moment of a sway portal ... 49
    2.9.2 Analysis of a six-storey frame ... 50
2.10 Conclusions ... 52
    References ... 52

Chapter 3. SECOND-ORDER ELASTIC ANALYSIS BY THE NEWTON–RAPHSON METHOD ... 55

3.1 Introduction ... 55
3.2 Bifurcation analysis versus load-deflection analysis ... 58
3.3 General solution methods for non-linear analysis ... 59
    3.3.1 The direct iterative method ... 60
    3.3.2 The pure incremental method ... 61
    3.3.3 Combined incremental–iterative method: the Newton–Raphson procedure ... 62
    3.3.4 The physical concept of the Newton–Raphson method ... 62
3.4 The tangent stiffness matrix ... 68
3.5 The secant stiffness matrix ... 70
3.6 Design application ... 72
3.7 Examples ... 73
    3.7.1 Elastic buckling load of a simple strut ... 73
    3.7.2 Second-order elastic analysis of a 4-storey moment-resistant frame ... 74
3.8 Conclusions ... 75
    References ... 75

Chapter 4. NON-LINEAR SOLUTION TECHNIQUES ... 77

4.1 Introduction ... 77
4.2 Genuinely large deflection analysis ... 79
4.3 Basic formulation ... 79
    4.3.1 The current stiffness parameter ... 80

	4.3.2	Displacement control method	81
	4.3.3	The constant work method	83
	4.3.4	The arc-length method	83
	4.3.5	The minimum residual displacement method	85
4.4	Numerical examples		87
	4.4.1	The William's toggle-frame	87
	4.4.2	The shallow arch	88
	4.4.3	The right angle frame	89
	4.4.4	The three-dimensional hexagonal frame	90
4.5	Conclusions		91
	References		91

Chapter 5. CONNECTION BEHAVIOUR AND MODELS ... 93

5.1	Introduction		93
5.2	Static monotonic loading tests		94
5.3	Connection behaviour		95
5.4	Classification of connection models		98
	5.4.1	Analytical models	98
	5.4.2	Mathematical models	99
	5.4.3	Mixed models	99
5.5	Formulation of connection models		100
	5.5.1	Linear model	100
	5.5.2	Multi-linear model	101
	5.5.3	Polynomial model	101
	5.5.4	Cubic B-spline model	103
	5.5.5	Power model	105
	5.5.6	Bounding-line model	107
	5.5.7	Exponential model	108
	5.5.8	Ramberg–Osgood model	111
	5.5.9	Richard–Abbott model	112
5.6	Connection spring element		113
5.7	Modified element stiffness matrix accounting for semi-rigid joints		115
5.8	Conclusions		117
	References		118

Chapter 6. NON-LINEAR STATIC ANALYSIS ALLOWING FOR PLASTIC HINGES AND SEMI-RIGID JOINTS ... 123

6.1	Abstract		123
6.2	Introduction		123
6.3	Literature review		125
6.4	Existing inelastic analysis methods		127
	6.4.1	Plastic hinge approach	128
	6.4.2	Plastic zone approach	137

6.5	Proposed refined-plastic hinge method		141
	6.5.1	The full and initial yield surfaces by section assemblage method	142
	6.5.2	Refined-plastic hinge method based on the section assemblage concept	145
	6.5.3	The elastic–plastic hinge method based on the section assemblage method	147
	6.5.4	Element stiffness formulation accounting for plasticity effect	148
	6.5.5	Checking of element stiffness in extreme conditions	149
	6.5.6	Correction of force point movement on full yield surface	152
6.6	Combined effect of joint flexibility and material yielding – the springs-in-series model		152
6.7	Numerical examples		159
	6.7.1	Linear analysis of space frame	159
	6.7.2	Effect of semi-rigid connections in truss analysis	160
	6.7.3	Buckling loads of semi-rigid simple portal frame	166
	6.7.4	Fixed-end beam	169
	6.7.5	Single storey braced frame	170
	6.7.6	Gable frame	174
	6.7.7	Inelastic response of simple semi-rigid jointed portal frame	176
	6.7.8	Inelastic limit load of a two-storey braced and unbraced semi-rigid frame with various support conditions	178
	6.7.9	Vogel six-storey frame	186
6.8	Conclusions		190
	References		191

Chapter 7. CYCLIC BEHAVIOUR OF FLEXIBLY CONNECTED ELASTIC STEEL FRAMES ... 195

7.1	Introduction		195
7.2	Research work on semi-rigid connections		197
7.3	Modelling of connection behaviour under cyclic loading		198
	7.3.1	Independent hardening method	199
	7.3.2	Kinematic hardening method	201
	7.3.3	Bounding surface method	201
7.4	Connection models used in this study		203
7.5	Hybrid element with connection springs		204
7.6	Numerical procedure for present non-linear transient analysis		206
	7.6.1	Assumptions	206
	7.6.2	Numerical solution method	206
	7.6.3	Derivation of shape function for beam–column element with flexible joints	207
	7.6.4	Formulation of the linear stiffness matrices	209
	7.6.5	Formulation of the geometric stiffness matrices	209
	7.6.6	Formulation of consistent mass matrix	210
	7.6.7	Formulation of viscous damping matrix	210

	7.6.8	Equations of motions	211
	7.6.9	Determination of resistant member forces	213
7.7		Procedure for present vibration analysis	216
7.8		Numerical examples	217
	7.8.1	Natural frequencies of beams with varying connection stiffness	217
	7.8.2	Vibration analysis of a cantilever column subjected to a varying axial load	218
	7.8.3	Vibration of two storey portals	218
	7.8.4	Linear dynamic analysis of six-storey frame	221
	7.8.5	Cantilever beam with uniformly distributed load	221
	7.8.6	Clamped–clamped beam subjected to impact point load	221
	7.8.7	Large deflection dynamic analysis of 45° curved cantilever beam	223
	7.8.8	Portal frame subjected to harmonic ground motion	226
	7.8.9	Single-bay two-storey frame for experimental verification	228
	7.8.10	Dynamic response of L-shaped frame	232
	7.8.11	Gravitational load effects on single-bay two-storey frame	237
	7.8.12	Buckling and vibration analysis of a portal	238
	7.8.13	Dynamic analysis of steel frames accounting for instability effect	241
	7.8.14	Static and dynamic analysis of the William's toggle frame with semi-rigid joints	243
	7.8.15	Analysis of the 3-dimensional hexagonal frame	244
	7.8.16	Large deflection analysis of 24 member dome frame with damping	246
	7.8.17	Dynamic behaviour of the Vogel two-bay six-storey frame	247
	7.8.18	Semi-rigid space frame under impulse force loading	252
	7.8.19	Seismic response of four-bay five-storey frame	253
7.9		Conclusions	259
		References	260

Chapter 8. TRANSIENT ANALYSIS OF INELASTIC STEEL STRUCTURES WITH FLEXIBLE CONNECTIONS **265**

8.1		Introduction	265
8.2		Review on cyclic plasticity models and methods	268
8.3		Strain-hardening rules	271
	8.3.1	Isotropic hardening rule	273
	8.3.2	Kinematic hardening rule	273
	8.3.3	Independent hardening rule	274
8.4		Existing cyclic plasticity models	274
	8.4.1	Elastic–perfectly plastic model	274
	8.4.2	Bi-linear strain-hardening model	274
	8.4.3	Multi-linear strain-hardening model	276
	8.4.4	Inversed Ramberg–Osgood model	276
	8.4.5	Strain rate dependent model	277
8.5		Proposed cyclic plasticity models used in this study	278
	8.5.1	Proposed elastic–perfectly plastic hinge model	278

		8.5.2	Proposed refined-plastic hinge model	279
8.6		Beam–column connection modelling		281
8.7		Numerical examples		283
		8.7.1	Elastic–plastic dynamic analysis of a simply supported beam	283
		8.7.2	Dynamic elastic–perfectly plastic response of Lee et al. steel arch	284
		8.7.3	Dynamic elastic–perfectly plastic response of Toridis–Khozeimeh frame	286
		8.7.4	Kam–Lin portal frame	290
		8.7.5	Rothert et al. frame	292
		8.7.6	Effect of yield stress on dynamic inelastic response	292
		8.7.7	Dynamic inelastic response of semi-rigid jointed steel frame	294
8.8		Conclusions		299
		References		300

Appendix A. 303

A.1	The 12×12 linear stiffness matrix, $[K_L]$	303
A.2	The 12×12 geometric stiffness matrix, $[K_G]$	305
A.3	The 12×12 consistent mass matrix, $[M]$	307

Appendix B. USER'S MANUAL OF GMNAF 311

B.1	Introduction	311
B.2	Data preparation	312
B.3	Activation of the program	314
B.4	Example	314
	References	323

LIST OF SYMBOLS **325**

AUTHOR INDEX **327**

SUBJECT INDEX **333**

Chapter 1

BEHAVIOUR, ANALYSIS AND DESIGN OF STEEL FRAMES

1.1. Introduction

The primary objective of a structure is to carry and transmit loads to foundations safely and to perform well in a serviceable condition. The old fashioned allowable stress design method controls the maximum stress of a structure below the allowable stress of the material, generally taken as the characteristic yield stress divided by a factor of safety. Additional checks will be imposed for checking against instability and buckling. In the more recently introduced and widely accepted limit state design concept, a structure is considered to reach a limit state when it is unable to fulfill the requirements for its functional uses. The inability of the structure or its components to meet these performance requirements can be considered as a structural failure at ultimate limit state and as non-structural failure at serviceability limit state, which comprises of a number of criteria related to safety and serviceability. During the service life of a structure, the structure should have an acceptable low probability of any kind of collapse or damage locally or globally and should not deflect to such an extent that the occupants feel uncomfortable or that the finishes or cladding are damaged, in addition, the fire resistance of the structure must allow sufficient time for the occupants to escape in case of fire and to prevent rapid spread of fire. The corrosion resistance should enable the structure to last for the designed life and the vibration should not be excessive, etc. These requirements are covered in several limit states detailed in the design codes and all these limit states are needed to be fulfilled for a generally satisfactory structure. This relationship can be stated in a general term as

$$\phi R \geqslant \lambda F, \tag{1.1}$$

in which ϕ is the material factor, R is the resistance of the structure, λ is the load factor and F is the external load.

Collapse load and the ultimate load-carrying capacity are related to the safety consideration for the design of a structure. Due to the importance of ultimate strength limit state, the safety margin or factor of safety is generally larger than other limit states and greater than 1. This book is mainly concerned with the limit analysis and design of a structure against collapse and under the action of static and cyclic loads. Obviously, the theory in this book can be applied equally well to the serviceability deflection limit state at a lower load level with a less severe degree of non-linearity.

A structure or its components are, in general, three dimensional. However, most structures and their components can be idealized as one or two dimensional for practical purposes. A plate or shell has lateral dimensions much larger than its thickness. The beam-column is one dimensional structural member with a longitudinal length much larger than its transverse dimensions. This includes the widely used structural elements such as I, channel, angle, hollow sections and others. These types of structural members are commonly used in steel and metal structures due to their structural efficiency, relatively light weight, ductility and ease of fabrication. They can be manufactured as hot-rolled, cold-formed or intruded for some types of alloy like aluminum. The cross sections of compressive members are generally thin-walled and seldom solid because of their larger flexural stiffness for the same weight of material. In general, the member is under the action of axial force, bending and torsional moments and shears. Most members are arranged to be under a dominant action from axial force and bending moments and essentially free from torsional moment, although this condition may not be attainable in all structures.

In engineering practice, a developer requesting for a project to be undertaken for a specific purpose will contact the architect. Normally, several options are then worked out. Structural feasibility will also be considered and the structural engineer will be consulted at this preliminary stage for some types of structures of considerable size. The structural engineer will estimate loading, and propose suitable framing systems. After consideration under the constraints of finance, environment, construction period, aesthetics and other relevant factors, an option will be chosen and a more detailed design will be worked out. The complete process may be iterated many times until all parties are satisfied with the final arrangement upon which the contract documents and specifications will be prepared. In the course of this option selection process, safety consideration plays an important role and may often affect the final solution. For example, in regions of seismic risk, the option of using steel is preferred and very often the management team requires a careful study on the safety of the structure under severe loads. Obviously the consideration of safety indirectly affects the architectural details.

In the past, structural engineers normally adopted linear analysis from which the result is employed for individual member design based on formulae in the design codes. In a linear analysis, the structural member and thus its composing structure is assumed to have a constant stiffness independent of the deformations and magnitude of member forces. Mathematically, this condition can be stipulated as

$$Ku = F, \tag{1.2}$$

in which K is the element stiffness matrix, u is the displacement vector and F is the force vector of a structural system. In a linear system, the stiffness K is independent of the force and displacement.

In the solution for the stiffness equation (1.2), the determination of external forces or loads and of the stiffness properties of the structural system is essential in getting the true behaviour of the structure. In the ultimate limit state design philosophy, they are obtained from a statistical basis.

1.2. Probability of failure

From a statistical point of view, it will be very uneconomical or even impossible to design a structure without any chance of failure. A structure is allowed to have a certain probability of failure, which is very low and in the order of 10^{-6}. To achieve this objective, the characteristic loads and strength are used in the design. The characteristic load is defined as the load with a certain low percentage being exceeded in the design life of a structure. Similarly, the characteristic stress is defined as the stress value with a certain low probability for the material to have yielding falling below the yield stress. Failure takes place when both these unfavourable conditions for the characteristic load and strength occur simultaneously.

For non-structural failure such as deflection limit state, the requirement is lower because the consequence of reaching the limit state is less serious. Because of this, its design requires an equally demanding level of accuracy since the chance of its occurrence is much higher than the limit state.

1.3. Loads

Applied forces acting on a structure can be external or internal. In general, loads are due to the dead weight of the structure and machines, live loads of the occupants and furnishes, wind and earthquake loads from the relative movement of air and earth and internal forces due to shrinkage and temperature change, foundation settlement, etc. In design, these loads should be combined in order to produce the most severe loading combination that may occur during the life span of a structure, with an acceptable low probability of actual loads exceeding these combined loads. Load factors are multiplied to the estimated loads and the magnitude for the factor varies from one country to another.

1.3.1. Dead loads

Dead loads are referred to as loads which do not vary with time during the life-span of the structure. These include the self weight of the structure, finishes and some types of permanent equipment such as lift motor. Density of material and estimation of object weight will be used for the computation of dead weight.

1.3.2. Live loads

Live loads are generally due to the occupants or objects that vary in position with time. In some types of live loads, the dynamic effects should be considered and a more detailed dynamic response of the structure should be conducted. In recent years, the wide use of delicate electronic equipment such as computers in offices has made serviceability vibration an important consideration in the design of floor beams which support this equipment. Suddenly applied loads on the floor may introduce vibration that may damage the equipment or information stored in them. It may cause a considerable loss of property and time.

1.3.3. Wind and earthquake loads

Wind loads are highly localized and national wind codes can provide a guideline in the determination of equivalent static wind pressure. Consideration will be given to the locations, height and surrounding environment of a building. The return period as well as the period taken for averaging the wind speed should also be determined for wind load intensity. In some countries, a fifty year return period with three second wind speed average is adopted. Different countries use a different period for averaging the wind speed and the calculated wind pressure is also dependent on this duration. In some cases for the design of tall or large buildings, a wind tunnel test may be required to determine the wind pressure at different locations of a building. In addition, dynamic response of a building may also be required to be simulated in a wind tunnel test.

Earthquake loads are due to ground acceleration and the inertia effect of the mass of a building. For a general analysis, the transient response of a structure is required to be traced and the strength, stability and also the ductility requirements are needed to be met. The checking of seismic resistance requires a rather different procedure from the conventional design method such as collapse load factor as this cannot be defined in the static case since the rate of this load will also affect structural safety. Instead, the stability and strength of a structure against time needs to be investigated and the response studied.

1.3.4. Internal forces

Internal forces may be due to foundation settlement, lack of fit, temperature change, creep and shrinkage of material and residual stress. The consideration of these effects requires the judgement of the engineer. The magnitude of these forces is difficult to be standardised and therefore, not commonly covered in detail in codes of practice. However, the maximum and minimum temperature may be specified in the codes for calculation of thermal strain. These effects can also be included in an analysis by appropriate modification of initial stress and displacement, etc.

1.3.5. Load factor and load combinations

Once the magnitude and directions of loads are determined, their combinations are required to be worked out. For conventional design, the worst combination should be determined from the dead, the live and the wind load. In Table 1.1, the loads factors used by the American Institute of Steel Construction (1986) and British Standard 5950 (1990) are listed for a comparison.

1.3.6. Collapse load factor

In the ultimate limit state design concept, the factor of safety can be defined as the ratio between the collapse and the design load factors. Load and resistance factors are multiplied to the material resistance and the load based on which checking is carried out. The load factor is then multiplied to the design load to give the permissible load of the structure until collapse. It can either be a local, a progressive or a global collapse.

Table 1.1
Load factors from AISC (1986) and BS5950 (1990) for different load types

Load combination	Dead load		Live load		Wind load	
	AISC	BS5950	AISC	BS5950	AISC	BS5950
1	1.4	1.4	0.0	1.6	0.0	0
2	1.2	1.2	1.6	1.2	0.0	1.2
3	1.2	1.4	0.5	0	1.3	1.4
4	−0.9	−1.0	0.0	0	1.3	1.4

In current design practice, the ultimate limit state design concept takes a conservative approximation on assuming the equality of load at the first plastic hinge to the ultimate design load. However, once yielding occurs, the stiffness of the structure changes and thus the original elastic assumption is no longer valid. The true collapse load after the formation of a series of plastic hinges is not considered. Further, even the variation in structural stiffness due to yielding, formation of plastic hinges and the presence of axial force in members are ignored in the analysis.

In recent years, the concept of prevention of progressive and disproportionate collapse has been proposed and enforced in practice in some countries. Identification of key structural elements, of which the failure causes the collapse of a large area, will be necessary and additional fictitious load on the element will be required to ensure the key member is capable of resisting additional and accidental loads.

As the validity of a linear analysis relies on the assumption that the member force is small and thus the material is elastic plus the assumption that, the deflection is small and the structure is stable, the theory is incorrect when the structure is heavily loaded so that it deforms considerably and the composing material yields. A stiffening and a softening system are selected for illustration of the error connected with linear analysis. The cantilever shown in Fig. 1.1 has been analysed and it can be seen that as the vertical deflection of the cantilever increases, the member is lengthened and thus a tensile force is developed, resulting in a pulling action of the member. This moves the tip of the cantilever inwards. In a linear analysis, there is no component in the longitudinal direction and this effect due to geometrical change cannot be observed. This geometrically non-linear effect will lead to the stiffening of the stiffness matrix for the complete structure when the tensile axial force is present. In situations where a structural member is under compression, a completely different result can be observed. Figure 1.2 illustrates the softening behaviour of another cantilever subjected to axial force at the end. The member stiffness deteriorates when the critical load is approached. Although the structure stiffens after this load, the large deflection in the post-buckling range is unattainable in most practical civil engineering structures and yielding occurs well before such a development of a bending strength in this range.

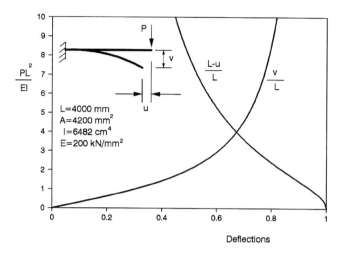

Fig. 1.1. Elastic large deflection analysis of a stiffening system.

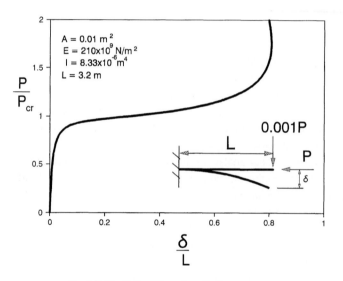

Fig. 1.2. Elastic buckling of a softening system.

1.4. Material properties

Steel has relatively uniform material properties when compared to concrete. The Young's modulus of elasticity is about 205,000 MPa with a maximum variation of about 2.5%. The idealised stress–strain curve for a steel bar is shown in Fig. 1.3. Before loaded beyond the elastic limit, the material fully recovers upon removal of external stress. For mild steel further loaded, the material exhibits a strain hardening behaviour. Further loading will cause the material to reach the ultimate stress and fracture.

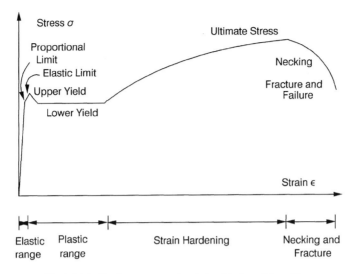

Fig. 1.3. Idealised stress vs. strain relationship for mild steel bar.

The strength of steel depends on the composition of the material. The 95% characteristic yield stress, defined as the the stress value at which 95% of the tested samples do not yield, is often used for design. Different steel grades have varied values for yield stress, but the same Young's modulus of elasticity. The characteristic yield stresses for common steel grades are 250, 350 and 450 MPa. Obviously, choosing a higher steel grade will be advantageous for structures with behaviour critically controlled by material yield, whilst the use of high grade steel has little or no advantage for structures with behaviour dominated by deflection limit state, instability or vibration. The uniaxial yield stress can further be used for prediction of material yielding under bidirectional direct and shear stresses. The yield criterion generally used for steel material is the Von Mises yield criterion derived from the maximum distortional energy theory, which is valid for metal material. In this Von Mises yield function, the following equation is employed:

$$\sigma_x^2 + \sigma_y^2 - \sigma_x\sigma_y + 3\tau_{xy}^2 = \sigma_{ys}^2, \tag{1.3}$$

in which σ_x and σ_y are the direct stress in x- and y-planes, τ_{xy} is the shear stress on the plane and σ_{ys} is the yield stress under uniaxial stress. It is of interest to note that by substituting zero stresses in the x- and y-directions, the yielding shear stress is $\sigma_{ys}/\sqrt{3}$.

In addition to higher stress versus weight ratio for steel when compared with its major rival, concrete, steel has an advantage in its high ductility and the ductile mode of failure. However, this is not often the case and to prevent brittle fracture, especially when the structure is loaded under low temperature, different steel grades of the same characteristic yield stress should be used for various wall thicknesses. For example, using the BS5950 (1990) system, Grade 43A steel should be used for relatively thin material, Grade 43B should be used for thicker material and Grade 43C, D or E should be used for even thicker material under cold environments.

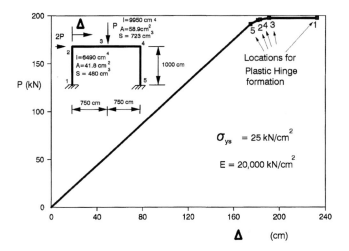

Fig. 1.4. Rigid plastic analysis of a simple portal.

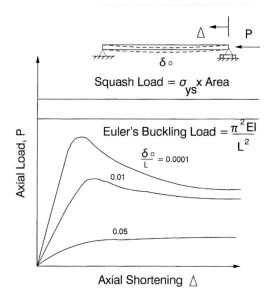

Fig. 1.5. Elasto-plastic buckling of a strut.

1.5. Plastic failure

In the second case for plastic analysis, a portal shown in Fig. 1.4 under a set of vertical loads and a small lateral disturbing force will reach its maximum load-carrying capacity at a vertical load of 197 kN at which a plastic mechanism is formed. In another example when a strut is subjected to a compressive axial load with small disturbance, as shown in Fig. 1.5, the member will exhibit an elasto-plastic buckling and neither the elastic buckling load nor the plastic squash load is close to the true failure load. It can be seen that judgement is necessary for the assessment of the importance of geometric and material non-linearities so that an approximate analysis method can be chosen for design and analysis. Owing to the improvement in material yield stress for steel in the past few decades and the availability of low-priced personal computers, slender steel structures are currently widely designed and constructed. The cost of computers, which was an important consideration before 1980s, has become a relatively unimportant factor for design to date and this trend will prevail in future with the advent in computer technology. Furthermore, the widely used ultimate limit state design philosophy places the factor of safety on the ultimate load capacity of the structure, of which the collapse load is required to be determined.

1.6. Instability of a structure

The design of a member under tensile axial load is relatively simple because the stiffness of a member is stiffened by the presence of axial force in the elastic range. The failure of the member is then controlled by material yielding. When a member is under the action of moment and/or compressive force, it may become unstable and failure may be due to buckling, material yield or their combinations, the so called elasto-plastic buckling.

In the design and analysis of steel structures, stability is an important consideration. An engineer should be fully aware of the possible buckling modes in order to account for this or to take appropriate action to prevent the occurrence of a particular type of buckling such as to provide bracing members to reduce the effective length. The phenomena and nature for various buckling modes are discussed in this section.

Instability refers to the phenomenon that a structure deforms non-proportionally and its stiffness reduces due to the change of geometry and an increasing load. This may be considered as a deteriorating stiffness phenomenon for the structure. There are basically several types of buckling modes which must be carefully considered before a safe design can be produced. These buckling modes for structural members can be classified as Euler buckling, torsional buckling, lateral–torsional buckling, local plate buckling and their combinations.

1.6.1. Euler buckling

Euler buckling is possibly the most widely encountered and studied buckling mode, both in theory and in practice. The frequent collapse cases for scaffolding (Chen et al., 1993) can be attributed partly to the buckling of members under axial forces. When an imperfect strut or column is subjected to an axial force, it has a tendency to deflect laterally along its length. For practical purposes, the bifurcation and the plastic collapse loads are only the upper bound solutions to the actual failure load. Figure 1.5 has already shown the load versus deflection path of an imperfect column and it can be seen that the member cannot sustain a load higher than the member squash load ($P_{sq} = \sigma_{ys} \times$ Area) or Euler's buckling load. The true load versus deflection relationship is dependent on the magnitude of initial imperfection. In some cases, the post-buckling path may be asymmetric to the

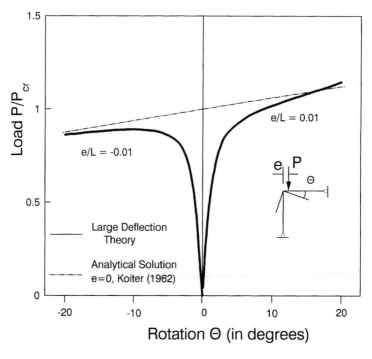

Fig. 1.6. Post-buckling behaviour of a right angle frame.

primary equilibrium path and the direction of initial imperfection will affect the behaviour of a structure, as indicated in Fig. 1.6 for the case of a right angle frame. The direction of load eccentricity has a significant influence over the pattern of the equilibrium path. When the eccentricity is positive, a stiffening behaviour is observed and vice versa. This behavioural phenomenon carries a physical implication that, for design against the worst possible condition, the direction as well as the magnitude of disturbing force and initial imperfection should be chosen in such a way so that the most severe effect results.

For the simplest case of a column under compressive axial load, the equation for Euler's buckling load can be obtained analytically as

$$P_{cr} = \frac{\pi^2 EI}{L_e^2}, \tag{1.4}$$

in which L_e is the effective length of a column, P is the axial force and π is a constant equal to 3.14159.

1.6.2. Lateral buckling of beams

When a member is under the predominant action of bending moment about its principal major axis, it may deflect and bend about its minor axis when the applied moment is close to its critical moment. The elastic buckling moment can be obtained by the solution of the

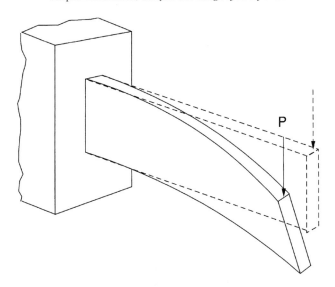

Fig. 1.7. Buckling of a cantilever of rectangular cross-section.

differential equilibrium equation (Trahair, 1965) or the energy equation using the principle of total potential energy (Chan and Kitipornchai, 1987a). The close form solution for the elastic buckling moment can then be obtained as

$$M_{cr} = m \frac{\pi}{L} \sqrt{EI_y GJ} \sqrt{1 + \frac{\pi^2 EI_\omega}{GJL^2}}, \qquad (1.5)$$

in which I_y is second moment of area about the minor principal axis, G is shear modulus of elasticity, J is torsional section constant, I_ω is warping section constant, L is the effective length of the beam and m is a coefficient to account for the effect of moment gradient given empirically by

$$m = \frac{1}{0.6 - 0.4\beta} < 2.5 \quad \text{and} \quad \beta = \frac{M_L}{M_S}, \qquad (1.6)$$

in which M_L and M_S are the larger and smaller nodal moments applied to the two ends of a beam, respectively.

For non-bifurcation type of analysis, a small disturbing force to initiate lateral deflection is necessary and the load versus deflection path can then be plotted using an incremental-iterative method which is discussed in the next Chapter. Figure 1.7 sketches the lateral buckling analysis of a cantilever beam of rectangular cross section.

1.6.3. Torsional buckling

A short thin-walled column under axial force may buckle and twist torsionally. This mode of buckling may be found in short and open section such as angle and cruciform,

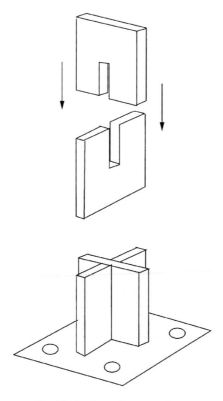

Fig. 1.8. Cruciform for support base.

which is used as the supporting base of space frames shown in Fig. 1.8. Short thin-walled columns under axial force may twist torsionally before reaching the squash load. Kitipornchai and Lee (1984) conducted experiments on the elasto-plastic buckling behaviour of angle with varying length and Chan and Kitipornchai (1987b) proposed a finite element approach to the elastic buckling analysis of angle struts. The elasto-plastic buckling load was also predicted by Chan (1989) using the finite element approach which compared well with the test results by Usami (1971). The elastic buckling load of a thin-walled member with asymmetric cross-section can be obtained as the root of the following cubic equation

$$(P - P_x)(P - P_y)(P - P_z)(r_p^2 + y_o^2 + z_o^2) - P y_o^2 (P - P_z) - P z_o^2 (P - P_y) = 0, \tag{1.7}$$

in which P_y and P_z are the Euler flexural buckling loads of the strut about the y- and the z-axes, P_x is the torsional buckling load about the shear centre axis, y_o and z_o are the coordinates of the centroid about the axes parallel to the principal axes and passing through the shear centre and r_p is the polar radius of gyration given by

$$r_p = \sqrt{\frac{I_x + I_y}{A}}, \tag{1.8}$$

in which I_y and I_z are, respectively, the second moments of area about the two principal axes respectively and A is the cross sectional area.

1.6.4. Local buckling

In the conventional theory of thin-walled beams, the cross section is assumed to remain undeformed when loaded to its ultimate strength. For the majority of hot-rolled sections, this assumption is valid and the result obtained on this basis is accurate. Using this assumption, theories developed for analysis of thin-walled beam-columns can be simplified and the time required for the analysis of this type of structures is greatly reduced.

When the thickness of the plate components making up a section is thin, the cross section of the member may deform and exhibit local plate buckling before the member yields or buckles in any other modes previously described. Figure 1.9 shows the local plate buckling simulated by the large deflection finite shell element computer program, NAF-SHELL (1993), for the top flange of a square hollow section (SHS) under moment. National codes for design of steel and metal structures have clauses to prevent this or to allow for a reduction in cross sectional capacity when the plate width versus thickness ratio exceeds a certain value, depending on the boundary condition of the plate being considered. Most hot-rolled sections are not susceptible to this type of buckling whilst built-up sections such as girders require careful consideration of the plate stability. In the design of structural elements susceptible to local buckling, the effective stress or the effective width concept is normally adopted. In the effective stress method, the permissible stress of the structural element is reduced to limit the permissible stress in the section. Codes such as BS5950, Part 1 (1990) adopt this approach for hot-rolled sections. For the effective width method, material lying outside the width limit of the composing plate

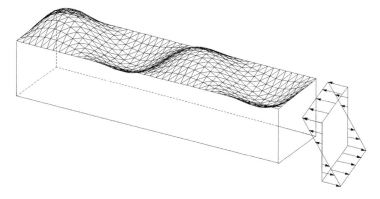

Fig. 1.9. Local buckling of top flange of a box section under moment.

elements making up the section is ignored. The overall resistance of the section is then lowered due to the ignorance of this material and therefore, the overall sectional capacity is reduced to cater for the effect of local plate buckling. In AS4100 (1990), this concept is used. Narayanan (1985) edited a series of recent research works in the area of plate buckling in various types of steel structures. A plate component may buckle by compressive edge loads or by shear loads acting along the edges of the plate. The lowest buckling load acting perpendicularly on the edge of a simple supported square plate is given by

$$P_{cr} = \frac{Et^3\pi^2}{12(1-v^2)b^2},\qquad(1.9)$$

and the shear buckling load by

$$P_s = \frac{9.35\pi^2 E}{12(1-v^2)b^2},\qquad(1.10)$$

in which b is the width of the square and v is the Poisson's ratio.

1.7. The P-δ effect

The P-δ effect is referred to as the second-order effect due to the deflection along a member and the axial force. It affects the state of stress as well as the stiffness of the member. Its careful consideration is important for buckling analysis and design of slender skeletal structures. To formulate the P-δ effect and to obtain the correct relationship between the bending coefficients under the action of axial force, resort can be made to the equilibrium equation for a beam-column element. Considering the beam-column element under an axial force as shown in Fig. 1.10, the differential equilibrium equation can be written as

$$EI\frac{d^2v}{dx^2} = -Pv + \frac{M_1 + M_2}{L}x - M_1,\qquad(1.11)$$

in which E is the Young's modulus of elasticity, I is the second moment of area, v is the lateral deflection in y-axis, x is the distance along the member, M_1 and M_2 are nodal

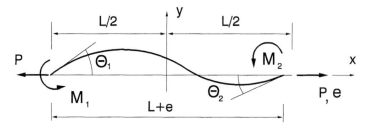

Fig. 1.10. The basic force vs. displacement relations in a member.

moments at the left and right ends, and P is the axial force. Note that positive nodal moment is assumed by the right hand rule using the z-axis pointing outside the plane of this paper:

$$v = \frac{M_1}{P}\left[\frac{\sin(\mu L - \mu x)}{\sin \mu L} - \frac{L-x}{L}\right] - \frac{M_2}{P}\left[\frac{\sin \mu x}{\sin \mu L} - \frac{x}{L}\right], \quad (1.12)$$

in which

$$\mu = \sqrt{\frac{P}{EI}}. \quad (1.13)$$

Differentiating Eq. (1.12) with respect to θ_1 and θ_2, and expressing the rotations at two ends as the nodal rotations as

$$\left.\frac{dv_1}{dx}\right|_{x=0} = \theta_1, \quad \left.\frac{dv_1}{dx}\right|_{x=L} = \theta_2,$$

we have the stiffness equation as

$$M_1 = \frac{EI}{L}(c_1\theta_1 + c_2\theta_2), \quad M_2 = \frac{EI}{L}(c_2\theta_1 + c_1\theta_2), \quad (1.14)$$

in which M_1, M_2, θ_1 and θ_2 are the nodal moments and rotations at the two ends of the member. E is Young's modulus of elasticity, L is the member length, I is the second moment of area about the bending axis, and c_1 and c_2 are the stiffness coefficients given by the followings:

$$c_1 = \frac{\phi(\sin\phi - \phi\cos\phi)}{2(1-\cos\phi) - \phi\sin\phi}, \quad (1.15)$$

$$c_2 = \frac{\phi(\phi - \sin\phi)}{2(1-\cos\phi) - \phi\sin\phi}, \quad (1.16)$$

in which $\phi = \sqrt{PL^2/(EI)}$.

Repeating the procedure for the tensile load case in which the sign of axial force, P, in Eq. (1.11) is reversed, we have the stability function for c_1 and c_2 as follows:

$$c_1 = \frac{\psi(\sinh\psi - \psi\cosh\psi)}{2(\cosh\psi - 1) - \psi\sinh\psi}, \quad (1.17)$$

$$c_2 = \frac{\psi(\psi - \sinh\psi)}{2(\cosh\psi - 1) - \psi\sinh\psi}, \quad (1.18)$$

in which $\psi = \sqrt{-PL^2/(EI)}$.

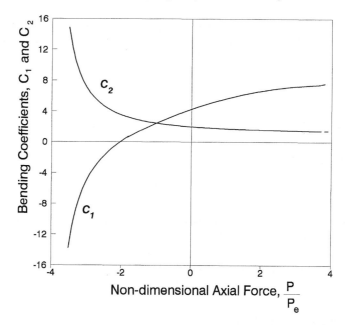

Fig. 1.11. Variation of bending coefficients against tensile and compressive axial force.

For the zero load case, Eq. (1.3) can be solved directly as in the case for linear analysis. The cubic displacement function for displacement will be obtained and the stability coefficients can be determined directly as

$$c_1 = 4, \qquad c_2 = 2. \tag{1.19}$$

The above procedure is an exact means of incorporating the P-δ effect into an element under the Timoshenko beam–column theory, which assumes that member is under predominant axial load and small end moment, since it considers correctly the second-order effect due to the element deflection.

Figure 1.11 shows the variation of axial force and the stability functions, c_1 and c_2. It can be seen that the conventional linear analysis of using c_1 and c_2 in Eq. (1.19) is only correct for the condition of zero axial load. The expressions in Eqs. (1.13) to (1.18) allow for the variation of the stiffness of a member in the presence of axial force and they can be used to revise the stiffness of the global stiffness matrix comprising hundreds of elements as well as predicting the buckling of a single member.

1.8. The P-Δ effect

When a structure deforms grossly, the original geometry can no longer be employed for the formulation of the transformation matrix simply because the coordinates have been varied. This effect may become important when the deflection and/or the conjugate force is large such as in the case of a building under a heavy mass at the roof and a lateral wind load. To include this effect in an analysis, we can update the geometry of a structure by the updated Lagrangian formulation as

$$x_{i+1} = x_i + u_i, \tag{1.20}$$

in which x_i and u_i are the coordinates and deflections at step i, respectively.

Based on the new or updated geometry, the transformation matrix is updated and the effect of large deflections is then included.

Alternatively, this effect can be included through element formulation so that the coordinate system remains unchanged for the complete loading history, generally termed as total Lagrangian formulation. With proper formulations, both these methods for including the effect of large deflections yield the same result (Zienkiewicz, 1977; Bathe, 1982).

The P-δ effect can also be transformed to the P-Δ effect by using several elements per member, at the expense of heavier computation time. Using more elements for a member, more nodes for geometry update can be introduced along the member and thus the effect of axial force can be included into the member stiffness.

1.9. Design of steel structures

In practical design of steel structures, not all the buckling modes are required to be included in an analysis, provided that their effects are prevented or insignificant. Beams with full restraints along the length do not require the determination of the effects of lateral instability. A column of I or box section and of sufficient torsional stiffness is not susceptible to torsional buckling. When the plate components are thicker than a certain breadth versus thickness ratio, local plate buckling will not occur before the structure reaches the plastic moment. In practical design of steel buildings, the second-order effect normally present is due to axial forces in columns or due to the vertical loads. The column strength is very much dependent on the member slenderness ratio and also the side sway of the whole building.

In the conventional linear design and analysis of steel structures, the linearized response of a structure is used to construct the bending moment diagram of the structure from which the strength and stability of each member is checked and designed. As many practical columns are of modest to high slenderness that invalidates the assumption of an analysis considering only material yielding, the effective length factor or the K-factor, is introduced to the design procedure for determining the permissible stress. Unfortunately, this slenderness depends on the rigidity of the connections as well as the rigidity of the connecting members to this column being designed and therefore a rather complicated formula is used in different national design codes.

In design of structures against extreme and dynamic loads, ductility, in addition to strength and stability, needs to be considered in the analysis and design. The ductility factor, defined as the ratio of the displacement at the instance of collapse to displacement at the onset of yielding, is determined. Framing is also aimed for the symmetrical and balanced layout in order to ensure stability and ductility in resisting dynamic forces. In dealing with this type of design and to obtain a better insight into the dynamic and collapse response of a structure, the analysis should be equipped with the ability to carry out inelastic and large deflection analysis.

1.10. Types of analysis

Although the actual behaviour of a structure is non-linear and complicated, different levels of analysis with various degrees of refinement can be used to design the structure. These analysis techniques require different degrees of complexity which, however, may not necessarily be directly related to the manual manipulation work involved due to the availability of powerful personal computers. To aid the designer and reader to fully appreciate the nature, limitation and accuracy of these methods, they are summarized below with a cross reference on the generalized load versus deflection path for a structure depicted in Fig. 1.12.

1.10.1. Linear analysis

Linear analysis, which has been used for decades, assumes the deflection is proportional to the applied force. Thus, the relationship between the applied load and the deflection at any point in a structure is a straight line, as shown in the line coinciding with the initial slope of any analysis in Fig. 1.12. A special feature or convenience about this method of

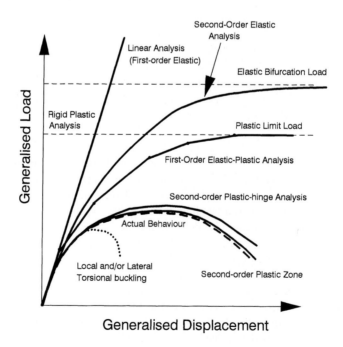

Fig. 1.12. General analysis types for framed structures.

analysis is on the validity of the principle of superposition of different load cases. Thus, one can superimpose the force diagram due to various loads in order to obtain the final force distribution in the structure. As this straight line by itself does not provide any information on stability and the true capacity of the structure, the designer must employ other means to check the safety of a structure such as the member capacity check based on an assumed K-factor for columns and to limit the maximum moments and forces in the structure. In order to account for any non-linear effect that has not been allowed for in a linear analysis, the amplification factor is used to increase the bending moment of a column obtained from a linear analysis. In the LFRD (1986) specification, the second-order elastic moment, M_u, can be estimated from a linear analysis by the following formula:

$$M_u = B_1 M_{nt} + B_2 M_{lt}, \tag{1.21}$$

in which M_{nt} and M_{lt} are the maximum first-order elastic moments in the member calculated on the assumptions of "lateral translation" and "without translation". B_1 and B_2 are the magnification factors corresponding respectively to the P-δ and the P-Δ effects and given in detail in the specification.

In BS5950 (1990), the slenderness ratio of a column in a multi-storey frame is determined from different charts for sway, non-sway and partial sway frames. Alternatively, the following factor can be used to amplify the moment in a sway frame.

$$\text{amplification factor} = \frac{\lambda_{cr}}{\lambda_{cr} - 1}, \tag{1.22}$$

in which λ_{cr} is the elastic buckling load factor which multiplied to the design or applied load vector gives the buckling load vector for the structural system. This factor can be obtained by an eigenvalue analysis detailed in next section.

It can be seen in Eq. (1.21) or (1.22) that the application of the formula becomes tedious when a large number of columns are to be designed and checked. In many cases, the effective length is difficult to assess by judgement. The situation will be more complicated when the member is slant or in cases where the member is under non-uniform axial force, as shown in Fig. 1.13. In many cases, the determination of the effective length factor is difficult. Furthermore, the stiffness of a slender member depends on the sign and magnitude of initial load in the member, but this effect of change of member stiffness due to external loads via the initial stress in members is ignored in a linear analysis.

1.10.2. Elastic bifurcation analysis

The buckling load factor, λ_{cr}, which causes the structural system to buckle elastically with the structural geometry assumed unchanged until buckling occurs, is termed as the elastic buckling factor. In the bifurcation type of analysis, an eigenvalue problem is formulated by a standard procedure as

$$|K_L + \lambda_{cr} K_G| = 0, \tag{1.23}$$

in which K_L and K_G are the linear and the geometric stiffness matrices, respectively.

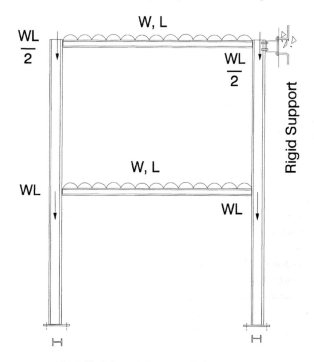

Fig. 1.13. Column under non-uniform axial load.

For a general stiffness matrix of a structural system with n degrees of freedom, there exist n roots for the bifurcation load and they represent n possible buckling modes. There exist a number of numerical methods for the solution of Eq. (1.23) and the computation of eigen-vectors which represent the buckling mode shape. The solution methods for the above problem are the methods of bisection, interpolation and the sub-space iteration schemes. A detailed discussion of these methods are given by Bathe (1982). Only the lowest or the lowest few modes are usually of practical interest. Manual methods based on the energy and the equilibrium considerations for the determination of critical loads have been presented by Timoshenko and Gere (1961), Chajes (1974) and Brush and Almroth (1975).

Although the method can only provide an upper bound solution to the stability of a frame, it can be used as an approximation to the instability stage. BS5950 (1990) requires the elastic buckling load factor to be larger than 4.6 for clad frames ignoring the stiffening effect from wall panels and 5.75 for uncladed frames or claded frames with inclusion of wall panel stiffening in analysis. The bifurcation analysis is generally used to determine the exact effective length factor K for a member in a structural system.

1.10.3. First order elastic–perfectly plastic analysis of rigid-jointed frames

In the plastic analysis of steel frames, Horne (1949) developed a kinematic energy approach for determining the plastic failure load of a steel frame, which was subsequently modified by others. This method ignores the buckling and the large deflection effect until collapse. The manual approach, however, becomes tedious when the structure size is large. Its application to a simple portal has been shown in Fig. 1.4. In this approach, a linear analysis is carried out first to search for the location for the formation of the first plastic hinge under a set of increasing loads acting upon a structure. Mathematically the load factor leading to the formation of the first plastic hinge analysis can be computed by the following simple equation

$$\Delta\lambda_j = \min\left(\left|\frac{{}_iM_p}{{}_iM}\right|\right), \quad i = 1,\ldots,n_p \quad \text{for the } j\text{-th cycle,} \tag{1.24}$$

in which ${}_iM$ is the moment due to an initial set of applied loads and at node i. ${}_iM_p$ is the plastic moment capacity of the section at node i. n_p is the total number of possible plastic hinge locations simulated by element springs. In the computer analysis, elements with end springs to simulate the softening of a section stiffness are used to model the members in a framed structure. When the possibility of formation of a plastic hinge along a member exists, several elements are needed to model each member. Once the location of the first plastic hinge is spotted, a hinge is inserted to this node by reducing the stiffness of end spring in the element to zero and the analysis is continued based on the modified structure with this new plastic hinge. The second incremental load factor, $\Delta\lambda_2$, is calculated by repeating the above process. This procedure is repeated until a mechanism is detected, which can be sensed by the semi-definite condition or zero determinant of the stiffness matrix and the final collapse load factor is determined as the sum of all the incremental load factors as

$$\lambda_p = \sum_{i=1}^{j} \Delta\lambda_i, \tag{1.25}$$

in which j is the total number of plastic hinges formed before a plastic mechanism is formed and λ_p is the plastic collapse load factor for the rigid-jointed frame. In this mechanism, deflections can occur without creation of strain energy. A very small force will cause infinite displacement.

The plastic and the elastic buckling factors can be combined to check the elastic and plastic stability of a frame. BS5950 (1990) specifies the range of these factors for steel frames in a very simple and empirical way and it can be used in conjunction with the elastic bifurcation load in Eq. (1.23) for a quick and approximate check for structural stability.

1.10.4. The Merchant–Rankine's formula for collapse load of rigid-jointed frames

Unless a structure is very slender or very stocky, neither the bifurcation nor the plastic load factors in Eqs. (1.23) and (1.25) are close to the true collapse load of a structure. In many practical structures, the geometric or the material non-linearity does not dominate its failure. Empirical formula has been developed to combine these two failure load factors to approximate the collapse load. Rankine (1863) used a reciprocal additive formula for computation of collapse load which was extended to frameworks by Merchant (1954) as

$$\frac{1}{\lambda_u} = \frac{1}{\lambda_p} + \frac{1}{\lambda_{cr}}, \qquad (1.26)$$

in which λ_u, λ_p and λ_{cr} are the ultimate, the plastic and the elastic buckling load factors, respectively.

Although the Merchant–Rankine formula is simple to use, it takes no account of many important factors such as initial imperfection, residual stress, the sway P-Δ effect and no rigid mathematical proof is yet available for verification of the formula. Although modified formulae are proposed to refine the equation for a better approximation of the collapse load through the elastic bifurcation load and the plastic collapse load, they are mostly empirical and lack a rational consideration of the effect of member initial imperfection and residual stress.

1.10.5. Second-order elastic analysis

A relatively more complicated approach for accurate analysis of structures involves the tracing of the equilibrium or the load versus deflection path. In principle, the effects of initial imperfection, the P-δ and the P-Δ effects can all be included in a rigorous analysis. These P-δ and P-Δ effects may be important for slender structures in practice and their existence is illustrated in Fig. 1.14. Their effects are due to deflections in the members

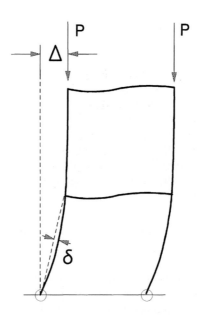

Fig. 1.14. The P-Δ and P-δ effects.

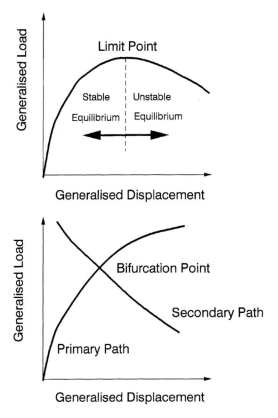

Fig. 1.15. Limit and bifurcation points.

and in the structures respectively and difficult to be rationally accounted for in a linear analysis for all cases.

In a non-linear analysis, one must distinguish between the limit and the bifurcation point which requires a different numerical treatment in tracing the equilibrium path by computer. For the stability theory involving large deflection effects, the critical point refers to the semi-definite condition for the tangent of the equilibrium path or the tangent stiffness matrix of the tangent stiffness matrix. This can either be a bifurcation point or a limit point as shown in Fig. 1.15. In a bifurcation point, the structure is assumed to be undeformed in all except the loaded directions. A limit point refers to the turning point for the equilibrium path of a structure. This can be considered as the transition point from stable to unstable equilibrium states or vice versa.

Second-order elastic analysis implies an analysis allowing for the second-order effects due to change of geometry and initial stress in members, which includes the P-δ and the P-Δ effects. Figure 1.14 shows these effects in a frame. As implied by its name, material yielding is not considered in the second-order elastic analysis. The checking of member strength relies on the application of design formula whilst the second-order effects such as the P-Δ and P-δ effects are included automatically in the analysis, the amplification factors in Eq. (1.22) or equivalent become unnecessary. As shown in the idealized load versus deflection response curves in Fig. (1.12), the path for second-order elastic analysis follows the exact equilibrium path before yielding. They deviate when the load is at a level causing material yield and thus the second-order elastic analysis over-predicts the collapse load of a structure. However, the member capacity check can be carried out by superposition of the computed nodal moments at the two ends of a member obtained directly from the second-order elastic analysis and the moment along the member due to member load and allowing for the P-δ effect. The process of determining the effective length of the member by a formula in a design code in order to calculate the second-order moments can, therefore, be skipped and the accuracy and saving in manipulation effort can be achieved.

1.10.6. Second-order inelastic analysis

If the effect of material yielding is included in the above second-order elastic analysis it becomes the second-order inelastic analysis. The inclusion of material non-linearity can either be carried out by the lump or by the fibre plasticity models. In the lump plasticity model, yielding is assumed to be concentrated in a small region of zero length, generally termed as the plastic hinge (Yau and Chan, 1994). In the fibre model, yielding is assumed to spread along the element length and across the section, and numerical integration is normally employed to evaluate the element stiffness of a partially yielded section (Chan, 1989). Generally speaking, the lump plasticity model is more efficient and easier to apply whilst the fibre model is more accurate and considered to be exact. However, the second-order inelastic analysis is not yet fully qualified for the requirement of an advanced analysis which can be used for direct design. The error for a second-order inelastic analysis may come from the approximation for the material model, residual stress and initial imperfection in the members but improvement can be achieved by refining the model. For example, a plastic hinge concept of reducing the stiffness at a particular location with zero length may not be able to completely represent the true yielding pattern of the structure. In many cases yielding is spread over such a distance that the plastic hinge cannot sufficiently capture the true behaviour of the frame. As a result, the second-order inelastic analysis may give an error in predicting the true collapse load. As shown in Fig. 1.12, the difference between the second-order inelastic analysis and the advanced analysis is expected. The error is difficult to quantify but is problem dependent. For example, if the yielding zone in a beam is distributed over a large area and only an element with a plastic hinge allowed to form at its ends is used to model each member, the idealization may become inaccurate and invalid.

1.10.7. Dynamic and cyclic analysis

Wind and seismic loads are time dependent and the response of a structure under this loading type is dynamic in nature. The widely used equivalent load approach to transform the dynamic force to a set of static forces in many cases cannot reflect the true behaviour of a structure. For instance, the phenomenon of resonance and ductility of a structure cannot be revealed in this static approach. The modal analysis and the spectrum analysis (Clough and Penzien, 1993) is an improved method to incorporating these dynamic effects into the analysis. A more natural and accurate analysis but expensive in terms of computer time, is to trace the complete equilibrium path, allowing for various sources of non-linearities such as the material yielding, large deflection and buckling. This so-called transient response analysis plots the load and deflection of the complete structure against time using various loadings or earthquake history. Depending on the severity of the applied load expressed as years of return period, environment, the structure is allowed to undergo a certain degree of damage and plasticity in structures is normally allowed, provided that it does not collapse. It is economically unjustified to design a structure to remain in its elastic stage under severe load, however, collapse should simultaneously be prevented even during a severe earthquake. For a less severe load, the structure should be repairable after an earthquake strike. It must be emphasised that, although the analysis method by itself is accurate, input information and the model must be precise and accurate in order to simulate the structural behaviour reliably. The sequence for formation of plastic hinges must be examined and controlled so that the structure remains stable. Energy absorption capacity must be adequate and fracture failure should be prevented. The Northridge earthquake in the USA in 1994 and the Hyogoken–Nanbu earthquake near Kobe city in Japan in 1995 revealed the design code at that period was still not fully sufficient to resist even a moderate earthquake attack. In the design of structures against extreme and dynamic loads, loading sequence is also an important parameter for analysis and ductility, in addition to strength and stiffness, and is required for consideration. Generally speaking, strength must be sufficient for stress reserve in a structure to resist loads before yielding. Stiffness is the rigidity of the structure against external loads and its inadequacy leads to buckling and excessive deformation. Ductility is the character of a structure to tolerate large deflection before fracture and its consideration is important in assessing the performance of a structure under dynamic loads. The ductility factor can only be evaluated accurately by a complete second-order plastic analysis.

1.10.8. Advanced analysis

Advanced analysis combines the effects of material yielding and geometrical change with sufficient accuracy for practical purposes so that isolated member capacity checks can be avoided. As such, Advanced analysis can be defined as any analysis method that does not require specification member capacity checks. It can be an improved or a refined second-order plastic-hinge analysis or direct plastic zone analysis. The effect of initial imperfections, residual stress etc. are accounted for in the analysis and a practical design can be directly carried out on the basis of the section capacity computed by this type of rigorous analysis. In other words, only the section capacity check is needed for member design. For an idealized model, the curves generated by this type of analysis follow closely to the actual behaviour of a structure, as indicated in Fig. 1.12. The geometrically non-linear or the second-order elastic analysis ignores the effects of material yielding and thus it can only follow the equilibrium path of a real structure before yielding. The second-order inelastic analysis includes the effects of material yielding but can be distinguished from the most sophisticated method of analysis, the Advanced analysis, by the fact that no detailed consideration on the accuracy of the plasticity model, residual stress, member initial imperfection and/or semi-rigid joint stiffness is included in the method. The most refined Advanced Method of Analysis includes all the second-order effects and also material properties and residual stress, initial member imperfection, erection procedure and interaction with foundations. At the present stage of development, out-of-plane buckling such as lateral torsional buckling should be prevented when using this type of analysis and design. This is due to the fact that these buckling modes may not be easily conducted in the global analysis for the complete structure and they have relatively less interaction with the global structure.

1.10.9. Objectives and structure of this book

Owing to the proposed uses of computer methods for rigorous, second-order and advanced analysis in various national design codes, a computer method and associated programs are urgently needed. The method for studying the second-order inelastic analysis of steel frames will be detailed in this book and explained with the aid of a computer program which has been applied successfully to a number of problems.

The underlying assumption for the used computer program is that all out-of-plane buckling is prevented by suitable bracing. In addition, local plate buckling is not considered. As previously mentioned, this buckling mode should be restrained by the addition of secondary bracing members and local plate buckling is prevented by controlling the plate width versus thickness ratio or by adding adequate stiffeners. Indeed, the slender section is not common in hot-rolled sections. Non-linearity due to the P-δ and the P-Δ effects as well as material yielding by the lump plasticity model is considered in the program. The reduction of stiffness due to these non-linear effects will affect the framing of the complete structure, i.e. the modification of a bracing system to prevent lateral drift will change the structural load path. This phenomenon is mainly due to the global influence of the effects when compared with the more localized buckling due to plate and torsional buckling of members. The general requirements for qualifying an analysis as *advanced* have been discussed in a number of recent publications and in the AS4100 (1990). This includes the prevention of any out-of-plane buckling by proper bracing.

In Chapter 2, a review of the linear method for analysis of skeletal structures will be given and methods for solution of a set of simultaneous equations, the concept of the stiffness method and techniques of accounting for semi-rigid connections in a linear analysis will be outlined. Chapter 3 gives an account of the second-order elastic method of analysis for steel frames where stability and large deflection are the major concern. The concept and procedure for forming the tangent and the secant stiffness matrices will be described in this Chapter. A thorough understanding of the method and the underlying concept for this type of incremental-iterative method of analysis is essential for the user of a non-linear computer program and also for the engineer who wants to gain a physical insight into, and theoretical appreciation of, the non-linear analysis method. The concept of the most widely used Newton–Raphson incremental-iterative method for non-linear analysis will also be described. For more advanced non-linear methods of analysis that have the capacity of traversing the limit point and also a faster convergence rate, refined methods based on the Newton–Raphson procedure are needed. The feature, limitation and implementation of these methods into a second-order analysis computer program are discussed so that the reader can be fully aware of the scope of their applications. The beam-column connection plays an important role in affecting the behaviour of a structure. It will also affect the very fundamental analysis output of a frame, the bending moment and the shear force diagrams. Chapter 5 discusses some of the more widely known connection types. Current state-of-the-art research and practice for design allowing for semi-rigid connections is also detailed. Requirements and procedure of incorporating the connection stiffness into an computer analysis program is reviewed and described. In Chapter 6 for second-order, static inelastic analysis of steel frames, the complete procedure allowing for yielding concentrated at nodes is detailed. In Chapter 7, static second-order analysis of semi-rigid steel frames is extended to non-linear cyclic analysis. The basis of the analysis method is similar to the static approach but allowing for the inertia and damping forces. Several examples will be given to illustrate the application, robustness and theory of the analysis method. Chapter 8 further considers the effect of material yield on the cyclic behaviour of a structure. A section with the example for the use of the analysis program GMNAF is explained in Appendix B. Only through such a practical analysis process, can readers develop an appreciation and concept on the computerised numerical procedure and theory in handling the design and analysis of structural forms. This process is believed to be invaluable to the career of a structural engineer in dealing with these types of structures and loadings.

References

American Institute of Steel Construction (1986): *Load and Resistance Factor Design, Specification for Structural Steel Buildings*, AISC, Chicago.

AS-4100 (1990): *Australian Standard for Steel Structures*, Sydney.

Bathe, K.J. (1982), *Finite Element Procedures in Engineering Analysis*, Prentice-Hall, Englewood Cliffs, NJ.

BS5950 (1990): *Structural Use of Steel in Building, Part 1*, British Standards Institution, UK.

Brush, D.O. and Almroth, B.O. (1975): *Buckling of Bars, Plates and Shells*, McGraw-Hill, New York.

Chajes, A. (1974): *Principle of Structural Stability Theory*, Civil Engineering and Engineering Mechanics Series, Prentice-Hall, Englewood Cliffs, NJ.

Chan, S.L. (1987): Nonlinear Analysis of Structures Composed of Thin-Walled Beam-Columns, Thesis submitted to the Department of Civil Engineering, University of Queensland, Australia for partial fulfillment of the Degree of Doctor of Philosophy.

Chan, S.L. (1989): Inelastic post-buckling analysis of tubular beam-columns and frames. *J. Eng. Struct.* **11**, 23–30.

Chan, S.L. and Kitipornchai, S. (1987a): Geometric nonlinear analysis of asymmetric thin-walled beam-columns. *J. Eng. Struct.* **9**, 243–254.

Chan, S.L. and Kitipornchai, S. (1987b): Nonlinear finite element analysis of angle and tee beam-columns. *J. Struct. Eng.* **113**(4), 721–739.

Chen, W.F. and Chan, S.L. (1994): Second-order inelastic analysis of steel frames by personal computers. *J. Struct. Eng.* **21**(2), 99–106.

Chen, W.F., Pan, A.D. and Yen, T. (1993): Construction Safety of Temporary Structures, Research Report CE-STR-93-26, Department of Structural Engineering, School of Civil Engineering, Purdue University, p. 33.

Clough, R.W. and Penzien, J. (1993): *Dynamics of Structures*, 2nd edn., Civil Engineering Series, McGraw-Hill, New York.

Horne, M.R. (1949): Contribution to "The design of steel frames" by J.F. Baker, *Struct. Engineer* **27**, 421.

Kitipornchai, S. and Lee, H.W. (1984): Inelastic Experiments on Angles and Tee Struts, Research Report CE54, Department of Civil Engineering, University of Queensland, Australia.

Merchant, W. (1954): The failure load of rigidly jointed frameworks as influenced by stability. *Struct. Engineer* **32** 185–190.
Narayanan, R. (1985): *Plated Structures – Stability and Strength*, Elsevier Applied Science, New York.
NAF-SHELL (1993): *Nonlinear Analysis of Shells*, User's Reference Manual.
Rankine, W.J.M. (1863): *A Manual of Civil Engineering*, 2nd edn., Charles Griffin, London.
Timoshenko, S.P. and Gere, J.M. (1961): *Theory of Elastic Stability*, 2nd edn., McGraw-Hill, New York.
Trahair, N.S. (1965): Stability of I-beam with elastic end restraints. *J. Inst. Eng. Australia* **38**, 157.
Usami, T. (1971): Restrained Single-Angle Columns Under Biaxial Bending, Dissertation presented to Washington University at St. Louis, MO, for partial fulfillment of the requirements for the degree of Doctor of Science.
Yau, C.Y. and Chan, S.L. (1994): Inelastic and stability analysis of flexibly connected steel frames by the spring-in-series model. *J. Struct. Eng.* 2803–2819.
Zienkiewicz, O.C. (1977): *The Finite Element Procedure*, 3rd edn., McGraw-Hill, New York.

Chapter 2

FIRST-ORDER LINEAR ELASTIC ANALYSIS FOR RIGID AND SEMI-RIGID FRAMES

2.1. Introduction

The purpose of engineering design is to produce a structure capable of withstanding the environmental loadings to which it may be subjected to. There is a two-stage operation in the design process: firstly, the forces acting on each structural element must be determined and, secondly, the load-carrying capacity of the structural element to these forces must be checked. The first stage involves an analysis of the stresses in the structure and the second involves member capacity checks as detailed by specifications or codes. Furthermore, once the structure is properly sized, analysis can be used to check the safety and serviceability. Structural analysis therefore serves to determine the structural response for given loads and expresses the response in terms of moment and force distribution leading to the internal moments and axial and shear forces and also the deformations. Re-design and modifications of the structure may become necessary when either the ultimate or the serviceability limit state is violated. Before an analysis can be carried out, the actual structure must first be idealized as a structural model ready for manual or computer calculation. This process involves a number of simplifications and idealizations, which may be acceptable in some cases but erroneous in others. It requires the experience and judgement of the structural engineer in assessing the acceptability of these simplifications and idealizations. The following are some of the typical assumptions made in an engineering analysis of framed structures.

(1) The members are line elements with only longitudinal dimension. They intersect at points with zero size.
(2) The material is homogeneous and isotropic. Although it is not complicated to include different material moduli for different members, they are not easy to quantify. For steel and other metals, the variation of material modulus is small so the assumption is valid.
(3) Warping and shear deformations are usually ignored in practice. For warping, the complexity lies in the assessment of the mechanism of transferring warping from one member to another (Baigent and Hancock, 1982; and Sharman, 1985) that it is not easy to generalize the warping transfer relation between members. Shear deformation can be included in the element formulation but it becomes unimportant for most members of moderate to high slenderness.
(4) For the first order linear elastic analysis used by many engineers and discussed in this Chapter, the deflection is proportional to the applied load and the structural stiffness

and stress are not affected by small changes in geometry and the presence of initial stresses.
(5) The contributions of secondary members are ignored in determining the global structural stiffness. These members, which also include bracing members, are not generally considered in actual analysis because of the difficulty in assessing their strength and stiffness and also of the small contribution they can normally offer. These non-structural elements include cladding, wall partitions and others.

2.2. Linear analysis

In spite of so many criticisms against a linear analysis, it is the most widely used method in practice and it has been used by the engineer for the design of the overwhelming majority of structures in human history. The method enjoys the advantages of being easy-to-understand, simplicity in computation and concept and saving in computational effort. The widely accepted principle of superimposition can also be applied in a linear analysis such that the response of the structure is equal to the sum of all the effects due to different load cases. Linear analysis is also particularly welcomed by the profession because of its simplicity. To date, most engineers still prefer a linear analysis for determination of stresses and deformations in a structure.

Another importance connected with linear analysis is the fact that the concept of a non-linear analysis cannot be dealt with without a thorough understanding of the method for linear analysis. Indeed, most non-linear analyses are carried out by a series of linearized analyses.

In conventional linear analysis, the deflection of a structure is assumed to be very small and the second-order effects due to geometrical change can be ignored. The stiffness of the structural members is also taken as constant and independent of the presence of axial force. Mathematically, the linear assumption for a structure ignores the P-δ and P-Δ effects due to the element and the global structural deflections as well as any material non-linearity including yielding and formation of plastic hinges. To check the load-carrying capacity of a member, the computed moments and forces acting on a member by a linear analysis must be checked against the design formulae considering material yield and instability as detailed by specifications or practice codes. This linear approximation may not be accurate in constructing the bending moment and force diagrams for a structure in some cases, especially for slender structures. The influence of the effects of instability and material yielding must be included at the later design stage which include checking of beam and strut buckling in the clauses of various national design codes of practice.

A number of manual and computer methods for linear analysis of skeletal structures have been developed over the past century. The slope deflection method provides a quick means of determining the moment in a beam. The iterative moment distribution method is another way of calculating the shear and the bending moment of a continuous beam of several spans or a simple portal. The more robust and general methods, i.e. the flexibility and the stiffness methods, are more suitable for computer applications. However, regardless of which method is used, three basic conditions must always be satisfied; the compatibility, the equilibrium and the material constitutive relationship.

Chapter 2. First-order linear elastic analysis for rigid and semi-rigid frames

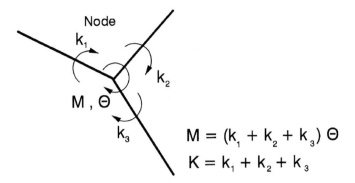

Fig. 2.1. The compatibility condition.

Compatibility refers to the condition where the members are connected without discontinuity. For the rigid-joint case, the members connected to a node share the same rotations and displacements. For pinned or semi-rigid connections, the displacements are common to the members connected to the node but rotations may not be identical for all connected members. However, the element rotations are still governed by the connection stiffness. In other words, the spring absorbs the relative rotation between the element and the node to which it is connected. In the stiffness matrix method of analysis, the assembling process of adding the contributions from the connected elements is in fact to impose the compatibility condition by assuming that the elements connected to a particular joint share the same rotations and displacements as shown in Fig. 2.1.

Equilibrium is a pre-requisite condition for a structure to stand without collapse. The external applied forces, together with the internal loads due to temperature, lack of fit, foundation settlement etc. must be balanced by the resistance of the structure and the reactions of the supports. The members in a structure transfer loads to foundations or supports. The summation of forces and moments for the whole system in any direction must be equal to zero. In addition, the equilibrium at nodes must also be satisfied via the solution of the set of simultaneous equilibrium equations. In some cases, forces and moments due to dynamic motion must also be accounted for in setting up the equilibrium equations. In the stiffness matrix method of analysis, the vanishing of any summed forces at the nodes about all the degrees of freedom must be fulfilled in order to meet the equilibrium requirement. The internal resistance equal to the product of the structural stiffness and the displacements must therefore, be equal to the applied force. As a result, for every node, the following condition must be satisfied:

$$\sum F_x = \sum F_y = \sum F_z = \sum M_x = \sum M_y = \sum M_z = 0, \qquad (2.1)$$

in which F_i and M_i are the force and the moment about the i axis, respectively.

The **material constitutive** or the stress versus strain relationship governs the stiffness and strength of the structure. When the material modulus is incorrect, the actual relationship between the stress and the strain is not equivalent to those values used in setting-up the stiffness of the element and the structure. Subsequently, the analysis results will be incorrect.

In the stiffness matrix method of analysis for framed structures, the compatibility and the equilibrium conditions are imposed in the solution of the system stiffness matrix equation. For linear analysis ignoring the second-order effect due to axial force, the solution of the equilibrium condition yields a cubic function and as a result, the cubic polynomial displacement function gives the exact solution in a *linear* analysis.

2.3. Coordinate systems

All structural analyses require a set of consistent coordinate systems to define the vector quantities. A properly defined coordinate system is essential for the development of a reliable, convenient and user-friendly analysis computer program. In some cases the loads are referred to in a global coordinate system such as the structural dead weight and the gravitational loads. In other cases, the loads are referenced by the local coordinate system which is attached to the element itself. This loading type includes the pressure load perpendicular to the element axis. Wind load can act on the projected area about either the global or the local axes, depending on the load case being considered.

2.3.1. The global coordinate system

The global coordinate system is a set of coordinate axes for the complete structure. The stiffness and applied forces for all the elements must be transformed to this coordinate system for summation. In this book, the right-hand screw rule is used to define the sign of moments and the right-hand rule is also employed for setting up the positive direction of the x, y and z axes which are taken to be orthogonal for sake of simplicity. The global coordinate system is shown in Fig. 2.2.

In the actual analysis, the structure should be placed in the global system with the nodal coordinates calculated by this global system. Although the orientation for the global axes can be arbitrary, it will be seen that a great saving in calculation of nodal coordinates can be achieved when it is properly set.

2.3.2. The local coordinate system

The element stiffness matrix is formulated in a set of local coordinate axes parallel to the principal axes for sake of convenience. When the element does not have an orientation the same as the global coordinate system, the element stiffness must be transformed to the global axes so that members with different orientations can be summed to balance the external loads.

The element local coordinate system for a general member is shown in Fig. 2.3. This set of coordinate systems is related to the global system by a series of transformations. To arrive at the final position of the element local axes shown in Fig. 2.2, the global axes can first be rotated by turning about the global y-axis as indicated in Fig. 2.2(a). The

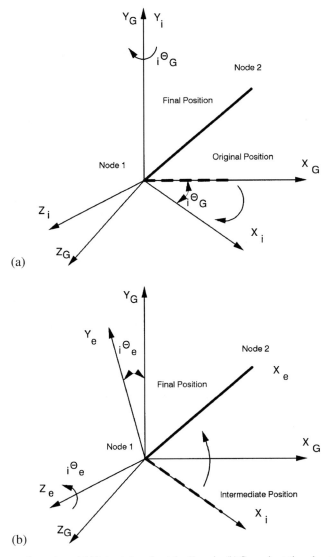

Fig. 2.2. The transformations. (a) First rotation about the Y_G axis; (b) Second rotation about the Z_i axis.

corresponding transformation matrix can be determined as

$$\begin{bmatrix} X_i \\ Y_i \\ Z_i \end{bmatrix} = \begin{bmatrix} \cos {}_i\theta_G & 0 & \sin {}_i\theta_G \\ 0 & 1 & 0 \\ \sin {}_i\theta_G & 0 & \cos {}_i\theta_G \end{bmatrix} \begin{bmatrix} X_G \\ Y_G \\ Z_G \end{bmatrix}, \quad [X_i] = [{}_iL_G][X_G], \tag{2.2}$$

in which ${}_i\theta_G$ is the angle between the x-component of the inclined element and the projected length of the element on the global x–z plane as shown in Fig. 2.2(a). 'i' stands for

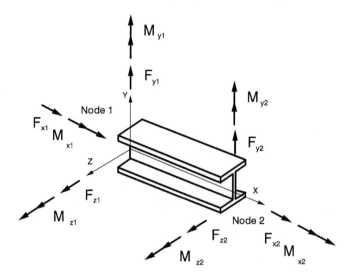

Fig. 2.3. The element local coordinate system.

intermediate position and 'G' stands for final position which can be expressed as

$$\cos {}_i\theta_G = \frac{C_x}{Q}, \qquad \sin {}_i\theta_G = \frac{C_y}{Q}, \qquad (2.3)$$

where C_x and C_y are the direction cosines of the element about the X_G- and Y_G-axes. Q is the project distance on the x–z plane of a unit length parallel to the element given by the following:

$$Q = \sqrt{C_x^2 + C_z^2}, \qquad (2.4)$$

and L is the length of the element given by

$$L = \sqrt{(x_2 - x_1)^2 + (y_2 - y_1)^2 + (z_2 - z_1)^2}, \qquad (2.5)$$

x_1, y_1, z_1, x_2, y_2 and z_2 are the coordinates of the two end nodes of the element (Fig. 2.2), respectively.

Substituting Eqs. (2.3) to (2.5) into (2.2), the transformation matrix relating i-axes to global G-axis can be expressed as

$$[{}_iL_G] = \begin{bmatrix} \dfrac{C_x}{\sqrt{C_x^2 + C_z^2}} & 0 & \dfrac{C_z}{\sqrt{C_x^2 + C_z^2}} \\ 0 & 1 & 0 \\ \dfrac{-C_z}{\sqrt{C_x^2 + C_z^2}} & 0 & \dfrac{C_x}{\sqrt{C_x^2 + C_z^2}} \end{bmatrix}. \qquad (2.6)$$

Chapter 2. First-order linear elastic analysis for rigid and semi-rigid frames

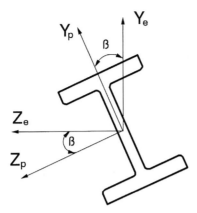

Fig. 2.4. The inclination of the principal axes.

To arrive at the final position, it is necessary to rotate an additional axis, the z_i axis which is the z-axis of the rotated element as shown in Fig. 2.2(b). The corresponding transformation matrix can be expressed as follows:

$$\begin{bmatrix} X_e \\ Y_e \\ Z_e \end{bmatrix} = \begin{bmatrix} \cos {}_e\theta_i & \sin {}_e\theta_i & 0 \\ -\sin {}_e\theta_i & \cos {}_e\theta_i & 0 \\ 0 & 0 & 1 \end{bmatrix} \begin{bmatrix} X_i \\ Y_i \\ Z_i \end{bmatrix}, \quad [X_e] = [{}_eL_i][X_i], \tag{2.7}$$

in which ${}_e\theta_i$ is the angle between the final position of the element and the x–z plane given by

$$\cos {}_e\theta_i = \sqrt{C_x^2 + C_z^2}, \qquad \sin {}_e\theta_i = C_y. \tag{2.8}$$

Substituting Eq. (2.8) into (2.7), the transformation matrix relating 'i' to the element local axes 'e' can be obtained as

$$[{}_eL_i] = \begin{bmatrix} \sqrt{C_x^2 + C_z^2} & C_y & 0 \\ -C_y & \sqrt{C_x^2 + C_z^2} & 0 \\ 0 & 0 & 1 \end{bmatrix}. \tag{2.9}$$

The two successive transformations cannot completely determine the orientation of an element. The analyst should also define the inclination of the principal axes of the element and an additional transformation matrix is therefore required. From Fig. 2.4, the mapping between the element and the principal axes can be expressed as follows:

$$\begin{bmatrix} X_p \\ Y_p \\ Z_p \end{bmatrix} = \begin{bmatrix} 1 & 0 & 0 \\ 0 & \cos\beta & \sin\beta \\ 0 & -\sin\beta & \cos\beta \end{bmatrix} \begin{bmatrix} X_e \\ Y_e \\ Z_e \end{bmatrix}, \quad [X_p] = [{}_pL_e][X_e], \tag{2.10}$$

in which β is the angle between the element local axes parallel to the global axes and the principal axes.

Multiplying the transformation matrices in Eqs. (2.6), (2.9) and (2.10), the final transformation matrix relating the element to the global coordinate axes can be evaluated as

$$[L'] = [_pL_e][_eL_i][_iL_G]$$

$$= \begin{bmatrix} C_x & C_y & C_z \\ \dfrac{-C_xC_y\cos\beta - C_z\sin\beta}{\sqrt{C_x^2+C_z^2}} & \sqrt{C_x^2+C_z^2}\cos\beta & \dfrac{-C_yC_z\cos\beta + C_x\sin\beta}{\sqrt{C_x^2+C_z^2}} \\ \dfrac{C_xC_y\sin\beta - C_z\cos\beta}{\sqrt{C_x^2+C_z^2}} & -\sqrt{C_x^2+C_z^2}\sin\beta & \dfrac{C_yC_z\sin\beta + C_x\cos\beta}{\sqrt{C_x^2+C_z^2}} \end{bmatrix}. \quad (2.11)$$

As each element possesses twelve degrees of freedom which can be grouped into four sets of vectors about the x, y and z axes, it becomes necessary to expand the transformation matrix in Eq. (2.12) to a 12×12 transformation matrix symbolically as

$$[L] = \begin{bmatrix} [L'] & 0 & 0 & 0 \\ 0 & [L'] & 0 & 0 \\ 0 & 0 & [L'] & 0 \\ 0 & 0 & 0 & [L'] \end{bmatrix}. \quad (2.12)$$

The element stiffness matrix in the global coordinate system can therefore be obtained through the application of the contragradient principal (Meek, 1971) as

$$[k] = [L]^T[k_e][L], \quad (2.13)$$

in which $[k_e]$ is the element stiffness matrix in local coordinates and $[k]$ is the stiffness matrix in the global coordinate system.

2.4. The principle of the total potential energy for element stiffness formulation

The principle of stationary potential energy or the virtual work can be applied to formulate the element stiffness matrix and the corresponding equilibrium equation. The total potential energy functional for a continuum is selected in this book because the energy functional can be expressed explicitly and the process is more familiar to most readers. It should be emphasized that, however, the two theorems will produce exactly the same outcome when applied properly to a problem and their choice is only a matter of convenience. The total potential energy functional, Π, is composed of two parts as

$$\Pi = U - V, \quad (2.14)$$

in which U is the strain energy and V is the external work done as

$$V = \sum F_i u_i \quad (2.15)$$

in which F_i and u_i are the applied forces and displacements, respectively.

The strain energy, U, can be written in terms of the stress (σ) and the strain tensor (ε) as

$$U = \frac{1}{2} \int_{\text{vol}} [\sigma]^T [\varepsilon] \, d\text{vol}$$
$$= \frac{1}{2} \int_{\text{vol}} [\varepsilon]^T [D][\varepsilon] \, d\text{vol} \qquad (2.16)$$

in which $[D]$ is the material Hookean's constant, vol is the volume of the continuum and T is the transpose operator.

The linear normal strain (ε_{ii}) of a continuum can be related to the displacement (Washizu, 1975) as

$$\varepsilon_{ii} = \frac{\partial u_i}{\partial x_i}, \quad i = x, y, z, \qquad (2.17)$$

in which u_i is the displacement along I-axis and $\partial(\cdot)/\partial x_i$ is the differential operator with respect to x.

The linear shear strain (ε_{ij}) can be written in terms of the displacements as

$$\varepsilon_{ij} = \frac{1}{2}\left[\frac{\partial u_i}{\partial x_j} + \frac{\partial u_j}{\partial x_i}\right], \quad i, j = x, y, z. \qquad (2.18)$$

For a the one-dimensional element as beam-column, not all the terms in a continuum need to be included in the formulation. Only the quantities on the plane perpendicular to the longitudinal x-axis are required to be considered. As a result, only the strain components, ε_{xx}, ε_{xy} and ε_{xz} are retained in the subsequent formulation. From the geometrical compatibility consideration for a beam-column, the displacement of an arbitrary point can be expressed in terms of the displacements and rotations about the principal coordinates as

$$u = u_c - y\frac{\partial v}{\partial x} - z\frac{\partial w}{\partial x}, \quad v = v_c - \theta_x z, \quad w = w_c + \theta_x y, \qquad (2.19)$$

where u, v, w, u_c, v_c and w_c are the displacements along the x, y and z axes of an arbitrary point and of the centroid and θ_x is the twist, respectively. The subscript 'c' refers to the centroid axis.

Substituting Eqs. (2.19) into (2.17) and (2.18) and then (2.16), we obtain the strain energy for a beam-column as

$$U = \frac{1}{2}\int_L \left(EA\left[\frac{\partial u}{\partial x}\right]^2 + EI_z\left[\frac{\partial^2 v}{\partial x^2}\right]^2 + EI_y\left[\frac{\partial^2 w}{\partial x^2}\right]^2 + GJ\left[\frac{\partial \theta_x}{\partial x}\right]^2\right) dx. \qquad (2.20)$$

It is necessary to assume a displacement function in order to express the displacement in terms of the nodal variables. Considering the differential equilibrium equation for a straight beam under nodal moments as follows, the cubic function is a solution for the differential equilibrium equation:

$$EI\frac{d^2v}{dx^2} = M_1\left(1 - \frac{x}{L}\right) + M_2\frac{x}{L}. \tag{2.21}$$

The cubic function contains four constants which can then be expressed in terms of the four nodal variables, namely, the two displacement and the two rotational degrees of freedom. The shape function can therefore be obtained as

$$v = a_0 + a_1 x + a_2 x^2 + a_3 x^3. \tag{2.22}$$

Making use of the boundary conditions in Eq. (2.23) for the axis passing through the two ends of an element, Eq. (2.24) for the shape function can be obtained (Fig. 2.5).

$$\begin{aligned} x = 0, \quad & v = 0, \quad \frac{\partial v}{\partial x} = \theta_{z1}, \\ x = L, \quad & v = 0, \quad \frac{\partial v}{\partial x} = \theta_{z2}; \end{aligned} \tag{2.23}$$

$$v = \left[x - \frac{2x^2}{L} + \frac{x^3}{L^2} \quad -\frac{x^2}{L} + \frac{x^3}{L^2} \right] [\theta_{z1}\ \theta_{z2}]^{\mathrm{T}}. \tag{2.24}$$

Similarly, lateral deflection along z can be expressed as

$$\begin{aligned} x = 0, \quad & w = 0, \quad \frac{\partial w}{\partial x} = -\theta_{y1}, \\ x = L, \quad & w = 0, \quad \frac{\partial w}{\partial x} = -\theta_{y2}; \end{aligned} \tag{2.25}$$

$$w = \left[x - \frac{2x^2}{L} + \frac{x^3}{L^2} \quad -\frac{x^2}{L} + \frac{x^3}{L^2} \right] [-\theta_{y1}\ -\theta_{y2}]^{\mathrm{T}}. \tag{2.26}$$

Note that the sign for nodal rotation is reversed for deflection in z axis because a positive slope $(\partial w/\partial x)$ is equal to $-\theta_y$ by the right-hand rule.

Since only two degrees of freedom exist for axial lengthening and torsion, their respective displacement functions can be assumed to be linear and the shape function for the displacement can be determined as

$$u = b_0 + b_1 x. \tag{2.27}$$

Making use of the boundary condition in Eq. (2.28), Eq. (2.29) for shape function of axial deformations can be obtained:

$$\begin{aligned} x = 0, \quad & u = u_1, \\ x = L, \quad & u = u_2; \end{aligned} \tag{2.28}$$

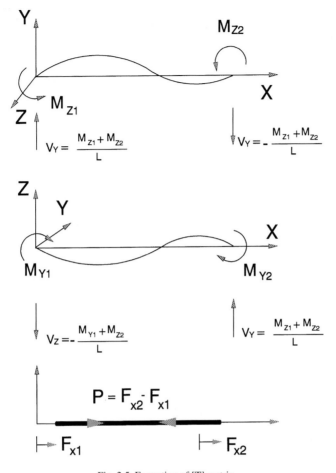

Fig. 2.5. Formation of [T] matrix.

$$u = \left[1 - \frac{x}{L}\right]u_1 + \frac{x}{L}u_2. \tag{2.29}$$

Similarly, the twist can be expressed as

$$\begin{aligned} x &= 0, & \theta &= \theta_{x1}, \\ x &= L, & \theta &= \theta_{x2}; \end{aligned} \tag{2.30}$$

$$\theta_x = \left[1 - \frac{x}{L}\right]\theta_{x1} + \frac{x}{L}\theta_{x2}. \tag{2.31}$$

Substituting the expression for displacements and rotations into the strain energy in Eq. (2.20), the strain energy functional can be expressed in terms of the displacement degrees of freedom. Making use of the principle of total potential energy that the first variation of the functional yields the equilibrium equations, the stiffness matrix can be derived as

$$\delta\Pi = \frac{\partial\Pi}{\partial x_i}\partial x_i = 0,$$
$$\frac{\partial^2 U}{\partial x_i \partial x_j}\delta x_j = k_{ij}\delta x_j = F_i. \tag{2.32}$$

By the energy principle, the stiffness coefficients can be obtained as

$$k_{ij} = \frac{\partial^2 U}{\partial u_i \partial u_j}. \tag{2.33}$$

The complete element stiffness equation can therefore be obtained by applying Eq. (2.33) to obtain the stiffness coefficients one by one and the final stiffness matrix can be written as

$$[k_b][u] = [P],$$

$$\begin{bmatrix} \frac{EA}{L} & 0 & 0 & 0 & 0 & 0 \\ 0 & \frac{4EI_y}{L} & 0 & 0 & \frac{2EI_y}{L} & 0 \\ 0 & 0 & \frac{4EI_z}{L} & 0 & 0 & \frac{2EI_z}{L} \\ 0 & 0 & 0 & \frac{GJ}{L} & 0 & 0 \\ 0 & \frac{2EI_y}{L} & 0 & 0 & \frac{4EI_y}{L} & 0 \\ 0 & 0 & \frac{2EI_z}{L} & 0 & 0 & \frac{4EI_z}{L} \end{bmatrix} \begin{bmatrix} \delta_x \\ \theta_{y1} \\ \theta_{z1} \\ \theta_x \\ \theta_{y2} \\ \theta_{z2} \end{bmatrix} = \begin{bmatrix} F_x \\ M_{y1} \\ M_{z1} \\ M_x \\ M_{y2} \\ M_{z2} \end{bmatrix}. \tag{2.34}$$

For a three-dimensional element with six degrees of freedom at each end, the total degrees of freedom will be twelve although these are not fully independent but equivalent to the six active forces and moments in Eq. (2.34). These twelve forces and moments can be related to these six internal and independent forces and moments by transformation considering the equilibrium of forces and moments in Fig. 2.5 or, equivalently, from geometrical compatibility and from displacements and rotations. This transformation matrix, $[T]$, can be expressed as

$$[F]_{L} = [T] \; [P], \tag{2.35}$$
$$\text{\scriptsize 12\times 1} \quad \text{\scriptsize 12\times 6} \; \text{\scriptsize 6\times 1}$$

in which

$$[T] = \begin{bmatrix} -1 & 0 & 0 & 0 & 0 & 0 \\ 0 & 0 & 1/L & 0 & 0 & 1/L \\ 0 & -1/L & 0 & 0 & -1/L & 0 \\ 0 & 0 & 0 & -1 & 0 & 0 \\ 0 & 1 & 0 & 0 & 0 & 0 \\ 0 & 0 & 1 & 0 & 0 & 0 \\ 1 & 0 & 0 & 0 & 0 & 0 \\ 0 & 0 & -1/L & 0 & 0 & -1/L \\ 0 & 1/L & 0 & 0 & 1/L & 0 \\ 0 & 0 & 0 & 1 & 0 & 0 \\ 0 & 0 & 0 & 0 & 1 & 0 \\ 0 & 0 & 0 & 0 & 0 & 1 \end{bmatrix}. \quad (2.36)$$

$[F]_L$ is the local element forces and moments $[F_{x1}, F_{y1}, F_{z1}, M_{x1}, M_{y1}, M_{z1}, F_{x2}, F_{y2}, F_{z2}, M_{x2}, M_{y2}, M_{z2}]^T$ (Fig. 2.3).

Combining the stiffness and the transformation matrices from Eqs. (2.34) to (2.36) the element stiffness matrices in the global coordinate system are obtained as

$$[k_e] = [T][k_b][T]^T, \quad (2.37)$$

in which $[k_e]$ is the element stiffness in the local coordinate having the dimensions of 12×12. This stiffness can then be substituted into Eq. (2.13) for the formation of the element stiffness matrix in the global coordinate system ready for assembling to form the global stiffness matrix.

2.5. Element stiffness for semi-rigid jointed members

In most analyses, the joints are assumed to be either pinned or rigid. In simple construction, beams are pin-joined to the columns and the lateral restraints are provided by other structural elements such as bracing and shear wall, as shown in Fig. 2.6. In rigid-jointed construction, the connections are assumed to be able to resist the moment induced and the stiffness warrants the analysis result in reasonably close agreement with the rigid frame analysis. The resistance against lateral forces can then be provided by the frame action in the structure. These assumptions have been made by engineers for decades to simplify the analysis procedure. More recently, personal computers of powerful capacity are available at low cost and a connection data bank has been set up (Kishi and Chen, 1986; and Chen and Toma, 1994). Models for simulating the joint stiffness are recommended in Eurocode-3 (1993) and are considered in American Load and Resistance Factor Design (1986). Other national design codes have also allowed the engineer to consider the effect of semi-rigid connections in their design.

The Load and Resistance Factor Design for steel buildings (1986) classifies joints as either semi-rigid or rigid (or partially or fully restraint construction). In Eurocode 3 for the

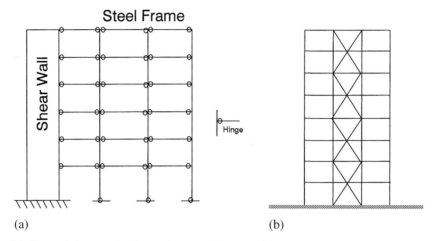

Fig. 2.6. Structural forms against lateral forces. (a) Simple construction laterally restrained by shear wall; (b) Braced steel frames.

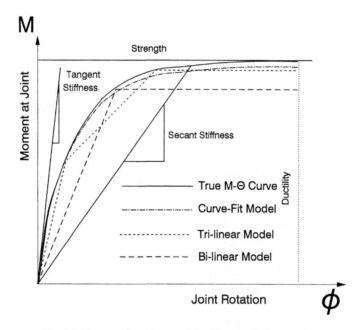

Fig. 2.7. The strength, stiffness and ductility of typical connection.

design of steel structures, a more detailed description on the requirement of beam-column connections is provided. In principle, a semi-rigid connection must have the required strength, stiffness and ductility. Figure 2.7 illustrates the generalized relationship between the three parameters in a typical moment versus rotation (M–θ) curve. The stiffness can

Chapter 2. *First-order linear elastic analysis for rigid and semi-rigid frames*

be directly input to the analysis program following a procedure detailed in the following section. The strength can be checked automatically in the analysis by comparing the moment at node and the moment at resistance of the connection whilst the rotation ductility can be ensured by comparing the induced relative rotation at joints and the ductility of connection used.

From the recommended model in Eurocode 3 and the assumptions made by a number of researchers [for example, Lui and Chen (1987)], a joint can be considered in an analysis as dimensionless with location at the intersection of the element centre lines. The spring element used for modelling the connection behaviour must be in equilibrium and satisfy the stiffness requirement for the rotations at its two ends. These conditions can be stipulated as follows:

$$M_e + M_i = 0, \tag{2.38}$$

in which M_e and M_i are the moments at the two ends of a connection. The corresponding external node is connected to the global node and the internal node is joined to the beam element.

The stiffness of the connection, S, can be related to relative rotations at the two ends of the connection spring as

$$S = \frac{M_e}{(\theta_e - \theta_i)} = \frac{M_i}{(\theta_i - \theta_e)}, \tag{2.39}$$

in which θ_i and θ_e are the conjugate rotations for the moments M_i and M_e.

Rewriting Eq. (2.39) above in matrix form, the stiffness matrix of a connection spring can be written as

$$\begin{bmatrix} S & -S \\ -S & S \end{bmatrix} \begin{bmatrix} \theta_e \\ \theta_i \end{bmatrix} = \begin{bmatrix} M_e \\ M_i \end{bmatrix}. \tag{2.40}$$

In the case of an element with connection springs at its ends, a hybrid element can be obtained by directly adding the two connection stiffnesses to the element bending stiffness in Eq. (2.34) as

$$\begin{bmatrix} S_1 & -S_1 & 0 & 0 \\ -S_1 & k_{11}+S_1 & k_{12} & 0 \\ 0 & k_{21} & k_{22}+S_2 & -S_2 \\ 0 & 0 & -S_2 & S_2 \end{bmatrix} \begin{bmatrix} \theta_{1e} \\ \theta_{1i} \\ \theta_{2i} \\ \theta_{2e} \end{bmatrix} = \begin{bmatrix} M_{1e} \\ M_{1i} \\ M_{2i} \\ M_{2e} \end{bmatrix}, \tag{2.41}$$

in which the first subscript refers to node 1 or node 2. The second subscript indicates whether the location for rotations and moments occurs at external or at internal node of the hybrid element (Fig. 2.8). k_{ij} is the stiffness coefficients of a prismatic beam. When i is equal to j, it is equal to $4EI/L$ and, when i is not equal to j, it is equal to $2EI/L$.

In order to reduce the degrees of freedom of a hybrid element to that for a conventional element for efficiency in computation and ease of computer implementation, it is necessary to statically condense the internal degrees of freedom of the stiffness expression in

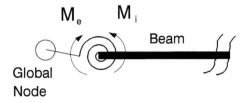

Fig. 2.8. Equilibrium at a joint.

Eq. (2.41). Considering the equations for the internal degrees of freedom and assuming moments are applied at the external nodes only, we can re-write the two equations for internal degrees of freedom in the following form:

$$\begin{bmatrix} k_{11}+S_1 & k_{12} \\ k_{21} & k_{22}+S_2 \end{bmatrix} \begin{bmatrix} \theta_{1i} \\ \theta_{2i} \end{bmatrix} + \begin{bmatrix} -S_1 & 0 \\ 0 & -S_2 \end{bmatrix} \begin{bmatrix} \theta_{1e} \\ \theta_{2e} \end{bmatrix} = \begin{bmatrix} 0 \\ 0 \end{bmatrix}, \qquad (2.42)$$

or

$$\begin{bmatrix} \theta_{1i} \\ \theta_{2i} \end{bmatrix} = \begin{bmatrix} k_{11}+S_1 & k_{12} \\ k_{21} & k_{22}+S_2 \end{bmatrix}^{-1} \begin{bmatrix} S_1 & 0 \\ 0 & S_2 \end{bmatrix} \begin{bmatrix} \theta_{1e} \\ \theta_{2e} \end{bmatrix}. \qquad (2.43)$$

Substituting Eq. (2.43) into (2.41), we obtain the stiffness expression of a beam element with both ends connected to a pair of springs as

$$[k_e][\theta_e] = [M_e], \qquad (2.44)$$

in which

$$[\theta_e] = [\theta_{1e} \quad \theta_{2e}]^T, \qquad [M_e] = [M_{1e} \quad M_{2e}]^T, \qquad (2.45)$$

$$[k_e] = \frac{1}{\beta} \begin{bmatrix} S_1\beta - S_1^2(S_2+k_{22}) & k_{21}S_1S_2 \\ k_{21}S_1S_2 & S_2\beta - S_2^2(S_1+k_{11}) \end{bmatrix}, \qquad (2.46)$$

and β is given by

$$\beta = \det \begin{vmatrix} k_{11}+S_1 & k_{12} \\ k_{21} & k_{22}+S_2 \end{vmatrix}, \qquad (2.47)$$

where S_1 and S_2 are the connection stiffness at nodes 1 and 2, respectively.

Equation (2.46), the stiffness coefficients of a semi-rigid jointed beam, is simple to program and represents a general stiffness of a semi-rigid beam (Fig. 2.9). When one spring stiffness is zero, it will be transformed to the stiffness matrix of a beam with one end pinned and the other end rigid. If the two spring stiffnesses are rigid and pinned respectively, the stiffness matrix will automatically be converted to the stiffness of a beam with corresponding rigid and pinned ends.

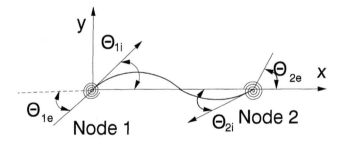

Fig. 2.9. The external and internal rotations.

In an analysis, the joint stiffness can be assumed as bilinear, trilinear or a curve as shown in Fig. 2.7. When the initial connection rotation is small, the initial slope of the moment versus rotation (M–θ) curve can be used. For joints where the rotation is significant, the secant stiffness of the joint can be used. Many practical connections have highly non-linear stiffness characteristics. The discussion on various joint models for these cases will be described in subsequent chapters for non-linear analysis.

2.6. Assembling of the stiffness matrix

The stiffness matrix should be first formed, transformed and assembled to form the stiffness matrix for the complete structure from which the displacements are solved. They will then be used for calculation of the element forces and moments for member capacity check and design. The assembling process is standard for the finite element analysis. However, there are several methods to store and solve the stiffness matrix. Before an efficient scheme can be derived or selected, the properties of the stiffness matrix must first be realized. For conservative systems widely encountered in the field of structural engineering, the stiffness matrix is symmetric and sparse. For linear analysis and pre-buckling non-linear analysis, the stiffness matrix is also positive definite. The size of the stiffness matrix can be very large, depending on the problems being studied. Under this condition, several approaches are available. These schemes include the full matrix, the band width scheme, the skyline or column storage scheme and the frontal solver. Each method has its advantages and short-comings and a proper choice of these methods depends very much on the nature of the problems being studied and the supporting computer environment and hardware available.

The *full matrix method* is referred to the setting-up of a $n \times n$ matrix to store the stiffness matrix of a structure of n degrees of freedom. This method wastes much computer memory and also leads to a heavy computing time. It is suitable for solving very small problems and for educational purpose at undergraduate level for demonstration of fundamental principles. Usually, manual method is used in connection with the approach and thus the problem size is limited.

The second widely used method is the *band-width technique* for saving the stiffness matrix. The maximum band width for the complete stiffness matrix is calculated in this method and, making use of the matrix symmetry, only the upper or the lower half of the stiffness matrix is stored. A stiffness matrix of $n \times n$ dimension can be stored in a $n \times h$ size rectangular matrix in which the h is the band width equal to the maximum column height in the stiffness matrix. The column height can be calculated as the maximum difference between any two nodes of the elements. Obviously, the numbering of the element nodes determines the column height. Algorithms to minimize the band width are also available in order to save memory and to reduce the number of algebraic equations involved. However, as computing time is needed for this minimization process, the final time saving in a linear analysis is difficult to estimate. For non-linear analysis where the stiffness matrix has to be solved many times in the tracing of equilibrium path whilst the minimization process is required to be activated only once, the saving in computing time may be justifiable.

The third method is the *skyline storage* or the *compact column scheme* in which the column height, instead of the maximum height for the complete matrix, is used to store the stiffness coefficients. All the non-zero terms are saved inside the column profile and matrix symmetry can also be made use of. The occupied memory by this scheme is always equal or less than the band width method and the minimization scheme is also available to further reduce the matrix size. This method is selected in this book for storing the global stiffness matrix. The modified Cholesky's method for factorizing the stiffness matrix is used in which the stiffness is split into the following form.

$$[K] = [U]^T [D_a][U] \tag{2.48}$$

in which $[U]$ is the upper triangular matrix with diagonal terms equal to unity and $[D_a]$ is the diagonal matrix of which the product is equal to the determinant of the stiffness matrix.

It should be noted that, the original Cholesky scheme of factorizing the stiffness into two parts (i.e. $[K] = [U]^T[U]$ being the upper triangular matrix and its transpose), is only suitable for linear and non-linear analysis in the stable equilibrium range. For tracing of unstable equilibrium paths where the determinant of the tangent stiffness matrix is indefinite and its determinant is negative, this original Cholesky scheme will break down as it attempts to square-root a negative number. In this case, the modified Cholesky scheme in Eq. (2.48) should be used and, for generality, it is adopted in this book. The currently used subroutine for factorisation and back-substitution for final solutions are adopted and simplified from the work of Felippa (1975) for solution of the stiffness matrix equation in the program analysis.

Another efficient method of solving the stiffness matrix equation is the *Frontal Solution Scheme* (Iron, 1970). The advantage of this method lies on the avoidance to store the complete stiffness matrix before the factorisation process is activated, which takes place during the assembling process. This method is particularly useful in the early computer stage when the in-core memory in a computer is very limited. However, it requires a relatively more complicated book-keeping process during the assembling and solution process. With the objective of clarity and the availability of powerful program compilers such as Lahey (1992) utilizing the hard-disc for memory storage, the aforementioned modified Cholesky method in conjunction with the skyline storage (Compact Column) scheme is used in this book.

A detailed discussion of these storage schemes and also the solution method for sparse and symmetric matrix is available in texts for numerical analysis (Bathe, 1982; and George and Liu, 1981) and will not be re-capitulated here. However, the three stage factorisation scheme used in the program GMNAF in the Appendix is summarized to ease readers in understanding the program operations.

The stiffness coefficients are saved in a single array instead of a square matrix for the stiffness. For example, the matrix shown below can be stored by a single array a_i as

$$\begin{bmatrix} a_1 & a_2 & a_4 & & & a_{12} \\ & a_3 & a_5 & a_7 & & a_{13} \\ & & a_6 & a_8 & & a_{14} \\ & & & a_9 & a_{10} & a_{15} \\ & & & & a_{11} & a_{16} \\ & & & & & a_{17} \end{bmatrix}. \tag{2.49}$$

Before the assembling process for the elements is activated, the profile of the global stiffness matrix such as the one in Eq. (2.49) must first be determined and saved for recording purposes. This can be easily done by using an array for saving the numbers of the coefficients appearing in the diagonal of the stiffness matrix, and the column height at i degree of freedom can then be simply equal to the difference between i and $i-1$ numbers for the diagonal coefficients. In the above example, the array storing the diagonal terms of a dimension equal to the size of total degrees of freedom can be written as

$$[LD] = [a_1 \quad a_3 \quad a_6 \quad a_9 \quad a_{11} \quad a_{17}]^T. \tag{2.50}$$

After assembling and with the information of forces, the displacement can be calculated by a three-stage process (Fellipa, 1975) as

$$\begin{array}{ll} \text{Factorisation} & [K] = [U]^T[D_a][U], \\ \text{Forward substitution} & [U]^T[Z] = [F], \\ \text{Scaling} & [y] = [D_a]^{-1}[Z], \\ \text{Back substitution} & [U][u] = [y]. \end{array} \tag{2.51}$$

With the displacement vector solved from Eq. (2.51), the deflections of the structure can be calculated and the element deflections, forces and moments can be computed for member capacity check discussed in the next section.

2.7. Member force determinations

An usual objective for a structural analysis is to compute the forces and moments in the members and thus to check their safety. With the solution for the displacements solved from Eq. (2.51), the forces and moments can be calculated by a direct multiplication of the element stiffness to the element nodal displacements as

$$[P] = [k_e][T][u_e]. \tag{2.52}$$

The twelve element nodal forces and moments can also be calculated by multiplying the transformation matrix $[T]^T$ to the basic element forces in Eq. (2.52) as

$$[F_e] = [T]^T[P]. \qquad (2.53)$$

2.8. Checking of equilibrium

It may be of interest to check whether or not the computed element forces and moments balance the applied external forces. This checking can be carried out by summing all the element forces and moments, transformed to global coordinates and compared with the external loads as

$$[F]_R = \sum_{\text{No. of element}} [L][F]_e,$$
$$[\delta F] = [F]_A - [F]_R, \qquad (2.54)$$

in which $[F]_R$ and $[F]_A$ are the resisting and the applied force vectors, respectively. $[\Delta F]$ is the unbalanced force vector respectively which is actually an error due to the finite precision or truncating error of the computer. If the magnitude of the error force vector is small, the accuracy of the results is sufficient, otherwise the analysis results contain a considerable error. This equilibrium error is suggested to be measured as the ratios of the error force to applied force norms as

$$\frac{[\Delta F]^T[\Delta F]}{[F]_A^T[F]_A} = \text{Accuracy} \leqslant \text{Tolerance}. \qquad (2.55)$$

For results of acceptable accuracy, Eq. (2.55) should be satisfied.

It should be noted that, in a non-linear analysis, the error may not be simply due to the truncation error, but also from the change of structural configurations and material properties, etc. In such cases, iteration may be necessary to dissipate the equilibrium error.

2.9. Examples

The computer program for linear analysis has been shown in the Appendix as a special case for the non-linear analysis program to be discussed in subsequent chapters. It can be seen that the switching of the analysis type can be activated by a parameter, NTYPE, in the program. The linear analysis is essentially a non-linear analysis with the number of iterations for equilibrium set to unity. However, the method of calculating the member lengthening by comparing the nodal coordinates in a non-linear analysis cannot be used in a linear case since no geometry update is carried out in a linear analysis.

2.9.1. The bending moment of a sway portal

A portal of width 16 m, height 6 m and under a vertical point load of 100 kN at beam mid-span and a lateral point load of 10 kN at column top has been analyzed with different values of connection stiffness (Fig. 2.10). The connections between the columns and the beams can be detailed as rigid by full welding, as pinned by angles web cleats and as semi-rigid by top and seat angles or flush end plate. For demonstrative purposes, the stiffness for the semi-rigid connection is related to the beam stiffness as $4EI_b/L_b$ in which I_b is the second moment of area of the beam and L_b is the length of the beam.

The bending moment diagrams for the three assumed beam connection types are plotted in Fig. 2.10. It can be seen clearly that the bending moment for the members are very much

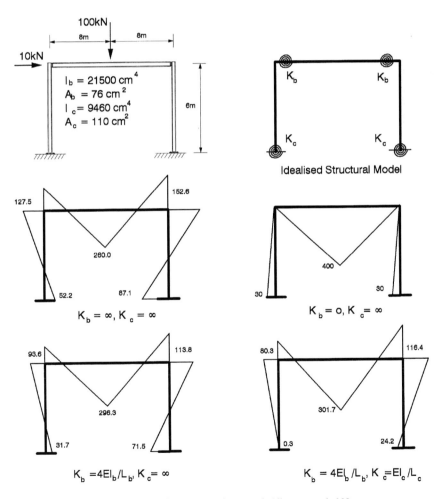

Fig. 2.10. The bending moments in a portal. All moments in kN m.

influenced by the connection details. For the column on the left hand side, even the sign for the bending moment at the column base can be changed by the connection details.

As a rigid base may not be possible in most practical cases due to the finite stiffness for the stanchion-to-foundation connection and may not be allowed in some design codes (BS5950, Part 1, 1990), the column base may be assumed to be semi-rigid of connection stiffness equal to EI/L. The case for the semi-rigid connection at stanchion-beam connection and at column-to-foundation connection is also studied and shown in Fig. 2.10. Again, the variation of the bending moment in the frame is significant. The response of the portal for other combinations of connection stiffness will be left to the readers as exercises.

2.9.2. Analysis of a six-storey frame

The six storey frame with rigid joints has been proposed by Vogel (1985) as a calibration example for checking the accuracy of an ultimate strength analysis. The frame is used here for comparing differences between the semi-rigid and the rigid connection assumption in terms of bending moment in beams and columns. The total height of the frame is 22.5 m and its width is 12 m. No bracing is provided to the frame. The sections used for the members are shown in Fig. 2.11, together with the material properties. The uniformly

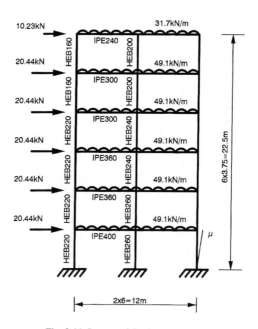

Fig. 2.11. Layout of the 6-storey frame.

Chapter 2. *First-order linear elastic analysis for rigid and semi-rigid frames* 51

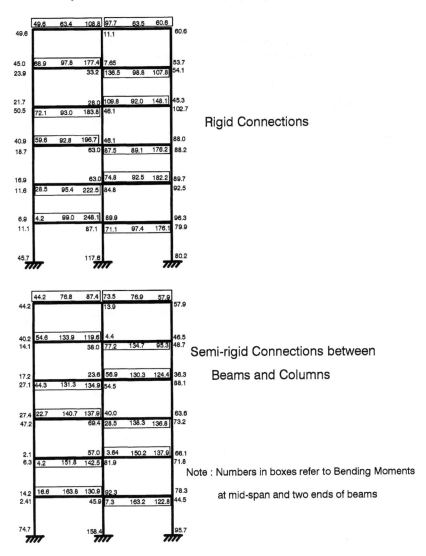

Fig. 2.12. Bending moments of the 6-storey frame by a linear analysis.

distributed load is modelled as a set of equivalent point loads. Four elements are used to model each beam and one element is used for the column. The connection stiffness used in the semi-rigid frame analysis is 12,430 kN m/rad which is close to the initial stiffness of a flush end plate connection. Only the beam-to-stanchion connection is assumed to be semi-rigid.

The bending moments at beams and columns are shown in Fig. 2.12. The discrepancy between the rigid and the semi-rigid joint cases is significant. For beams in semi-rigid joint case, the mid-span moment is larger when compared to the rigid joint case. However, the end moment may be smaller and therefore, more economical since the part of the bending moment at two ends of a beam has been distributed to its mid-span. Nevertheless, the top storey and right hand corner deflection for the rigidly jointed case is 1/308 of the storey height when compared with the deflection in the semi-rigid joint case of 1/150 of storey height. This magnitude may already violate the national code of practice requirements and result in damage to other building elements such as cladding and finishes. In practice, the deflection can be reduced by the use of bracing members or core/shear walls and this should make the semi-rigid connection more readily adopted in actual design.

2.10. Conclusions

The basic concept and formulation for linear elastic analysis is introduced in this chapter. The analysis procedure is simple and can be very effective in analyzing steel frames with various connection stiffnesses. The bending moment diagrams constructed by the method show that the difference between a rigid and a semi-rigid frame is significant. An analysis allowing for the semi-rigid connections involves only a few additional data defining the type and stiffness of the connections at the two ends of a beam and the computing time for such an analysis is also minimal and acceptable when using a powerful personal computer.

Eurocode 3 provides a guideline for the design of steel frames allowing for semi-rigid connections. Checking of connections concentrates on strength, stiffness and ductility and can be carried out easily by computer analysis. However, though the connections between beams and columns have been tested and reported, their studies are focused on the connections between the major principal axes of beams and stanchions. Beams connected to the web of a stanchion, which is bent about its minor principal axis, appears to have not received sufficient attention by researchers. This prohibits the direct analysis of a three dimensional semi-rigid frame unless an additional assumption is made with regard to the stiffness of this connection type or to idealize the structure as a series of planar frames tied to one other.

References

Baigent, A.H. and Hancock, G.J. (1982): Structural analysis of assemblages of thin-walled members. *Eng. Struct.* **4**(3), 207–216.
Bathe, K.J. (1982): *Finite Element Procedures in Engineering Analysis*, Prentice-Hall, Englewood Cliffs, NJ.
British Standards Institution (1990): BS5950, Structural Use of Steelwork in Buildings. Part 1, Code of practice for design in simple and continuous construction: hot rolled sections.
Chen, W.F. and Toma, S. (1994): *Advanced Analysis of Steel Frames; Theory, Software and Applications*, CRC Press, Boca Raton, FL.
Eurocode 3 (1993): *Design of Steel Structures*, Draft for Development.
Felippa, C.A. (1975): Solution of linear equations with skyline-stored symmetric matrix. *Comput. Struct.* **5**(1), 13–29.
George, A. and Liu, J.W. (1981): *Computer Solution of Large Sparse Positive Definite System*, Prentice Hall Series in Computational Mathematics, Englewood Cliffs, NJ.
Iron, B.M. (1970): A frontal solution program for finite element analysis. *Int. J. Num. Methods Eng.* **2**, 5–32.

Kishi, N. and Chen, W.F. (1986): Data base of steel beam-to-column connections, Structural Engineering Research Report CE-STR-86-26, School of Civil Engineering, Purdue University, West Lafayette, IN.

Lahey F77L-EM/32 (1992): *Fortran Program Compiler*, Program Reference, Lahey Computer System.

Load and Resistance Factor Design (LRFD) (1986): Specification for Structural Steel Building, American Institute of Steel Construction (AISC), Chicago, IL.

Lui, E.M. and Chen, W.F. (1987): Effects of joint flexibility on the behavior of steel frames. *Comput. Struct.* **26**(5), 719–723.

Meek, J.L. (1971): *Matrix Structural Analysis*, McGraw-Hill, New York.

Sharman, P.W. (1985): Analysis of structures with thin-walled open sections. *Int. J. Mech. Sci.* **27**(10), 665–677.

Washizu, K. (1975): *Variational Methods in Elasticity and Plasticity*, 2nd edition, Pergamon Press, New York.

Vogel, U. (1985): Calibrating frames. *Stahlbau* **54**, 295–301.

Chapter 3

SECOND-ORDER ELASTIC ANALYSIS BY THE NEWTON–RAPHSON METHOD

3.1. Introduction

During linear analysis of a structure, the force is assumed linearly proportional to the displacement and the principle of superimposition can be applied to obtain the final bending moment and force diagrams. This feature leads to much convenience in practical analysis and design. The non-linear response of a structure, i.e. buckling, will be checked by the design formulae in national codes or standards. This superimposition technique cannot be applied to a non-linear analysis because the structural response is affected by the interaction between loads and deformations and they cannot, therefore, be isolated from one another.

Generally speaking, linear analysis over-simplifies the actual behaviour of a structure, especially at its limit state. For some particular types of structure including cable-stayed masts and scaffolding, the degree of non-linearity is so severe that linear analysis cannot be used as a reliable tool for understanding the actual behavior of the structure. For general structures when the load is sufficiently large and close to the ultimate capacity of a structure, the linear assumption no longer holds.

For geometrically non-linear problems, two sources of non-linearities are considered. They are the P-Δ and the P-δ effects discussed in Chapter 1. Their consideration in an analysis requires special procedures at formulation and programming levels. The P-Δ effect is due to geometrical change of the structure and the P-δ effect is the result of the change of member stiffness in the presence of axial force in the member and the deflection along the member length.

Whilst numerical solution methods traversing the limit point will be discussed in detail in the next Chapter, it is considered to be necessary to describe the scheme for a general non-linear analysis based on the Newton–Raphson procedure in order to allow the reader to fully and easily understand the procedure of a second-order and non-linear analysis. The Newton–Raphson method of analysis also gives the structural response exactly at the input load level, taken as the ultimate or serviceability design loads, which is the most frequently sought response for practical design. Other advanced solution methods in next Chapter do not possess this feature. In fact, the understanding of this simple non-linear procedure is invaluable in visualizing how a more general non-linear analysis works.

For static non-linear analysis, the presence of forces or loads change the structural geometry and member stiffness. Before the solution for the displacement in the equilibrium equations is available, it is not possible to give exactly the displacement and, thus, the

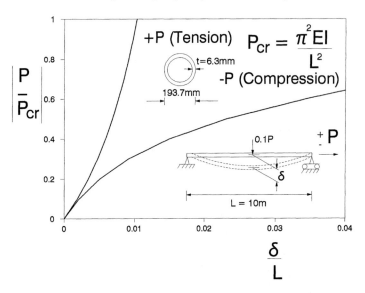

Fig. 3.1. Central deflection of a beam-column under tensile and compressive axial force.

final geometry which is distorted by the displacements. As the element stiffness matrix is required to be transformed to global coordinates, this unknown structural geometry will result in the error for the element stiffness which, in the global coordinate system, depends on the transformation matrix $[L]$ in Eq. (2.12), Chapter 2. This type of non-linearity is defined as the $P\text{-}\Delta$ effect.

Another source of non-linearity is due to the change in element stiffness in the presence of axial force and the deflection along an element, defined as the $P\text{-}\delta$ effect. The variation of element stiffness coefficients have been plotted in Fig. 1.11, Chapter 1. It can be seen from the Fig. 1.11 that the coefficients vary with the direction as well as the magnitude of the axial load. Due to variation of element stiffness as indicated in Fig. 3.1, a member is more difficult to bend when it is in tension than when it is in compression.

To accurately model the behaviour of a structure allowing for all sources of non-linearity is complicated. Even when only the buckling effects are considered, the load versus deflection path allowing for all types of buckling modes such as Euler's column buckling, local plate buckling, lateral-torsional beam buckling and torsional buckling is not easy to plot. In theory, one can employ a refined and rigorous finite shell element computer program for analysis of a beam or even a frame to cater for all kinds of instability effects. However, the approach is difficult to use in practice due to the excessive computer time and data preparation effort. In addition, the assumption for various boundary conditions is tedious to model in the finite shell element mesh and this complex approach does not carry strong physical meanings to the design engineer for his interpretation. On the other hand, when using the simplified computer approach based on the stiffness frame analysis method, some of the buckling modes may be missed or convergence may not be obtained. In design codes for steel structures, this type of lateral or local plate buckling is not considered in the analysis part, but accounted for in the member capacity check such as using bracing to prevent lateral-torsional buckling and adopting compact sections to ensure no early local buckling. Once these buckling modes are prevented, their influence to the global structural behaviour can be ignored.

For members under high axial force, it is more difficult to completely eliminate the effects of instability by increasing the member size or using closely spaced bracings. This is due to economical reasons and structural and architectural functional requirements. The P-Δ effect resulting from building sway can be reduced by bracing, but cannot be completely eliminated. Columns and struts are commonly subjected to heavy axial compression, although it is better, but not always possible, to revert the member force from compression to tension using innovative framing systems such as hanging columns. Whilst design of columns under tensile load is relatively simple, more effort is needed for design of columns and struts under compression, either wholly in compression for columns or partially in compression for beams. In current practice, the instability of struts and columns is prevented by checking the member strength against the Perry–Robertson formula, which, unfortunately, requires an approximation of effective length by engineering judgement from other manual methods. This effective length is assessed by the design engineer and is only an approximate and convenient means of estimating the buckling resistance of a compressive member. In some cases it may be difficult to accurately assess the effective length factor. For example, the bow-shaped truss shown in Fig. 3.2 carrying glazing panels is supported at both the top and bottom. When under suction wind pressure, the curved back chord is under compression and the determination of the effective length for the chord buckling out of plane is not easy by manual inspection. This task, however, becomes trivial when using the second-order analysis (Chan, Sun and Koon, 1999).

The purpose of this Chapter is to introduce the method for second-order elastic analysis of framed structures. The illustration through discussion in this Chapter, and examples presented herewith, should provide a clear picture of the fundamental principles of the non-linear numerical analysis procedure for a structural system. This forms the basis for further understanding of the more advanced non-linear analysis method detailed in the next Chapter. The results will automatically include second-order effects due to change of geometry and variation of element stiffness in the presence of axial force. When allowing for this second-order effect, the calculated deflections and forces and moments in the frame will be more accurate than the linear analysis. Empirical methods of amplifying the moment in the presence of axial force used in various national codes becomes unnecessary and the effective length is also not required to be assumed.

In the context of buckling analysis, two fundamental assumptions are normally made on the inclusion of the buckling effect; a perfectly straight column under purely concentric axial force will exhibit a bifurcation type of behaviour and it does not deform laterally until the bifurcation load is reached at which the column deflects laterally infinitely at no increase of load. The vanishing of the stiffness of the structure at this instance is due to the cancellation of the linear stiffness matrix by the geometric stiffness matrix. This approach to the bifurcation type of analysis is hardly encountered in practice, but represents a simplified and idealized buckling analysis (see Fig. 1.12). In the second approach allowing for change of geometry, the structural geometry is continuously updated and the load versus deflection response of the structure is traced. This method reflects more closely the actual behaviour of a structure and is termed as the load versus deflection analysis.

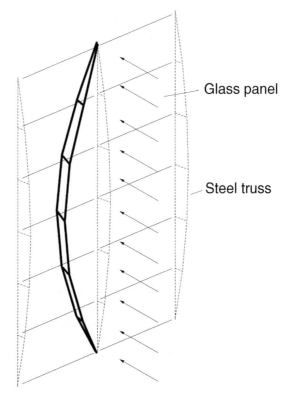

**Suction wind making back chord
in compression. What is its effective length ?**

Fig. 3.2. Buckled shape of a bow-shaped truss.

As it can be used to study a wide range of non-linear problems including geometric and material non-linearities, this approach is used and described in this book.

3.2. Bifurcation analysis versus load–deflection analysis

Bifuraction analysis is based on a mathematical idealization that the structure deforms only in the loaded directions until the bifurcation load is reached. As a result, the element stiffness is only a function of force in the structural members and other basic material and geometrical properties before loaded. The tangent stiffness relating incremental displacements to incremental loads will be independent of the displacement and the structure is assumed to deform only in the loaded direction so the effect of geometrical change is unimportant. Under these assumptions, the structure will be termed as perfect and the loads are exactly concentric without components in other directions. The condition for the determination of the bifuraction load will then be given by the following characteristic equation:

$$|K_L + \lambda_{cr} K_G| = 0, \tag{3.1}$$

in which K_L is the linear stiffness matrix, K_G is the geometric stiffness matrix and λ_{cr} is the load factor causing the determination of Eq. (3.1) to vanish.

The validity of the solution for the buckling load of a structure is based on the assumption that the structure is perfect. For the case of a column, it is then perfectly straight and the axial force is ideally concentric. Strictly speaking, Eq. (3.1) is a mathematical idealization that can hardly be achieved in practice. All structures contain a certain degree of initial imperfection and the loads are not perfectly concentric so that deflections, however small, would occur in directions other than the loading plane. In general, the result from bifurcation analysis should not be used directly for practical design and it only provides an upper bound solution to the elastic buckling load of a structure. The advantages of this approach are the shorter computational time and simpler formulation. Solution time for the bifurcation load, λ_{cr}, in Eq. (3.1) is a fraction of the load versus deflection plot described in the next section. Determinant search techniques such as the methods of bisection, interpolation and subspace iteration can be used to solve Eq. (3.1) and the text by Bathe (1982) gives a detailed account of various numerical procedures for the solution of eigenvalue problems in the finite element context.

To predict more accurately the response of a structure under load, the load versus deflection response analysis is needed. In this approach, the equilibrium path for the structure is traced and the stresses and deflections are continuously updated and monitored. Material yielding can be considered or suppressed in the analysis by using the tangent modulus. This approach, however, consumes longer computational time and may involve divergence problems and more complicated formulation and programming efforts.

The major formulation difference between the load versus deflection analysis and the bifurcation analysis is the inclusion of large displacement effects in the stiffness equation. The influences due to initial stress and deformed geometry of the structure are allowed for in the stiffness equation. When using the current value of Young's modulus to cater for material yielding, the approach can further be employed for the prediction of collapse load of a structure allowing for material non-linearity. There is no doubt that this method is more accurate in predicting the actual behaviour of a structure. Complicated formulation and programming effort and heavier computer time are becoming less important factors due to the availability of low-priced and fast-speed computers. In view of this, this book will focus on this load versus deflection approach.

3.3. General solution methods for non-linear analysis

In general, the difficulty for a non-linear analysis is a result of the inter-dependence of the stiffness, the load and the displacement as follows:

$$[F] = [K(F, u)][u]. \tag{3.2}$$

In contrast to a linear analysis, the stiffness matrix in a non-linear analysis depends on the displacement as well as the force. The applied force will create displacement which alters the geometry and thus the structural resistance against external loads. Using the force and an initial stiffness matrix for computation, the displacement is obtained which, together with the applied force, changes the stiffness matrix and invalidates the results for the previous analysis. In order to arrive at a solution which satisfies Eq. (3.2), a linearized incremental approach, a trial and error process or a combined incremental-iterative scheme is needed. Generally speaking, a direct iterative method, a pure incremental method or a predictor-corrector procedure can be used for the solution of Eq. (3.2).

3.3.1. The direct iterative method

In this method, the stiffness is assumed and, based on the applied force, the displacement is computed which is then used to re-calculate the stiffness and to re-compute the new displacement. This process is continued until the input data is very close to the output result in which case convergence is achieved. The graphical representation of this process is shown in Fig. 3.3. Mathematically, these steps can be summarized as follows:

$$[u]_{i+1} = [K_s(u_i)]^{-1}[F]_{i+1}. \qquad (3.3)$$

This process is continued until the next and the last displacement are equal or very close to each other in which case the stiffness has been correctly assumed and the dis-

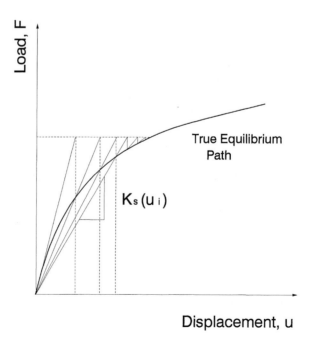

Fig. 3.3. The direct secant iterative method.

placements and resisting forces have been correctly calculated. This condition can be stipulated as $u_{i+1} \approx u_i$. The technique of using only the secant stiffness relations was adopted by Hancock (1974), in conjunction with the Southwell plot, for determination of bifurcation load of beams. Chen and Atsuta (1977) used the direct secant iterative scheme with the influence coefficient method for second-order analysis of tubular members. A reasonable estimate of the deflected shape must be assumed otherwise convergence may not be achieved in this method. Papadrakakis (1981) employed the dynamic relaxation scheme for the second-order and large deflection analysis of trusses. In this method, an appropriate selection of time step, damping coefficients and mass coefficients is needed. All these schemes require only a set of secant stiffness relations relating the total forces and total displacements for their solutions and their convergence rate is generally slow. However, the method has an advantage of simplicity in using only the secant stiffness and the complex formulation of the tangent stiffness matrix is not needed.

3.3.2. The pure incremental method

To approximate a curve, one naturally thinks of using a series of segments. The smaller the size for these segments, the closer the approximation will be, but a heavier computational effort is required. The pure incremental method is based on a similar concept of splitting the total load by a series of small load increments, as shown in Fig. 3.4. The procedure can be summarized mathematically as

$$
\begin{aligned}
&[\Delta u]_{i+1} = [K(u_i)_\text{T}]^{-1}[\Delta F]_{i+1}, \\
&[u]_{i+1} = [u]_i + [\Delta u]_{i+1}, \\
&[F]_{i+1} = [F]_i + [\Delta F]_{i+1}.
\end{aligned}
\tag{3.4}
$$

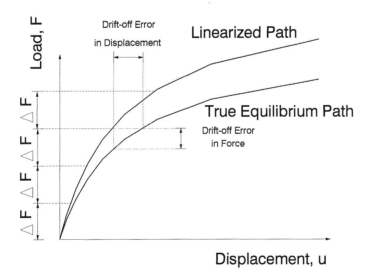

Fig. 3.4. Pure incremental method with constant load increments.

In contrast to the direct iterative method, this procedure requires only the tangent stiffness and the secant stiffness is not needed. This tangent stiffness must be able to accurately relate the incremental displacement to the incremental force and the accuracy of the method depends on the accuracy of the tangent stiffness as well as the size of the load step. This procedure, in principle, can trace the equilibrium path in pre- and post-buckling ranges. The increase or decrease of load can be monitored by the sign of the determinant of the tangent stiffness matrix. It is simpler than the combined incremental-iterative scheme described below since the iterative stage is skipped. However, the method requires the load step to be sufficiently small for an accurate solution of highly non-linear problems and the numerical drift-off error is unknown. Two trial runs with different load size are normally required to compare the discrepancy for the solution paths due to a reduction in load size. This convergence checking is obviously expensive, tedious and inconvenient for routine uses.

3.3.3. Combined incremental–iterative method: the Newton–Raphson procedure

The secant iterative method is more accurate since it satisfies the equilibrium condition whilst the pure incremental method progresses along the equilibrium path without the divergence problem. Their combined use, therefore, is a sensible choice in improving the computational efficiency and accuracy of a non-linear analysis. In this scheme, the tangent stiffness is used as a predictor for estimation of the displacement increment due to a small but finite force. The secant stiffness is then used as a corrector of equilibrium. If the error is too large or unacceptable, a further iteration is activated. This relatively complicated scheme requires more effort in computer programming but, generally speaking, the improvement in computational efficiency and accuracy well deserves this additional work. Under the Newton–Raphson scheme, the tangent stiffness can be reformed at every iteration as the original or conventional version (Fig. 3.5(a)) and, when the tangent stiffness is reformed at only the first iteration, it is termed the modified Newton–Raphson method (Fig. 3.5(b)).

In this book, the above scheme is used and discussed in full detail. For the most frequently sought solution related to the analysis of a structure, the load control Newton–Raphson method with constant load control is used to check the element or member forces and moments and to determine the displacement for the structure at the ultimate and serviceability design load. This is the typical and most common objective of an analysis in a design office. Additional check on the collapse behaviour and load may be followed when the scheme of the structure is finalised so that the ultimate and collapse behaviour can be evaluated. The corresponding more advanced analysis method for traversing limit points will be discussed in Chapter 4.

3.3.4. The physical concept of the Newton–Raphson method

It is important for anyone involved in non-linear analysis to fully understand the Newton–Raphson procedure in order to carry out a correct non-linear analysis reflecting the true response of a structure. This understanding becomes more important if the reader wishes to code or to modify a computer program for non-linear analysis to suit his particular need.

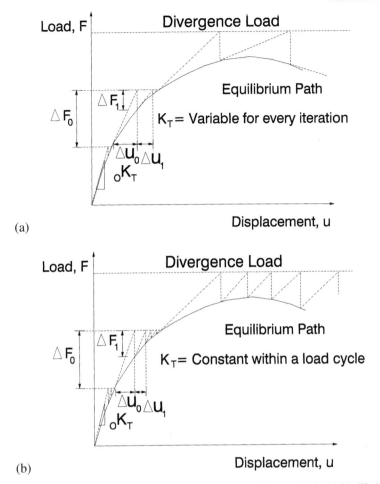

Fig. 3.5. The Newton–Raphson method. (a) Conventional Newton–Raphson method; (b) Modified Newton–Raphson method.

To obtain an insight into the non-linear phenomenon and the Newton–Raphson numerical method, it is best to make reference to a physical problem and to study how it behaves and the scheme used in tracing the equilibrium path. Consider a slender beam-column with both ends pin-supported and of very small bending stiffness but a large axial stiffness as shown in Fig. 3.6, being subjected to a point load at mid-span. For simplicity, the material making up the column is assumed to have an infinite yield stress so that only geometrical non-linearity is considered. This assumption simplifies significantly the problem for the purpose of clarity. At the onset of load application, the stiffness of the member against the point load is due to the bending stiffness and it is therefore weak. Using the first estimated central deflection, the new length of the member can be computed and the tension force in the member is determined. The tension due to the change in length may be

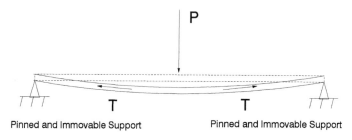

Fig. 3.6. Cantenary action of a largely deformed slender beam-column.

so large that the vertical component exceeds the applied load and the equilibrium condition is violated. This condition is unacceptable and therefore the unbalanced force, equal to the difference between the applied and resistant force, will be substituted back into the analysis for computation of a corrective deflection increment. In the iterative process, a deflection increment due to the unbalanced force will be added to the nodal coordinates of the structure for correction of the structural geometry in response to the unbalanced force and the value for total deflection. Based on this information, the condition of equilibrium is checked again and a further unbalanced force is calculated. The iterative procedure is continued until the unbalanced force is very small in which equilibrium is considered to have been achieved.

It should be noted that, at the first instance upon the application of the set of loads, the stiffness of the structure is formed on the basis of the *undeformed* geometry of the structure and the displacement is solved, as represented by $_0k_T$ in Fig. 3.5. In the current example, this stiffness comes solely from the bending stiffness and the axial stiffness has no contribution as the member is perpendicular to the applied point load. If the structural behavior is linear, or very close to it, the first prediction will be sufficient and can then be used for force, displacement, moment and finally stress computation. For small deflection problems, the set of forces and moments in the structure after transformation will balance the applied force so that equilibrium can be satisfied. However, when the structure behaves non-linearly, the force and moment resistance calculated from the displacement cannot balance the applied loads. Therefore, the displacements for the complete structures should finally be calculated as

$$[\Delta u]_i = [K_T]^{-1} [\Delta F]_i, \tag{3.5}$$

in which $[\Delta u]_i$ is the displacement increment due to the load increment $[\Delta F]_i$ and $[K_T]$ is the tangent stiffness. $[\Delta F]_i$ is the incremental load at the initial iteration (i.e. $i = 0$) or the unbalanced force for the subsequent iteration onwards (i.e. $i \geqslant 1$). Note that for the conventional Newton–Raphson method, the tangent stiffness is reformed every iteration and, for the modified Newton–Raphson method, it is only reformed in the first iteration. The conventional Newton–Raphson method normally takes fewer iterations for convergence but the computer time for each iteration in the modified Newton–Raphson method is shorter since the solution and formation of the tangent stiffness matrix is a time-consuming process. Furthermore, the full Newton–Raphson method generally has a higher chance of convergence. The combined use of these two approaches of reforming the tangent stiffness for every, say, three iterations, takes advantage of these two methods. However, the most economical number of iterations for reforming of tangent stiffness is problem dependent and cannot be over-generalized.

The incremental element nodal displacement will be then calculated by extracting corresponding displacement degrees of freedom and transforming to the element local axis as

$$[\Delta_e u]_i = [T][L][\Delta \bar{u}]_i, \qquad (3.6)$$

in which $[\Delta_e u]_i$ is the element nodal displacement vector of dimension 6 in the element local axis, $[\Delta \bar{u}]_i$ is the extracted displacement vector for the two nodes of an element in the global coordinate axis and $[L]$ is the global to local transformation matrix.

The element forces and moments will then be determined by the element stiffness equation as

$$[\Delta_e R]_i = [k_e][\Delta_e u]_i, \qquad (3.7)$$

in which $[\Delta_e R]_i$ is the incremental change of element forces and $[k_e]$ is the element stiffness matrix. The total or accumulated forces for the element can be further updated as

$$[_e R]_i = \sum_i [\Delta_e R]_i$$
$$= [_e R]_{i-1} + [\Delta_e R]_i. \qquad (3.8)$$

Transforming the element forces and moments from the local to the global axes, the resistance of the complete structure can be assembled and computed as

$$[R]_i = \sum_{NELE} [_e R]_i, \qquad (3.9)$$

in which *NELE* is the number of elements in the system.

Comparing the resistance against the applied load, the following equilibrium equation for unbalanced forces, $[\Delta F]_{i+1}$ is generated:

$$[K_T][\Delta u]_{i+1} = [\Delta F]_{i+1} = [F]_i - [R]_i. \qquad (3.10)$$

The total displacement will be accumulated and updated simply as

$$[u]_{i+1} = [u]_i + [\Delta u]_{i+1}. \qquad (3.11)$$

Graphically, the error is illustrated in Fig. 3.5 as $[\Delta F]_i$. To eliminate the error, it is necessary to revise the displacement in the structure and to re-check the equilibrium condition. In the Newton–Raphson method, the error in force, $[\Delta F]$, is calculated in Eq. (3.10) and then substituted back into the analysis for displacement increment determination. The unbalanced displacement, $[\Delta u]_{i+1}$, at this instance will then be added to the last total displacement to obtain the updated displacement, $[u]_{i+1}$ in Eq. (3.11).

As shown in Fig. 3.5, this method assumes a small load increment parallel to the applied load and the corresponding displacement increment is then computed and accumulated to the total displacement and updating of geometry. Based on the updated and new geometry and the forces, the resistance of the structure is then determined and compared with the applied load. When this unbalanced force or equilibrium error is too large, it will be substituted back into the incremental equilibrium Eq. (3.10) for calculation of another displacement increment and the process is repeated until the equilibrium error is sufficiently small. This convergence check is carried out using the following conditions:

$$[\Delta u]^T[\Delta u] < TOLER * [u]^T[u],$$
$$[\Delta F]^T[\Delta F] < TOLER * [F]^T[F]. \qquad (3.12)$$

After satisfaction of equilibrium, another new load increment is imposed to the structure and the new configuration is sought by repeating the previous process.

The flow chart for the iterative process shown in Fig. 3.7 represents a standard procedure used in second-order analysis. Although the concept for this Newton–Raphson analysis procedure is essentially the same when used by different researchers, the element stiffness, i.e. the expression for the tangent stiffness matrix and the secant stiffness and the method of calculating the element displacements and rotations vary considerably among these works. For example, the element stiffness matrix derived by the stability function (Oran, 1973a, 1973b), the cubic Hermite function (Jennings, 1968), the self equilibrium polynomial function (Chan and Zhou, 1994), the weak formulation element (Kondoh and Atluri, 1986), the higher order element (Izzuddin and Elnashai, 1993) and many others are all different, not only in the method of derivation but also their accuracy and efficiency. A further study may reveal that even under the name of stability function, there may still be different versions. Majid (1972), Livesley (1964), Chen and Lui (1991) and Oran (1973a, 1973b) used varied forms for the analyses.

In calculating element deflections, a number of formulations are available which include the total secant stiffness method used by Chen and Lui (1991) and Chan and Kitipornchai (1987), the joint orientation method by Oran (1973a, 1973b) and the incremental secant stiffness method by Chan (1989). The advantages and limitations of these formulations have been discussed by Chan (1992).

As the objective of this book is not to investigate the characteristics, merits and limitations of element and formulation, only the methods used by the authors and in the computer program GMNAF are described. The procedure was found to be computational stable, robust, reliable and efficient for most engineering structures. This not only saves time for the reader in understanding a non-linear analysis, but also makes the discussion clearer.

Step 1	Initialisation of variables and parameters.
Step 2	Input first incremental load, boundary condition, connectivity, structural geometry and material properties.
Step 3	Formation of the element tangent stiffness matrix $[_e k_T]$.
Step 4	Transformation from local to global axis $$[k_T] = [L]^T [_e k_T][L].$$
Step 5	Assembling to form the tangent stiffness matrix $[K_T]$ $$[K_T] = \sum_{i=1}^{NELE} [k_T].$$
Step 6	Solving for displacement increment $$[\Delta u]_{i+1} = [K_T]^{-1} [\Delta F]_{i+1}.$$
Step 7	Accumulation of displacement increment to total displacement $$[u]_{i+1} = [u]_i + [\Delta u]_{i+1}.$$
Step 8	Updating of geometry $$[x]_{i+1} = [x]_i + [\Delta u]_{i+1}.$$
Step 9	Calculation of element displacements by extracting from the global displacement vector and transforming to the local element coordinate axes $$[\Delta u_e]_{i+1} = [T][L][\Delta u]_{i+1}.$$
Step 10	Computation of element force $$[\Delta_e R]_{i+1} = [k_e][\Delta u_e]_{i+1}.$$
Step 11	Accumulate the element force vector to obtain the final element force vector and transform to global coordinate for resistance force $$[_e R]_{i+1} = [_e R]_i + [\Delta_e R]_{i+1},$$ $$[R]_{i+1} = [L]^T [T]^T [_e R]_{i+1}.$$
Step 12	Comparison with the applied forces, the unbalanced forces are determined $$[\Delta F]_{i+1} = [F]_{i+1} - [R]_{i+1}.$$
Step 13	Check for equilibrium. If satisfied, apply next load increment or stop. Otherwise go to Step 3 for conventional Newton–Raphson method or Step 6 for modified Newton–Raphson method. Check: $$[\Delta F]^T [\Delta F] < TOLER * [R]^T [R],$$ $$[\Delta u]^T [\Delta u] < TOLER * [u]^T [u].$$

Fig. 3.7. Flow chart for second-order analysis using the Newton–Raphson procedure.

3.4. The tangent stiffness matrix

It becomes apparent from the flow chart in Fig. 3.7 that the tangent stiffness matrix must be formulated and used for prediction of displacement increments. The expressions in Eqs. (3.6) to (3.12) are based on predicted displacement increments for the computation of resisting forces. The determination or, more exactly, the estimation of these incremental displacements will be carried out by another set of relations between incremental forces and displacements, namely the tangent stiffness matrix. Although it can be said that inaccurate predictions of displacements may not directly lead to an error for the solution in a path-independent problems, the gross error may result in numerical divergence or slow convergence. In addition, the tangent stiffness will be used directly for prediction of bifurcation load in a bifurcation type of analysis in which case the accuracy of the tangent stiffness matrix directly determines the accuracy of the buckling loads. In recent studies for exact location of limit points (Chan, 1993; Fujii and Okazawa, 1997) the accuracy of the tangent stiffness becomes essential.

In principle, the tangent stiffness should be formed on the basis of the most updated information for the structure and also the work done due to incremental change in displacements, both the first and the higher order terms, and initial force in the element should be included in its formulation. By taking the second variation of the energy functional, it can be seen that the tangent relationship between the incremental force and displacement is related by the geometric stiffness matrix, the large displacement matrix and the linear stiffness matrix (Zienkiewicz, 1977) as

$$k_T = k_L + k_G + k_O, \tag{3.13}$$

in which k_T is the tangent stiffness matrix, k_L is the linear stiffness matrix, k_G is the geometric stiffness matrix and k_O is the large displacement matrix.

The geometric stiffness matrix accounts for the work done by the initial stress or force and the incremental displacements. It can be obtained as the product of the initial stress and the second-order displacement as

$$k_G = \frac{P}{2}\left[\int_0^L \left(\frac{\partial v}{\partial x}\right)^2 + \left(\frac{\partial w}{\partial x}\right)^2 dx\right]. \tag{3.14}$$

The large displacement matrix caters for the change in geometry. This can be evaluated explicitly (Mallet and Marcal, 1968; Clarke and Hancock, 1990), or, implicitly by continuous updating of the structural geometry (Chan, 1993; Meek and Tan 1984; and Lui and Chen, 1986). The linear stiffness matrix has been explained previously in Chapter 2. Combining these three matrices, written in a combined form and transformed to the global coordinate system, we have the following expression:

$$K_T = \sum_{j=1}^{NELE} k_T = \sum_{j=1}^{NELE} [L]([T][k_e][T]^T + [N])[L]^T, \tag{3.15}$$

Chapter 3. Second-order elastic analysis by the Newton–Raphson method

in which $[T]$ is a transformation matrix relating the six element forces and moments to the twelve global forces and moments and given in Eq. (2.36). $[N]$ is the matrix to account for the work done by the initial forces and the nodal displacements given by the following expression:

$$N_{ij} = \frac{P}{2} \int_L \left(\frac{\partial v}{\partial x}\right)^2 dx. \tag{3.16}$$

Substituting

$$v = \left(1 - \frac{x}{L}\right)v_1 + \frac{x}{L}v_2,$$

one obtains

$$N_{ij} = \begin{cases} P/L & \text{for } i = j, \\ -P/L & \text{for } i \neq j. \end{cases} \tag{3.17}$$

Repeating the procedure for displacement in the z-axis, the $[N]$ matrix is finally obtained as follows:

$$\begin{bmatrix}
0 & 0 & 0 & 0 & 0 & 0 & 0 & 0 & 0 & 0 & 0 & 0 \\
  & P/L & 0 & 0 & 0 & 0 & 0 & -P/L & 0 & 0 & 0 & 0 \\
  &   & P/L & 0 & 0 & 0 & 0 & 0 & -P/L & 0 & 0 & 0 \\
  &   &   & 0 & 0 & 0 & 0 & 0 & 0 & 0 & 0 & 0 \\
  &   &   &   & 0 & 0 & 0 & 0 & 0 & 0 & 0 & 0 \\
  &   &   &   &   & 0 & 0 & 0 & 0 & 0 & 0 & 0 \\
  &   &   &   &   &   & 0 & 0 & 0 & 0 & 0 & 0 \\
  &   &   &   &   &   &   & P/L & 0 & 0 & 0 & 0 \\
  &   &   &   &   &   &   &   & P/L & 0 & 0 & 0 \\
  & \text{SYMMETRIC} &   &   &   &   &   &   &   & 0 & 0 & 0 \\
  &   &   &   &   &   &   &   &   &   & 0 & 0 \\
  &   &   &   &   &   &   &   &   &   &   & 0
\end{bmatrix}. \tag{3.18}$$

The work done by the initial force and displacement along the element due to nodal rotations has been included in Eq. (3.14). The displacement of a generic point along the element length due to these nodal displacements can be simply expressed as

$$v_N = v_1\left(1 - \frac{x}{L}\right) + v_2\frac{x}{L},$$
$$w_N = w_1\left(1 - \frac{x}{L}\right) + w_2\frac{x}{L}, \tag{3.19}$$

in which v_N, w_N are the lateral deflections of a generic point and v_1, v_2, w_1 and w_2 are the nodal lateral displacements along the y- and the z-axes, respectively.

3.5. The secant stiffness matrix

To illustrate the use of the numerical method, it is necessary to first derive the tangent stiffness and the secant stiffness matrices. Whilst the semi-rigid jointed element has been formulated previously in Chapter 2, this Section demonstrates the formation of the rigid jointed cubic Hermite beam-column element for sake of simplicity. The objective is to illustrate to the readers the fundamental concept and application of a general numerical procedure for second order non-linear analysis. In addition, for a general study of non-linear problems, the cubic element is sufficient only when the degree of non-linearity is mild or two or more elements will be required to model a member. Bathe and Bolouri (1979) recommended the simple cubic element for large deflection analysis but Chan and Zhou (1994) noted that the element is not of sufficient accuracy when an element is used to model a member which has been practiced by the engineer for decades. For present studies which involve formation of plastic hinges and instability, this simple element shows a good performance in the elastic and inelastic ranges when using two or more elements to model a member.

The energy functional of a beam-column element with reference to the coordinate axes passing through its end nodes and about the principal axes as shown in Fig. 2.3 can be written as

$$\Pi = \frac{1}{2}\int_L EI_y \left(\frac{\partial^2 w}{\partial x^2}\right)^2 dx + \frac{1}{2}\int_L EI_z \left(\frac{\partial^2 v}{\partial x^2}\right)^2 dx$$
$$- M_{y1}\theta_{y1} - M_{y2}\theta_{y2} - M_{z1}\theta_{z1} - M_{z2}\theta_{z2}$$
$$+ \frac{P}{2}\left[\int_L \left(\frac{\partial v}{\partial x}\right)^2 dx + \int_L \left(\frac{\partial w}{\partial x}\right)^2 dx\right]. \tag{3.20}$$

The first variation of the energy functional yields the equilibrium equations. Operating on Eq. (3.20), we have

$$\Delta M_{y1} = \left(\frac{4EI_y}{L} + \frac{4PL}{30}\right)\Delta\theta_{y1} + \left(\frac{2EI_y}{L} - \frac{PL}{30}\right)\Delta\theta_{y2},$$
$$\Delta M_{y2} = \left(\frac{2EI_y}{L} - \frac{PL}{30}\right)\Delta\theta_{y1} + \left(\frac{4EI_y}{L} + \frac{4PL}{30}\right)\Delta\theta_{y2},$$
$$\Delta M_{z1} = \left(\frac{4EI_z}{L} + \frac{4PL}{30}\right)\Delta\theta_{z1} + \left(\frac{2EI_z}{L} - \frac{PL}{30}\right)\Delta\theta_{z2}, \tag{3.21}$$
$$\Delta M_{z2} = \left(\frac{2EI_z}{L} - \frac{PL}{30}\right)\Delta\theta_{z1} + \left(\frac{4EI_z}{L} + \frac{4PL}{30}\right)\Delta\theta_{z2},$$

in which Δ represents the incremental form of the quantities. P is the axial force being positive for tension and negative for compression.

The axial force can be obtained directly from the change in element length as

$$u_x = u_a + u_b, \tag{3.22}$$

in which u_x, u_a and u_b are the total, the bending and the bowing shortenings respectively. u_a can be obtained by subtracting the deformed length by the initial length of the element calculated from the coordinates of the element nodes. u_b is the shortening due to bowing given by

$$u_b = \frac{1}{2}\int_0^L\left[\left(\frac{\partial w}{\partial x}\right)^2 + \left(\frac{\partial v}{\partial x}\right)^2\right]dx$$

$$= \frac{L(2\theta_{y1}^2 - \theta_{y1}\theta_{y2} + 2\theta_{y2}^2)}{30} + \frac{L(2\theta_{z1}^2 - \theta_{z1}\theta_{z2} + 2\theta_{z2}^2)}{30}. \quad (3.23)$$

The incremental form for Eq. (3.23) can be obtained by taking a variation on the equations as

$$\Delta u_x = L_{i+1} - L_i$$
$$+ \frac{L}{30}[4\Delta\theta_{y1}\theta_{y1} - \Delta\theta_{y2}\theta_{y1} - \Delta\theta_{y1}\theta_{y2} + 4\Delta\theta_{y2}\theta_{y2}$$
$$+ 4\Delta\theta_{z1}\theta_{z1} - \Delta\theta_{z2}\theta_{z1} - \Delta\theta_{z1}\theta_{z2} + 4\Delta\theta_{z2}\theta_{z2}.] \quad (3.24)$$

The torsional stiffness can be calculated by directly substituting the torsional twist to the torsional stiffness as

$$\Delta M_x = \left(\frac{GJ + Pr_0^2}{L}\right)(\Delta\theta_{x2} - \Delta\theta_{x1}). \quad (3.25)$$

Combining Eqs. (3.21) to (3.25), the stiffness matrix can be obtained for a beam-column element with six degrees of freedom. The resistance for the element can then be evaluated in an incremental manner as in Eqs. (3.7) and (3.8).

The above Newton–Raphson type of non-linear analysis procedure iterates at a constant load. It is suitable for conventional design where stresses, deflections and response at the design load level are required. The method, however, cannot obtain a solution in some cases when the load level is above the limit point of the structure. Iteration at such a level higher than the limit load will lead to divergence and, in some cases, to a response range beyond the limit point. The overlook of this behavior can be prevented by using sufficiently small load increments which, lead to additional computational time. As a general guide, the load increment should be taken as about 10% of the estimated elastic buckling load so that 10 load increments should be adequate to arrive at the buckling load level. However, from experience (except for highly non-linear structures such as pre-tensioned trusses), using the design load as the first increment does not usually have divergence problems when the section moment is controlled to be under plastic moment of the section for practical design. Nevertheless, the choice of proper load size is problem dependent and requires experience and a trial and error process is normally required.

3.6. Design application

The Newton–Raphson method gives the response of a structure allowing for various second-order effects and at a specified design load level, normally taken as the design load. It can, therefore, be used directly for practical design. To achieve this aim, a second-order analysis will be carried out with allowance for the simulation of the second-order effects such as member initial imperfection, building out-of-plumbness and notional force for which different national codes and standards have their own specified requirements. When these effects are properly simulated, the widely used first-plastic-hinge design can be conducted by the present method. The more complex second-order inelastic analysis is not generally accepted as a routine design tool in the design office, except for design against extreme loads such as seismic forces in which elastic design is costly. Therefore, based on the first-plastic-hinge design philosophy, only the section capacity of each member will be checked. The individual member check, which generally involves an uncertainty in the value for effective length, is not required. The comparison for the procedures of conventional and the proposed design approaches is shown in Fig. 3.8.

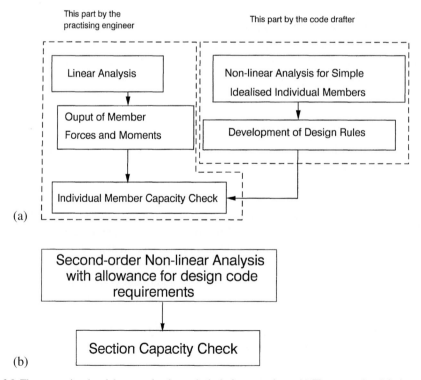

Fig. 3.8. The conventional and the second-order analysis design procedures. (a) The conventional design procedure; (b) The second-order analysis design procedure.

3.7. Examples

Two examples have been selected to demonstrate the application of the Newton–Raphson method in the elastic range of a structure. In the first problem, limitations of the Newton–Raphson method of divergence near buckling are illustrated, showing that numerical failure may not necessarily be an indication of true structural failure. In the second problem, the P-Δ and P-δ effects are studied for a low-rise unbraced steel frame against the result of a linear analysis. This demonstrates the error for a linear analysis in estimating the deflection, moment and axial force of a low-rise steel building at serviceability and ultimate load levels.

3.7.1. Elastic buckling load of a simple strut

In the first example, the simple strut shown in Fig. 3.9 with initial imperfection of 0.1% of its length is under an axial force. The length of the member is 10 m, the outer diameter of the section is 193.7 mm and the wall thickness is 6.3 mm. In the problem, 10% of the Euler's load is imposed as the incremental load and the full Newton–Raphson scheme is used such that the tangent stiffness is reformed every iteration. Four elements have been used to model the complete member.

The maximum converged load is 80% of the Euler's buckling load, demonstrating the limitation of the load control Newton–Raphson method in handling this simple problem. It is not possible, therefore, to rely on the divergence load as the true buckling load of a structure since it may not be sufficiently accurate. The converged buckling load depends on many factors not solely related to the structural properties. This includes the load size used, whether one is using the Newton–Raphson method or its modified form which diverges earlier in most cases, and the convergence tolerance. In addition, the degree of non-

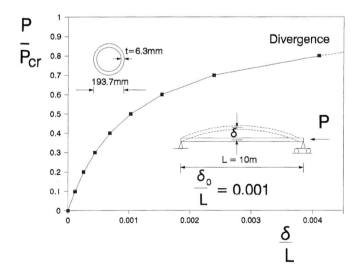

Fig. 3.9. The Newton–Raphson method for a simple strut.

3.7.2. Second-order elastic analysis of a 4-storey moment-resistant frame

In this example, the 4-storey, 5 bays moment-resistant steel frame shown in Fig. 3.10 is under a vertical unfactored working load of 40 kN/m and a horizontal working wind load of 30 kN/m acting on the side column. The total width of the structure is 35 m and the height is 14 m. All joints between members are assumed rigid and connections of columns to supports are pinned. For simplicity, one group of column size and another group of beam size are used. The columns are 168 cm^2 in area and 22,530 cm^4 in second moment of area. For beams, the cross-sectional area is 129 cm^2 and the second moment of area is 61,520 cm^4. The frame is idealized as two-dimensional. The beams are laterally restrained so that lateral-torsional buckling will not occur and the columns are loaded about its major principal axis.

The deflected shape and the results of analyses using the linear assumption and the second-order analysis are shown in Fig. 3.10. The deflections at serviceability load levels for linear and second-order elastic analysis are 40.3 mm and 44.7 mm, respectively. This represents a 11% difference. For ultimate limit state with a load factor of 1.2 for simultaneous action from wind, live and dead loads, the most heavily loaded column, being the second column from the left, indicated in Fig. 3.10, is under an axial force of 1459 kN

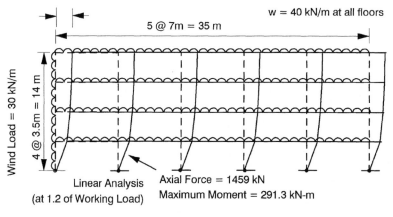

Fig. 3.10. Second-order non-linear analysis of an unbraced 4-storey steel building.

and moment 291.3 kN m when using the linear analysis. The computed axial force and moment by the second-order elastic analysis for this column are 1461 kN and 335.6 kN m. For the axial force, the results between linear and second-order non-linear analysis results is small, but the discrepancy in moment is significant and about 15%.

When using the linear analysis in practical design, the engineer would allow for the second-order effect via some additional reduction in column strength using the concept of effective length or K-factor, which are difficult to assess accurately. This procedure is more complicated, tedious and inaccurate since, in general, the presence of axial forces in members affects their stiffness which changes the distribution of force and moment in a structural system. This effect cannot be simulated accurately in a linear analysis.

It can be seen that the present second-order elastic analysis may not fulfill all the requirements for the "Advanced Analysis" which refers to an analysis capturing the behavior of a structure so that individual member checks are not necessary. However, when the effects of member initial imperfection, out-of-plumbness for the frame etc. are modelled in the analysis through the use of a series of segments for modelling initially curved members and/or notional forces to model any initial imperfection for frames and members, the analysis output can then be directly used for section capacity check instead of the conventional approach of designing members individually. It can be seen that these considerations can be easily incorporated into the present method by assigning appropriate nodal coordinates and/or with application of notional forces.

3.8. Conclusions

This Chapter introduces the fundamental concept and computational scheme of non-linear analysis. Several basic approaches of iteration using only the secant stiffness, of pure incremental approach using only the tangent stiffness and the generally most versatile, effective and efficient combined incremental-iterative scheme based on the Newton–Raphson procedure are discussed. The simple second-order elastic analysis of frames and beam-columns has been employed for demonstration of the application of the load control Newton–Raphson method. Divergence when the method approaches the buckling load is demonstrated in a simple example, and determination of deflection, axial force and moment in a low-rise building frame is carried out by the method. Through these examples, the concept of the most general non-linear analysis method commonly used in conjunction with the finite element or matrix stiffness method of analysis is introduced. The unique feature of this Newton–Raphson method over other methods is its ability to estimate the structural response at a specified load level, usually taken as the serviceability or ultimate design load. This is the most frequently sought solution for practical design. The program contained in Appendix B is capable of conducting the described analysis.

References

Bathe, K.J. (1982): *Finite Element Procedures in Engineering Analysis*, Prentice-Hall, Englewood Cliffs, NJ.
Bathe, K.J. and Bolourchi, S. (1979): Large deflection analysis of three dimensional beam structures. *Int. J. Num. Methods Eng.* **14**, 961–986.

Chan, S.L. (1989): Inelastic post-buckling analysis of tubular beam-columns and frames. *J. Eng. Struct.* **11**, 23–30.

Chan, S.L. (1993): A nonlinear numerical method for accurate determination of limit and bifurcation points. *Int. J. Num. Methods Eng.* **36**, 2779–2790.

Chan, S.L. (1992): Large deflection kinematic formulations for three dimensional framed structures. *Comput. Methods Appl. Mech. Eng.* **95**, 17–36.

Chan, S.L. (1989): Inelastic post-buckling analysis of tubular beam-columns and frames. *Eng. Struct.* **11**, 23–30.

Chan, S.L., Koon, C.M. and Sun, S. (1999): Design of glazed trusses by buckling analysis. In: *Glass in Buildings*, Conference Proceedings, Bath, UK, 31 March – 1 April.

Chan, S.L. and Kitipornchai, S. (1987): Geometric nonlinear analysis of asymmetric thin-walled beam-columns. *Eng. Struct.* **9**, 243–254.

Chan, S.L. and Zhou, Z.H. (1994): A pointwise equilibrating polynomial (PEP) element for nonlinear analysis of frames. *J. Struct. Eng. ASCE* **120**(6), 1703–1717.

Chen and Atsuta (1977): Space behaviour and design. In: *Theory of Beam-Columns*, Vol. 2, Ch. 12, McGraw-Hill, New York.

Chen, W.F. and Lui, E.M. (1991): *Stability Design of Steel Frames*, CRC Press, Boca Raton, FL.

Clarke, M.J. and Hancock, G.J. (1990): Finite element nonlinear analysis of stressed-arch frames. *J. Struct. Eng. ASCE* **117**(10), 2819–2837.

Fujii, F. and Okazawa, S. (1997): Pinpointing bifurcation points and branch-switching. *J. Eng. Mech.* **123**(13), 179–189.

Hancock, G.J. (1974): A matrix non-linear analysis of elastic thin-walled beams. *Civil Eng. Trans. Inst. Eng. Australia* **CE16**(2), 168–173.

Izzuddin, B.A. and Elnashai, A.S. (1993): Eulerian formulation for large displacement analysis of space frames. *J. Eng. Mech. ASCE* **119**(3), 549–569.

Jennings, A. (1968): Frame analysis including change of geometry. *J. Struct. Div. ASCE* **94**(ST3), 627–637.

Kondoh, K. and Atluri, S.N. (1986): A simplified finite element method for large deformation, post-buckling analysis of frame structures using explicitly derived tangent stiffness matrices. *Int. J. Num. Methods Eng.* **23**, 69–90.

Livesley, R.K. (1964): *Matrix Methods of Structural Analysis*, Pergamon, Oxford.

Lui, E.M. and Chen, W.F. (1986): Analysis and behaviour of flexibly jointed frames. *Eng. Struct.* **8**, 107–118.

Majid (1972): *Non-Linear Structures*, Butterworth.

Mallet, R.H. and Marcal, P.V. (1968): Finite element analysis of nonlinear structures. *J. Struct. Div. ASCE* **94**(ST9), 2081–2105.

Meek, J.L. and Tan, H.S. (1984): Geometrically nonlinear analysis of space frames by an incremental-iterative technique. *Comput. Methods Appl. Mech. Eng.* **47**, 261–282.

Oran, C. (1973a): Tangent stiffness in plane frames. *J. Struct. Div. ASCE* **99**(ST6), 973–985.

Oran, C. (1973b): Tangent stiffness in space frames. *J. Struct. Div. ASCE* **99**(ST6), 987–1001.

Papadrakakis, M. (1981): Post-buckling analysis of spatial structures by vector iteration methods. *Comput. Struct.* **14**(5–6), 393–402.

Zienkiewicz, O.C. (1977): *The Finite Element Method*, 3rd edition, McGraw-Hill, New York.

Chapter 4

NON-LINEAR SOLUTION TECHNIQUES

4.1. Introduction

The load control Newton–Raphson method has been discussed in Chapter 3. It is powerful in providing the response of a structure at a specified set of loads. However, the method diverges when the solution point is close to the limit point and taking this as the failure load of a structure may not be correct. In many occasions, the analyst cannot distinguish whether it is a structural instability or a numerical failure which can be due to many factors including the load step being too large or the tangent stiffness not providing a close estimate to the solution.

In the design of some practical structures such as shallow domes and rafters, snap-through buckling needs to be considered. Furthermore, the ultimate capacity for most metal structures always requires the consideration of the geometrically non-linear effect. The inability of the load control Newton–Raphson method to converge near the limit point is a handicap to any computer program for non-linear analysis of structures.

Basically, the difficulties associated with tracing up to and beyond the limit point is due to the ill-conditioning of the tangent stiffness matrix. When the determinant of the tangent stiffness is exactly zero, the computer program will have over-flow error in the factorization of the tangent stiffness matrix. Fortunately, this condition is very seldom encountered in a second-order analysis of structures. Even when using an algorithm for locating the limit or bifurcation point where factorisation of tangent stiffness matrix is carried out in the vicinity of the critical point, this condition of exact semi-definite stiffness matrix is not encountered (Chan, 1993). However, when the load-control Newton–Raphson method is used in tracing the equilibrium path near the limit point, the equilibrium error in Eq. (3.12) increases with the number of iterations until numerical over-flow. This divergent problem is quite common in a non-linear analysis.

The pure incremental method described in the last chapter is theoretically capable of handling this problem of tracing the equilibrium path near and beyond the limit point since it does not iterate. Computational time, however, is expensive and uncertain in the magnitude of drift-off error. Furthermore, whilst a smaller load step will reduce the error, its value is difficult to determine but problem-dependent.

In Chapter 3, it was demonstrated that the load control Newton–Raphson method diverges at 80% of the elastic buckling load of an imperfect column. To prevent this divergence, a number of solution schemes are available with the aim of tracing the equilibrium path near and beyond the limit point. Earlier work by Argyis (1965) for second-order analysis of trusses and frames adopted the pure incremental approach by splitting the total load by a series of smaller incremental loads. Wright and Gaylord (1968) and Sharifi and Popov (1973) employed the artificial spring method to prevent the singularity of the stiffness matrix in snap-through problems. This method adds an artificial spring to the loaded degree of freedom so that the tangent stiffness matrix does not become ill-conditioned. The value of spring stiffness, however, is difficult to assess and it will also destroy the symmetry of the tangent stiffness matrix. The degree of freedom to be argumented is also difficult to determine for a large structure.

Whilst convergence may not be achieved near the limit point when using the Newton–Raphson method, it is possible to suppress iteration in the vicinity of the limit point. The switching of the combined incremental-iteration Newton–Raphson and pure incremental methods is suggested to be controlled by the current stiffness parameter proposed by Bergan (1981). Although this method improves the general performance of the solution scheme, the deficiency of the pure incremental method is not yet completely eliminated. The drift-off error near the limit point is still unknown.

In a typical limit point problem, the load is reduced with an increase in deflection. Using deflection as a controlling parameter will be an alternative to achieve convergence. Batoz and Dhatt (1979) developed a displacement control technique for handling the snap-through analysis by incrementing the displacement instead of the load. However, when the structure exhibits a snap-back instead of a snap-through type of behaviour, the displacement control breaks down for the same reason as the load-control Newton–Raphson method. Sabir and Lock (1972) combined the load control and the displacement control methods in a single analysis for elastic shells. They used the constant load method until the slope of the tangent stiffness is nearly horizontal, then they used the displacement control method. The switching of these two methods is determined by the slope of the tangent stiffness.

Many solution algorithms combine methods for treating the problem at different conditions in order to avoid limitations. However, a single method capable of handling the analysis under different adverse conditions is preferred for consistency and full automation. Riks (1979) and Wempner (1971) introduced an additional equation to the original stiffness equations for the arc-length control of tracing the equilibrium path. This additional constraint equation, however, destroyed the symmetry of the stiffness matrix which is highly undesirable in the finite element analysis based on the symmetric stiffness matrix. The technique was later improved by Crisfield (1981) and Ramm (1981) through the re-installation of the symmetrical property of the tangent stiffness matrix by an iterative process, which is nevertheless inevitable for a non-linear analysis. Bathe and Dvorkin (1983) argued that the constraint of keeping the work done constant in a load increment is a more logical choice in place of the arc length constraint. This concept was also used by Yang and McGuire (1984). Chan (1988) noted that all these constraints of constant arc-length or the constant work done in a load cycle are not the actual objective of an analysis. Instead, the aim of an iterative process is to dissipate the unbalanced equilibrium error and, therefore, a scheme, the minimum residual displacement method, of minimizing the unbalanced error in each iteration was developed. The expression for this method was found to be the simplest and the most logical, and it does not involve solution of quadratic equations in the arc-length method.

A comprehensive review of these common numerical methods was presented by Clarke and Hancock (1990) and Chan and Ho (1990). Although it can be said that no a single method is superior to others in all cases, the example chosen by these researchers contains the basic features of snap-through and snap-back behaviour and their findings are also valuable for selection of a particular method in an analysis.

This Chapter reviews the formulation and logic for some of the more commonly accepted and used non-linear solution methods. Examples are given to demonstrate the application of these methods in snap-through and snap-back, two and three dimensional problems.

4.2. Genuinely large deflection analysis

Unlike the problems discussed in Chapter 3, the post-buckling analysis of structures involves not only geometrical non-linearity, but also large deflection analysis. For geometrically non-linear analysis of structures with moderately large deflection, we normally assume the rotations are finite but small of, say, less than 15 degrees. Under this assumption, the rotations can be treated as vectors and simplicity and convenience can be gained since they are additive. This assumption is valid for the practical problems solved by the Newton–Raphson method in the last chapter since excessively large rotations are uncommon in civil engineering structures.

However, for genuinely large rotation problems in three dimensional space where the rotations are so large that they cannot be accumulated as vectors, the formulation becomes more complicated and other more rigorous methods like the joint orientation method (Oran, 1973) or the more simple incremental secant stiffness method (Chan, 1988) can be used. This phenomenon of non-vectorial property for rotations can be demonstrated simply by rotating a plane about the x-axis and then the y-axis each by a right angle each. Different orientations will be obtained if we rotate the plane about the y-axis first and then the x-axis by the same angle. For two dimensional models, this problem does not exist since only one rotation is present at any single node. An evaluation of these formulations is given by Chan (1992) and will not be discussed here. The incremental secant stiffness approach is used in this book because of its simplicity, robustness and subsequent suitability for educational purposes. Although it is limited by the rotations not being greater than 15 degrees within a particular load step, the constraint is seldom encountered in practical or even research problems.

4.3. Basic formulation

It can be seen that many of these methods can be derived from the same approach of super-imposition of the original incremental equilibrium equation and the constraint equation. The incremental equilibrium equation can be written as

$$[\Delta F] = [K_T][\Delta u], \tag{4.1}$$

in which $[\Delta F]$ is the unbalanced force, $[K_T]$ is the tangent stiffness matrix and $[\Delta u]$ is the unbalanced displacement.

The constraint equation with force vector parallel to the applied load vector can be written as

$$\Delta \lambda [\Delta \overline{F}] = \Delta \lambda [K_T][\Delta \bar{u}], \tag{4.2}$$

in which $\Delta \lambda$ is a load corrector factor for imposition of the constraint condition, $[\Delta \overline{F}]$ is the force vector parallel to the applied load vector and of arbitrary length and $[\Delta \bar{u}]$ is the conjugated displacement vector to $[\Delta \overline{F}]$.

Adding Eqs. (4.1) and (4.2), we have the resulting incremental equilibrium equation as

$$([\Delta F] + \Delta \lambda [\Delta \overline{F}]) = [K_T]([\Delta u] + \Delta \lambda [\Delta \bar{u}]). \tag{4.3}$$

In the following discussion, the left subscript represents the degree of freedom, the right subscript is the iteration number within a load cycle and the right super-script refers to the number of the load cycle. When confusion is unlikely, these scripts are omitted for clarity.

With the determined correction load factor discussed in the next section, the load and displacement in each iteration is updated as

$$[F]_{i+1} = [F]_i + \Delta \lambda_{i+1}[\Delta \overline{F}], \tag{4.4}$$

$$[u]_{i+1} = [u]_i + [\Delta u]_{i+1} + \Delta \lambda_{i+1}[\Delta \bar{u}], \tag{4.5}$$

in which the subscript 'i' refers to the i-th iteration number within a load cycle.

Commonly used numerical procedures for incremental-iterative non-linear finite element analysis are in fact formulated on the basis of Eq. (4.3) with their difference on the load corrector factor $\Delta \lambda$. Obviously, there can be many versions for numerical methods in this line of development. The load control Newton–Raphson method can also be taken as a special case of putting $\Delta \lambda$ to zero. These methods can be used with reformulation of the tangent stiffness at every iteration or otherwise with the merits and disadvantages similar to the discussion of conventional and modified Newton–Raphson methods in Chapter 3. Below are some of these methods proven to be reliable and robust in most non-linear problems.

4.3.1. The current stiffness parameter

The relative slope of the tangent along the equilibrium path is always used as a yardstick to check the state of stiffness of a structure. In some methods, when this relative stiffness is less than a certain value of its original stiffness, the method will be converted to a pure incremental method and when the stiffness is large, the iterative procedure can be activated since divergence is unlikely. Bergan (1981) proposed the current stiffness parameter for such a measurement of stiffness and sensing of the approach of the limit point along the load versus deflection path.

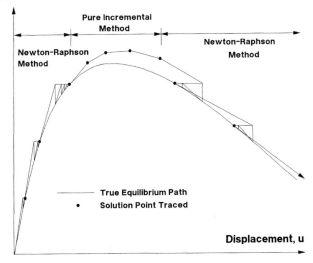

Fig. 4.1. The method of current stiffness parameter.

The current stiffness parameter is determined by the ratio of the current slope to the initial slope as

$$S_p = \frac{(dF/du)^k}{(dF/du)^1} = \frac{[\Delta \overline{F}]^T[\Delta \bar{u}]_1^1}{[\Delta \overline{F}]^T[\Delta \bar{u}]_1^k}, \qquad (4.6)$$

in which the right super-script 'k' refers to the k-th load cycle and the right subscript to the number of iteration within a load cycle.

The current stiffness parameter in Eq. (4.6) is scalar and it has a value of unity in the first load cycle. For a stiffening system, the parameter will be larger than unity and for a softening system, the value is less than one. For post-buckling equilibrium paths, the parameter will be negative. In the current stiffness parameter method, when this parameter is less than a certain value, iteration is suppressed and when it is larger than this specified value, iteration is resumed. The graphical representation of the method is indicated in Fig. 4.1.

4.3.2. Displacement control method

The displacement control method by Batoz and Dhatt (1979) is possibly the earliest general numerical method which can satisfy equilibrium conditions at every solution. This method simultaneously possesses the capacity of traversing the limit point without destroying the symmetrical property of the tangent stiffness matrix. A single degree of freedom is chosen to be the steering displacement degree of freedom for control of the advance of the solution for equilibrium path, and the magnitude for each displacement

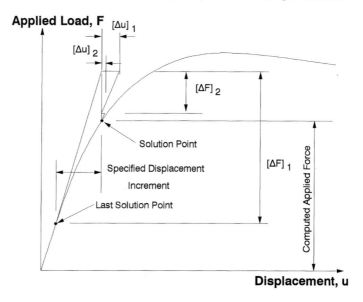

Fig. 4.2. The displacement control method.

increment must be decided. Similar to the load control Newton–Raphson method which diverges at a limit load, the selection is essential since it must increase along the complete equilibrium path otherwise divergence will occur. For example, the method breaks down in a snap-back problem since no solution is available for a displacement increment. A graphical representation of the method is shown in Fig. 4.2.

We assume m-th degree of freedom has been chosen as the steering displacement degree of freedom. In the first iteration for each load increment, the displacement increment is designated as $\Delta_m u_1$ and substituted into Eq. (4.5), we have the corrector factor as

$$\Delta \lambda_1 = -\frac{\Delta_m u_1}{\Delta_m \bar{u}_1}. \tag{4.7}$$

For second onward iteration within a load cycle, u_{i+1} will be equal to u_i for no displacement increment at m-th or the steering degree of freedom, Eq. (4.5) can therefore be re-written as

$$\Delta \lambda_i = -\frac{\Delta_m u_i}{\Delta_m \bar{u}_i} \quad \text{for } i \geqslant 2. \tag{4.8}$$

It can be seen from Eqs. (4.7) and (4.8) that the behaviour of the steering degree of freedom controls the performance of the solution method. Its selection must be undertaken with great care and may not be obvious for large scaled structures with many degrees of freedom. In addition, the method is not fully automatic but requires selection of steering displacement by the analyst.

4.3.3. The constant work method

The concept for the constant work method proposed by Bathe and Dvorkin (1983) and Yang and McGuire (1984) is to enforce a constant or a certain fixed value of work done in each load cycle. This implies that when the displacement increases, the force is reduced in order to keep the work done constant. During iteration, the force level will be reduced with an increase in displacement in order to keep the work done within a load cycle constant. In contrast to the displacement control method, this approach considers all degrees of freedom and, therefore, the selection of a steering freedom is not needed. The work done in each load cycle can be calculated as the input load increment and the displacement in the first load cycle. The work for the subsequent load increment and the conjugate displacement is equal to this constant work, ΔW, as

$$\Delta W = [F]_1^T [u]_1 = \Delta \lambda_1 [\Delta \overline{F}]^T [\Delta \bar{u}],$$
$$\Delta \lambda_1 = \frac{\Delta W}{[\Delta \overline{F}]^T [\Delta \bar{u}]}. \tag{4.9}$$

For the second iteration onwards in each load cycle, the work done by the residual displacement and force will be zero. From Eqs. (4.4) and (4.5), we have

$$\Delta \lambda_i [\Delta \overline{F}]^T ([\Delta u]_i + \Delta \lambda_i [\Delta \bar{u}]_i) = 0,$$
$$\Delta \lambda_i = -\frac{[\Delta \overline{F}]^T [\Delta u]_i}{[\Delta \overline{F}]^T [\Delta \bar{u}]_i}. \tag{4.10}$$

The graphical representation of this constant work method is shown in Fig. 4.3. It is interesting to note that, in a single degree of freedom, Eq. (4.9) will convert to the same expression for the displacement degree of freedom with the size of displacement increment controlled by the constant work done condition. It can therefore be deduced that for a structure with negative work generated by reducing load increment and decreasing displacement, the method will also fail. In other words, for a structural system with snap-back response over multiple degrees of freedom, the method may not function.

4.3.4. The arc-length method

The arc-length method is based on using the complete displacement vector for the control of the advance of the equilibrium path. The arc distance, taken as the dot product of the displacement vectors, is fixed in a particular load cycle. The graphical scheme for the method of controlling the arc-distance of displacements is shown in Fig. 4.4. Denoting the arc-distance as S, the constraint equation in the first iteration of a load cycle from Eq. (4.5) can be obtained as

$$\Delta \lambda_1 = \frac{S}{\sqrt{[\Delta \bar{u}]_1^T [\Delta \bar{u}]_1}}. \tag{4.11}$$

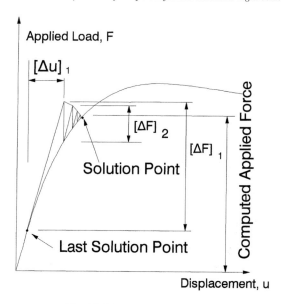

Fig. 4.3. The constant work method.

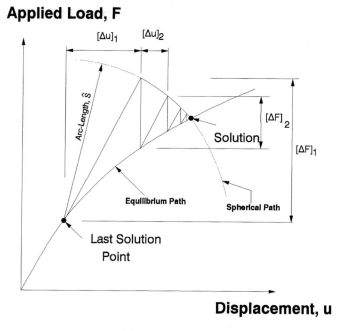

Fig. 4.4. The arc-length method.

For the second iteration onwards, the arc distance S is kept constant. Thus we have

$$\left([u]_{i-1} + [\Delta u]_i + \Delta\lambda_i[\Delta \bar{u}]_i\right)^T \left([u]_{i-1} + [\Delta u]_i + \Delta\lambda_i[\Delta \bar{u}]_i\right) = S^2. \tag{4.12}$$

Expanding, we have

$$\alpha_1 \Delta\lambda_i^2 + \alpha_2 \Delta\lambda_i + \alpha_3 = 0, \tag{4.13}$$

in which

$$\begin{aligned}
\alpha_1 &= [\Delta\bar{u}]^T[\Delta\bar{u}], \\
\alpha_2 &= 2\left([u]_{i-1} + [\Delta u]_i\right)^T[\Delta\bar{u}], \\
\alpha_3 &= \left([u]_{i-1} + [\Delta u]_i\right)^T\left([u]_{i-1} + [\Delta u]_i\right) - S^2.
\end{aligned} \tag{4.14}$$

Upon solution of the two roots in Eq. (4.13), the one maintaining a positive angle between the origin and the updated incremental displacement vector is chosen. If both of these roots meet this condition, the one closer to the following linear solution as follows will be selected:

$$\Delta\lambda_i = -\alpha_3/\alpha_2. \tag{4.15}$$

It will be inconvenient for the users to specify an arc-distance for a particular problem. Instead, the initial arc-distance can be computed from the dot-product of the displacement vectors in the first iteration as

$$S^1 = \sqrt{[\Delta u]_1^{1T}[\Delta u]_1^1}. \tag{4.16}$$

In order to use a smaller arc-distance in a highly non-linear range and a larger arc-distance for a relatively linear range, the arc-distance is varied according to the number of iterations in the last cycle as

$$S^j = S^{j-1}\sqrt{\left(\frac{I_d}{I^{j-1}}\right)}. \tag{4.17}$$

The arc-length method has been very widely used to-date. The essential part of the scheme, however, the iterative procedure, involves the solution of a quadratic equation in Eq. (4.13). Furthermore, the root can be imaginary and the selection of an appropriate root from the two roots in Eq. (4.13) is complicated.

4.3.5. The minimum residual displacement method

The ultimate objective of an iterative process is to eliminate the residual displacement or the unbalanced forces, instead of meeting the constant work done or the constant arc-distance. A logical method is, therefore, aimed at minimizing the error in each iteration.

Chan (1988) proposed a simple technique to search for a direction leading to the minimum value for the displacement error expressed as $\Delta\lambda_i[\Delta\bar{u}] + [\Delta u]_i$. This can be achieved by differentiating the residual displacement with respect to the parameter, $\Delta\lambda_i$, as follows:

$$\frac{\partial \lfloor (\Delta\lambda_i[\Delta\bar{u}] + [\Delta u]_i)^{\mathrm{T}}(\Delta\lambda_i[\Delta\bar{u}] + [\Delta u]_i) \rfloor}{\partial \Delta\lambda_i} = 0. \tag{4.18}$$

Simplifying, we have

$$\Delta\lambda_i = -\frac{[\Delta u]_i^{\mathrm{T}}[\Delta\bar{u}]}{[\Delta\bar{u}]^{\mathrm{T}}[\Delta\bar{u}]}. \tag{4.19}$$

For the size of load increment, the simple scaling of the load factor is used as follows:

$$\Delta\lambda_1^k = \Delta\lambda_1^1 S_p^\gamma \quad \text{and} \quad 2 > \frac{\Delta\lambda_1^k}{\Delta\lambda_1^1} > 0.1, \tag{4.20}$$

in which γ is a parameter to control the load size and the second condition ensures the incremental load size is not too large or too small.

It can be seen that the essential part for convergence mainly lies on the iterative part whilst the load size should be prevented from being too large or too small. The Minimum Residual Displacement method offers one of the simplest and most efficient scheme with solid mathematical proof on its convergence. Graphically this scheme arrives at a solution point shown in Fig. 4.5.

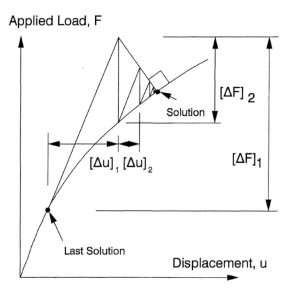

Fig. 4.5. The minimum residual displacement method.

4.4. Numerical examples

To test the proficiency of the described numerical methods, four elastic structures with snapping behaviour are studied. All these problems belong to the class of elastic and large deflection analysis. In the first problem, the well-known William's toggle frame is analyzed using different boundary conditions. In the second problem, the shallow arch with snap-back response is studied. In the third example, the equilibrium path of a right angle portal under a point load with positive and negative load eccentricity is plotted. This problem further quantifies the error by using a single and two elements to model each member. The last and the fourth example are the three-dimensional analysis of the hexagonal frame. All these problems are solved by the minimum residual displacement method as the iterative scheme, with arc-distance control as the load incremental size control, which was experienced by the authors to be the best numerical procedure for simplicity and robustness.

4.4.1. The William's toggle-frame

The shallow William's toggle frame is one of the earliest problems for demonstrating the snap-through buckling behaviour of a structure. Williams (1964) conducted an experiment on the bucking response of a shallow two-bar structure shown in Fig. 4.6. The structure is of rise 0.368 inches and total length 25.872 inches. The axial rigidity (EA) is 185.5 kips and the flexural rigidity (EI) is 9.27 ksi. Three boundary conditions have been assumed; rigid supports and rigid connections, pinned support connections with rigid joint between members and the pinned truss case with members pinned at both ends. For the first two analyses, two elements are used to model each inclined member and for the last case of truss analysis, each member is modelled by a single element with ignorance of all rotational degrees of freedom.

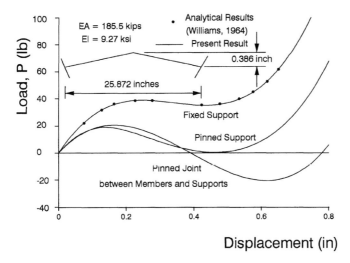

Fig. 4.6. Buckling of toggle frame.

The vertical downward load against the vertical tip downward displacement is plotted in Fig. 4.6. These curves involve a different degree of non-linearity and the proposed method coded in the computer program GMNAF has no difficulty in traversing these limit points and in handling the analysis with different boundary conditions. The curve for the rigid joint case is compared very well with the experimental results by Williams (1964). This shows the validity of the proposed method in dealing with this class of snap-through non-linear problem. Rafters and shallow trusses under downward loads may exhibit this type of snap-though buckling. The validity of the method in this example shows its application to the design and checking of the structure when load factors and member initial imperfections are considered properly.

4.4.2. The shallow arch

The shallow arch with ends pinned and subjected to a point load with an eccentricity of 200 mm is shown in Fig. 4.7. The axial rigidity (EA) is 2×10^6 N/mm^2, the flexural

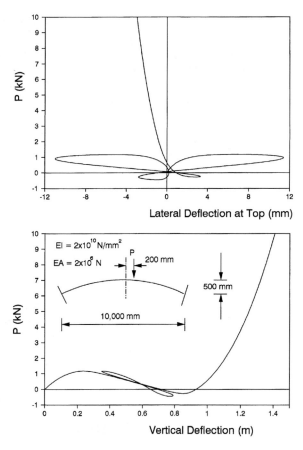

Fig. 4.7. Post-buckling analysis of shallow circular arch.

rigidity (EI) is 1×10^{10} N/mm^2 and the horizontal length is 10 m with a rise of 0.5 m. This example exhibits snap-through and snap-back buckling response and, therefore, it is a good example for evaluation of the performance of various non-linear solution methods.

Load versus deflection curves for vertical movement of the mid-span node and the deflection curve for horizontal deflection at the mid-span node are shown in Fig. 4.7. It can be seen that the structure exhibits a snap-through as well as snap-back behaviour for the loaded degree of freedom, implying that both the load control and displacement control methods cannot obtain convergence for the complete equilibrium path. When using the Minimum Residual Displacement method, the curve can be plotted without a divergent problem.

4.4.3. The right angle frame

The right angled frame shown in Fig. 4.8 is under an eccentric vertical point load. The cross-sectional area is 600 in^2, the member length is 1000 inches, the second moment of area is 60,000 in^4 and the Young's modulus of elasticity is 10,000 psi. The load eccentricity, being essential for converting the structural system to an imperfect system, is taken as 0.01 of its length and both directions of eccentricity have been considered. The bifurcation load is 8.331 kips which is further used to normalize the vertical load. Analytical solutions previously given by Koiter (1962) are also plotted in the same figure. One and two elements have been used to model each member for comparison of the accuracy of using a coarse model.

The load versus rotation plot of the portal by using different number of elements and with positive and negative eccentricity can be seen in Fig. 4.8. The error is noted to be

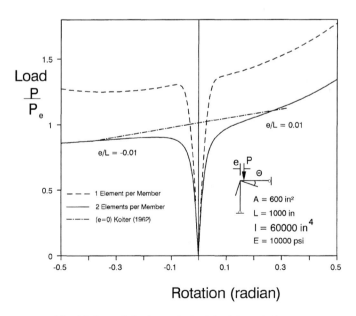

Fig. 4.8. Large deflection analysis of the right angle frame.

significant, both in the buckling load prediction and in the equilibrium path. The finer model of using two elements per member is close to the analytical solution while the use of one element per member predicts a buckling load with an error more than 20% in both load cases. This large error is due to the axial force in the member being close to the buckling load of the member. Thus, when this condition of members under large axial load is likely, more Hermite cubic elements or a refined element such as the Pointwise Equilibrium Polynomial Element (Chan and Zhou, 1994), the tangent stiffness matrix for large deflection spatial element (Kondoh and Atluri, 1986) and the element for large deflection analysis (Izzuddin, Elnashai and Dowling, 1990) should be used to model each member.

4.4.4. The three-dimensional hexagonal frame

The three dimensional shallow space frame of hexagonal shape is shown in Fig. 4.9. The axial rigidity (EA) of all members is 217,261 kips and their flexural rigidity (EI) is 8,796 ksi. It is subjected to a point load at top. For the case of roller support, a single and two elements have been used to model each member. Further, the pinned support case is analysis using a single element per member since the axial load in member is less in this case. The results of analysis are plotted using 1 and 2 elements per member.

As shown in Fig. 4.9, the results for the two boundary cases are different and the case for pinned supports is stiffer. The error for using the single and two elements to model each member is minimal, due to the axial force in members being quite small when compared to the Euler's buckling load for the members. The assumption for the small error of using a single cubic Hermite element per member should be adopted with caution and, when the axial force is large, more cubic elements or the refined element, the Pointwise Equilibrium Element by Chan and Zhou (1994), should be used.

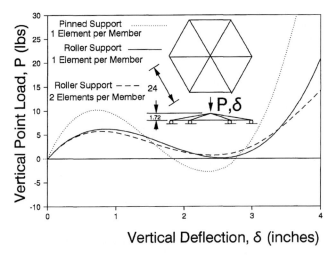

Fig. 4.9. Post-buckling analysis of three-dimensional hexagonal frame.

4.5. Conclusions

The advanced non-linear methods capable of traversing the limit points are discussed in this Chapter. The limitations and robustness of these solution schemes which include the method of current stiffness parameter, the displacement control method, the constant work method, the arc-length method and the minimum residual displacement method are described. The cubic element and the simple incremental secant stiffness procedure are used in the analysis. Four examples for snap-through and post-buckling analysis are solved and solutions are obtained without the divergence problem. Through these problems, the use of one cubic Hermite element to model each member is demonstrated to be inadequate when the axial force in the member is large. The use of several cubic elements or the Pointwise Equilibrium Element should be used. This condition is quite common in civil engineering structures whilst extreme large rotations invaliding the vectorial assumption for rotations are unusual. The theory is developed along this line of consideration and therefore both simplicity and accuracy are preserved.

References

Argyris, J.H. (1965): Continua and discontinua. In: *Proc. 1st Conf. on Matrix Methods in Structural Mechanics*, Wright-Patterson AFB, OH, pp. 11–189.

Bathe, K.J. and Dvorkin, E.N. (1983): On the automatic solution of non-linear finite element equations. *Comput. Struct.* **17**(5–6), 871–879.

Batoz, J.L. and Dhatt, G. (1979): Incremental displacement algorithms for non-linear problems. *Int. J. Num. Methods Eng.* **14**, 1262–1267.

Bergan, P.G. (1981): Strategies for tracing the nonlinear response near limit points. In: W. Wunderlich, E. Stein and K.J. Bathem, eds., *Nonlinear Finite Element Analysis in Structural Mechanics*, Springer-Verlag, Berlin, pp. 63–69.

Chan, S.L. (1988): Geometric and material nonlinear analysis of beam-columns and frames using the minimum residual displacement method. *Int. J. Num. Methods Eng.* **26**, 2657–2669.

Chan, S.L. (1992): Large deflection kinematic formulations for three dimensional framed structures. *Comput. Methods Appl. Mech. Eng.* **95**, 17–36.

Chan, S.L. (1993): A nonlinear numerical method for accurate determination of limit and bifurcation points. *Int. J. Num. Methods Eng.* **36**, 2779–2790.

Chan, S.L. and Ho, G.W.M. (1990): Comparative study on the nonlinear numerical algorithms. In: *3rd Int. Conf. on Advances in Numerical Methods in Engineering, Theory and Applications*, University College Swansea, UK, January 7–11, pp. 552–565.

Chan, S.L. and Zhou, Z.H. (1994): A pointwise equilibrating polynomial (PEP) element for nonlinear analysis of frames. *J. Struct. Eng. ASCE* **120**(6), 1703–1717.

Clarke, M.J. and Hancock, G.J. (1990): A study of incremental-iterative strategies for nonlinear analysis. Research Report R587, School of Civil and Mining Engineering, University of Sydney.

Crisfield, M.A. (1981): A faster incremental/iterative solution procedure that handles snap-through. *Comput. Struct.* **13**, 55–62.

Izzuddin, B.A., Elnashai, A.S. and Dowling, P.J. (1990): Large displacement nonlinear dynamic analysis of space frames. In: W.R. Kratzig et al., eds., *European Conf. on Structural Dynamics, Eurodyn'90*, Bochum, Germany, A.A. Balkema, Rotterdam, pp. 491–496.

Koiter, W.T. (1962): Post-buckling analysis of a simple twp-bar frame. In: B. Broberg, J. Hult and F. Niordson, eds., *Recent Progress in Applied Mechanics, the Folke Odquist Volume*, Admquist and Wiksell, Stockholm, Sweden, pp. 337–354.

Kondoh, K. and Atluri, S.N. (1986): A simplified finite element method for large deformation, post-buckling analysis of large frame structures, using explicitly derived tangent stiffness matrices. *Int. J. Num. Methods Eng.* **23**, 69–90.

Oran, C. (1973): Tangent stiffness in space frames. *J. Struct. Div. ASCE* **99**(ST6), 987–1001.

Sabir, A.B. and Lock, A.C. (1972): The application of finite elements to the large deflection geometrically nonlinear behaviours of cylindrical shells. In: C.A. Brebbia and H. Tottenham, eds., *Variational Methods in Engineering*, Southampton University Press, pp. 7/66–7/75.

Sharifi, P. and Popov, E.P. (1973): Nonlinear finite element analysis of sandwich shells of revolution. *Am. Inst. Aeronaut. Astronaut. J.* **11**, 715–722.

Ramm, E. (1981): Strategies for tracing the nonlinear response near limit points. In: W. Wunderlich, E. Stein and K.J. Bathe, eds., *Nonlinear Finite Element Analysis in Structural Mechanics*, Springer-Verlag, Berlin, pp. 63–89.

Riks, E. (1979): An incremental approach to the solution of snapping and buckling problems. *Int. J. Solid Struct.* **15**, 529–551.

Wempner, G.A. (1971): Discrete approximations related to nonlinear theories of solids. *Int. J. Solids Struct.* **7**, 1581–1599.

Williams, F.W. (1964): An approach to the nonlinear behaviour of the members of a rigid joint plane framework with finite deflections. *Quart. J. Mech. Appl. Math.* **17**, 451–469.

Wright, E.W. and Gaylord, E.H. (1968): Analysis of unbraced multi-storey steel rigid frames. *J. Struct. Div. ASCE* **ST5**(94), 1143–1163.

Yang, Y.B. and McGuire, W. (1984): A work control method for geometrically nonlinear analysis. In: *Proc. Int. Conf. in Advances in Numerical Methods in Engineering*, Swansea, UK, pp. 913–921.

Chapter 5

CONNECTION BEHAVIOUR AND MODELS

5.1. Introduction

In conventional analysis and design of steel structures, beam–column joints are usually assumed to be either perfectly rigid or ideally pinned for simplicity. This implies that the angle between adjoining members remains unchanged for the rigid joint case and no moment will be transferred between adjoining members for the pinned joint case. The rigid assumption implies the connection stiffness relative to the beam and column stiffness is very large whilst the pinned joint assumption considers the joint stiffness to be small when compared with the connected member stiffness. In design of steel frames, two major assumptions are made with the condition of sway and non-sway behaviour with which the connection stiffness plays different role. In the simple construction, the lateral stiffness is provided by other structural systems such as core and shear walls and an effective bracing system such that the beams are simply supported on stanchions. In this case the connections are assumed frictionless pinned and the beams and stanchions only resist vertical loads. In the second assumption, the lateral stiffness is provided by the moment action in the frames. This moment frame assumption assumes rigid connections between members. The first approach of simple construction normally results in a larger size beams and requires the use of another structural system in resisting the lateral loads whilst the second approach requires more heavily detailed connections between members. A more realistic and possibly more economical design is to allow for a certain degree of connection stiffness at connections since extra connection detailing cost can be eliminated whilst the nominal connection stiffness can be made use of.

Focusing on the behaviour of connections, experiments (Popov and Stephen, 1970; Popov, 1983; Nader and Astaneh, 1991) demonstrated that connections behave non-linearly in a manner between the two extremes of fully rigid and frictionless pinned connections. This means that a finite degree of joint flexibility exists at connections. Connections in practice are, therefore, semi-rigid. Apart from the geometrically non-linear effect, joint flexibility is known to be a significant and major source of non-linearity in the structural behaviour of steel frames under static or dynamic loading. Recently, the influence of semi-rigid connections on realistic structural response has been recognized, and provision for semi-rigid connections has been given in several national steel design codes including the American Load and Resistance Factored Design (LRFD) (1986), British Standard BS5950 (1990), Eurocode 3 (1992) and Australian AS4100 (1990). For example, the LRFD code (1986) published by the American Institute of Steel Construction (AISC) permits two basic types of construction: Type FR (fully restrained) construction which is commonly designated as rigid frame, and Type PR (partially restrained) construction which allows for semi-rigid connection stiffness. Although the LRFD code permits a partial restraint to be utilized, it gives little guidance to engineers on the use of finite connection stiffness. In the inelastic and large deflection analysis of steel frames using the Eurocode 3 (1993), the moment resistance, connection stiffness and rotational capacity are required to be checked for a particular connection. The last check for ductility is only needed for inelastic and large deflection case involving the tracing of equilibrium path beyond the elastic limit. In this Chapter, the behaviour of connections and several commonly used connection models are described and this is essential for latter studies of frames with semi-rigid connections.

5.2. Static monotonic loading tests

Extensive research work on static monotonic loading tests for different types of beam-column connections had been carried out in the past few decades. Table 5.1 summarizes the data available for each type of connections. The experimental results give valuable data for understanding the behaviour of various connection types and for modelling the connection component in the analysis.

Since the earliest studies of rotational stiffness of beam-column connections by Wilson and Moore (1917), hundreds of tests have been conducted to establish the relationship between moments and relative rotations of beam-column connections. Prior to 1950, riveted connections were tested by Young and Jackson (1934) and Rathbun (1936). In parallel with the interest in using high-strength bolts as structural fasteners, this connection type was tested by Bell et al. (1958). Subsequently, the behaviour of header plate connections was investigated in twenty tests by Sommer (1969).

Table 5.1
Available experimental $M-\phi_c$ data of various connection types

Connection type	Reference	Fastener	Column axis restrained	No. of tests
Single web cleat	Lipson (1977)	Bolts	Major	43
	Lipson and Antonio (1980)	Bolts	Major	33
Double web cleats	Rathbun (1936)	Rivets	Minor	7
	Thompson et al. (1970)	Bolts	Major	24
	Bjorhovde (1984)	Bolts	Major	10
Header plate	Sommer (1969)	Bolts	Major	16
	Kennedy and Hafez (1986)	Bolts	Major	8
Top and seat angle cleats	Hechtman and Johnston (1947)	Rivets	14 Major 5 Minor	19
	Azizinamini et al. (1985)	Rivets	Major	20
Flush end plate	Zoetemeijer and Kolstein (1975)	Bolts	Major	12
	Ostrander (1970)	Bolts	Major	24
Extended end-plate	Sherbourne (1961)	Bolts	Major	4
	Bailey (1970)	Bolts	Major	13

Extended end-plate and flush end-plate connections have been used extensively since the late 1960s and extended end-plate connections are designed to transfer considerable moments from beams to columns. Flush end-plate and extended end-plate connections for more rigid connections were tested by Ostrander (1970) and Johnstone and Walpole (1981), respectively.

Davison et al. (1987) performed a series of tests on a variety of beam-to-column connections including the web-cleat, flange cleat, combined seating cleat and web cleats, flush end-plate and extended end-plate connections. An informative database of connection moment–rotation M–ϕ_c curves was collected. More recently, Moore et al. (1993a, 1993b) carried out tests on five full-scale steel frames. The connections under the test included the flush end-plate connection, the extended end-plate connection and the flange cleat connection. Complete records of deformation of members and joints were reported. The results were also compared with the current British design code BS5950 Part 1 (BS5950, 1990), and extensions and modifications to the existing code have been proposed for better prediction of analysis results accounting joint stiffness.

In mathematical modeling of joint stiffness, a number of models were proposed to represent the moment versus rotation curves of the tested connections. In 1975, Frye and Morris collected a total of 145 test results during the period 1936 to 1970. Seven types of connection test data were classified, covering the range from the weakest single web angle connection to the stiffest T-stud connection. They also modelled the moment–rotation relationship by a polynomial function which was previously adopted by Sommer (1969). More recently, Ang and Morris (1984) collected a further 32 experimental results from the period 1934 to 1976, which were grouped into five types, namely the single web angle, the double web angle, the header plate, the top and seat angle and the strap angle connections. The more accurate Ramberg–Osgood function (Ramberg and Osgood, 1943) was used to represent the standardized moment–rotation relationship for the five commonly used connection types. In addition, Goverdhan (1984), gathered a total of 230 test results of moment–rotation curves which were digitized to form a database of connection behaviour. Nethercot (1985) conducted a literature survey for the period from 1951 to 1985. The connection test data and the corresponding curve representations were reviewed. Kishi and Chen (1986a; 1986b; 1987a, 1987b) further extended the collection to include over 300 test results which are classified into seven connection types. Namely, these are the single/double web-angle, top- and seat-angle with/without double web-angle, extended/flush end-plate and header plate connections. A computerized data bank program called the Steel Connection Data Bank (SCDB) was developed on steel beam-to-column connections to store the data base systematically. Using the system of tabulation and plotting in the SCDB program, the moment–rotation characteristics and the appropriate analytical connection model for the seven commonly used connection types can be easily obtained and used in the analysis.

5.3. Connection behaviour

In construction of steel framed buildings, beam-to-column connections are widely used. Figure 5.1 shows types of commonly used connections. Strictly speaking, the widely used

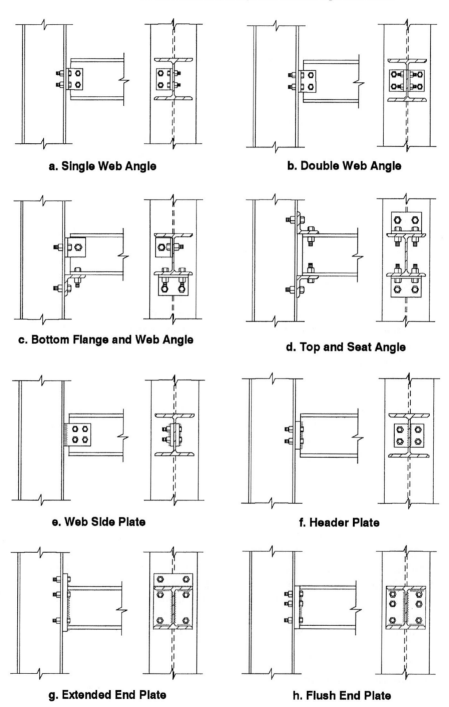

Fig. 5.1. Types of beam-to-column connections.

assumption of perfectly rigid and frictionless pinned beam–column connections is practically unattainable. Numerous experiments have shown that practical connections behave non-linearly due to gradual yielding of connection plates and cleats, bolts and so on. The properties of connections are complex and uncertainties in connection behaviour are common. The effects of frame non-linearity further complicate the problem. The practical sources of complexity includes geometric imperfections, residual stress due to welding, stress concentration and local secondary effects such as panel zone deformation and

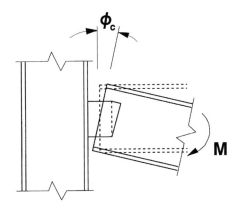

Fig. 5.2. Rotational deformation of a connection.

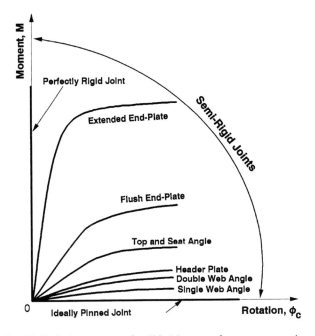

Fig. 5.3. Typical moment–rotation (M–ϕ_c) curves of common connections.

strain hardening. For most connections, however, the axial and shear deformations are usually small compared with the flexural deformation. For simplicity, only the rotational behaviour of connections due to flexural action is considered in this book. Figure 5.2 shows the rotational deformation of a connection.

The non-linear behaviour of a connection can be represented by a moment–rotation (M–ϕ_c) curve which can generally be obtained from test results. Typical M–ϕ_c curves of several commonly used connections are shown in Fig. 5.3. As illustrated from the figure, the two extreme cases are the ideally pinned (i.e. the horizontal line) and perfectly rigid (i.e. the vertical line) cases. Theoretically, those connections between the extreme cases are classified as semi-rigid joints. In order to incorporate the M–ϕ_c curves more systematically and efficiently into a frame analysis computer program, the moment versus rotation relationship is usually modelled by a mathematical function. A good mathematical function should be simple, of few parameters, easy in determination of these parameters, physically meaningful, numerical stable, containing no negative first-derivative, and capable of representing a wide range of connection types.

5.4. Classification of connection models

Generally speaking, the connection behaviour can be simplified as a set of moment versus rotation, M–ϕ_c, relationships. Mathematically, these relations can be expressed in a general form of

$$M = f(\phi_c), \tag{5.1a}$$

or, inversely,

$$\phi_c = g(M), \tag{5.1b}$$

in which f and g are some mathematical functions; M is the moment at the connection; and ϕ_c is the slip angle (i.e. the rotation of the connection) equal to the difference between the angles at the two ends of a connection.

Basically, the classification of the moment–rotation relationships of joints can be broadly divided into three main types: analytical, mathematical and mixed models. In the analytical models, the M–ϕ_c relationship is based on the physical characteristics of a connection. In the mathematical models, on the other hand, the relationship is expressed by a mathematical function in which the parameters are determined by the curve-fit of test results. Lastly, the mixed models combine the analytical and the mathematical models. These three types of models are described in the following.

5.4.1. Analytical models

Analytical models are derived to predict the joint stiffness on the basis of the geometrical properties and arrangements of connection components. With the assumptions on the deformation mechanism of the connection components for a particular connection type, the mechanical behaviour of the joint can be predicted by some methods such as the finite element method. Hence, the deformations of the components and the resisting moment of the joint can be determined and the moment–rotation relationship for this connection type can be obtained. Parametric studies are usually conducted on the effects of various geometric and force variables related to the components of the connections. Practical ranges of these variables are then analyzed to produce data for the actual analysis. However, the cost and time involved is generally unacceptable for practical applications. This is because each type of connections or different configuration of joint components requires a new set of formulation for the $M-\phi_c$ relationship. Further, inherent uncertainty of the joints may significantly affect the joint stiffness computed by the models. Besides, additional data handling procedures are required to incorporate the analytical results into the actual semi-rigid frame analysis. Examples of adopting the analytical models include Youssef-Agha and Aktan (1989) for the top and seat angle connection type, and Shi et al. (1996) for the end-plate connection type.

5.4.2. Mathematical models

Currently, the most commonly used approach for determining the moment–rotation relationships of connections is to curve-fit the experimental data with simple expressions. These expressions are called the mathematical models which directly relate the moment and rotation of connections by means of mathematical functions with some curve-fitting constants as in Eq. (5.1a) or (5.1b). The curve-fitting constants are determined from the experimental data. When these constants are obtained, the $M-\phi_c$ relationships can be explicitly expressed and directly used in the frame analysis. The mathematical models are, therefore, simpler than the aforementioned analytical models. Examples includes the Richard–Abbott model (Richard and Abbott, 1975), the power model (Colson and Louveau, 1983), the Chen–Lui exponential model (Lui and Chen, 1986) and the bounding-line model (Al-Bermani et al., 1994). Since extensive tests on various connection types have been conducted in the past few decades, much $M-\phi_c$ data is available for calculating the required parameters or constants of the mathematical models by the curve-fitting techniques. A good mathematical model should be simple, and requiring few parameters with physical meanings. Further, it should always guarantee to generate a smooth curve with positive first-derivative or connection stiffness. Besides, the model is desirable to cover a wide range of various connection types.

5.4.3. Mixed models

Mixed models are combinations of the analytical and the mathematical models. In the formulation of the mixed models, the functions are expressed in terms of both curve-fitting constants and geometric parameters. In order to develop a general expression for all connections with similar arrangement of components of a given joint type, the functions are usually standardized for convenience. The curve-fitting constants are determined by the curve-fitting technique while the geometric parameters are based on the geometry of the connection components. The mixed models require few parameters comparable to those of the mathematical models. Besides, similar to the analytical models, they also retain the geometrical parameters which relates with the physical characteristics of connections. This feature is not possessed by the curve-fitting mathematical model. In general, these models can be used to calculate the initial stiffness of particular types of connections and predict the non-linear behaviour of connections. Examples include the polynomial model (Frye and Morris, 1975) and the Ramberg–Osgood model (Ang and Morris, 1984).

5.5. Formulation of connection models

There are several commonly used mathematical and mixed models for representing the M–ϕ_c curves of connections. Namely, they are the linear model (Arbabi, 1982; Kawashima and Fujimoto, 1984; Chan, 1994), the bilinear model (Sivakumaran, 1988; Youssef-Agha, 1989), the trilinear model (Stelmack et al., 1986; Gerstle, 1988), the polynomial model (Frye and Morris, 1975), the cubic B-spline model (Cox, 1972; Jones et al., 1980), the bounding-line model (Al-Bermani et al., 1994; Zhu et al., 1995), the power model (Batho and Lash, 1936; Krishnamurthy et al., 1979; Colson and Louveau, 1983; Kishi and Chen, 1987a; King and Chen, 1993), the Ramberg–Osgood model (Ramberg and Osgood, 1943; Shi and Atluri, 1989), the Richard–Abbott model (Richard and Abbott, 1975; Gao and Haldar, 1995), the Chen–Lui exponential model (Lui and Chen, 1988), and others. The linear and bilinear models are simple to use but they may be too coarse and produce drastic change in stiffness which is undesirable in terms of accuracy and computational stability. The polynomial model is able to provide a better approximation, but could give undesired negative connection stiffness. The cubic B-spline model can represent the results accurately but requires many parameters. The Ramberg–Osgood (1943) and the Richard–Abbott (1975) models require three and four parameters respectively and give a fairly good fit. The Chen–Lui (1988) exponential model can provide an excellent fit and at least six parameters are used in modelling. The characteristics of these commonly used connection models are described in the following sub-sections.

5.5.1. Linear model

This is the simplest connection model and requires only one parameter defining the stiffness of a connection. Owing to its simplicity, the linear semi-rigid connection model has been widely used in the early stages of developing analysis methods in semi-rigid joints (Batho and co-workers, 1931, 1934, 1936; Rathbun, 1936; Baker, 1934; Monforton and Wu, 1963) and in the bifurcation and vibration analysis of semi-rigid frames (Chan, 1994). The moment–rotation function can be written as

$$M = S_c^o \phi_c, \tag{5.2}$$

in which S_c^o is a constant value of the initial stiffness of a connection. The value of S_c^o can be obtained from experiments. Mathematically, a simple function of S_c^o can be expressed in terms of the beam stiffness. Lightfoot and LeMessurier (1974) assumed the connection stiffness as

$$S_c^o = \lambda \frac{4EI}{L}, \qquad (5.3)$$

in which EI and L are the flexural rigidity and length of the beam respectively; and λ is the rigidity index which is proposed to indicate the degree of connection flexibility. The value of λ ranges from zero for the ideally pin-ended joint case and is infinite for the perfectly rigid-jointed connection case.

Alternatively, Eq. (5.3) can further be rationalized by the adoption of a fixity factor η first suggested by Romstad and Subramanian (1970) and Yu and Shanmugam (1986) as

$$S_c^o = \frac{\eta}{1-\eta}\left(\frac{4EI}{L}\right). \qquad (5.4)$$

The fixity factor η is equal to zero for pinned joints and unity for rigid joints.

The linear model is simple to use because the initial stiffness of connections is easily obtained from experiments and is constant throughout the analysis without the requirement for updating the connection stiffnesses. However, as illustrated in Fig. 5.4(a), it is not accurate in the large deflection range and its is considered that they should only be used in linear, vibration and bifurcation analysis where the deflections are small.

5.5.2. Multi-linear model

In the linear model described above, the linear assumption is suitable only in a low loading range such as the severability check for lateral deflection. However, in a large deflection analysis, the degradation of the stiffness of connections should be accounted for. The bi-linear model (Melchers and Kaur, 1982; Romstad and Subramanian, 1970; Lionberger, 1967; Lionberger and Weaver, 1967; Sugimoto and Chen, 1982; Lui and Chen, 1983; Maxwell et al., 1981; Tarpy and Cardinal, 1981) and the multi-linear model (Moncarz and Gerstle, 1981; Vinnakota, 1983; Razzaq, 1983; Stelmack et al., 1986; Gerstle, 1988) have been proposed for improving the accuracy of the analysis. As shown in Fig. 5.4(a), both the bi-linear and the multi-linear models perform better than the linear model. However, they may have the deficiency of slope discontinuity which is undesirable in numerical analysis and modeling.

5.5.3. Polynomial model

To provide a smoother moment–rotation curve, Frye and Morris (1975) proposed an odd-power polynomial function, which was based on a procedure by Sommer (1969). It has the form as

$$\phi_c = C_1(KM)^1 + C_2(KM)^3 + C_3(KM)^5, \qquad (5.5a)$$

in which K is a standardization parameter which is a function of the significant geometrical parameters such as adjoining member size, plate thickness, etc.; and C_1, C_2 and C_3 are

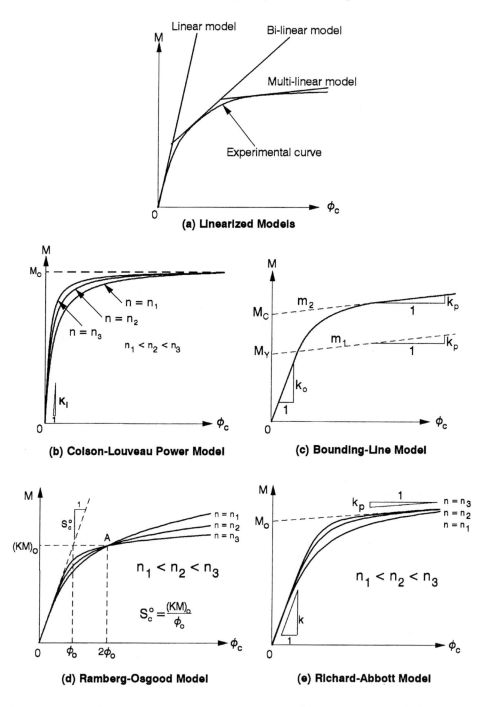

Fig. 5.4. Forms of connection models.

curve-fitting constants. The slope of the curve, which is the tangent connection stiffness, S_c, is, therefore, given by

$$S_c = \frac{dM}{d\phi_c} = \frac{1}{C_1 K + 3C_2 K (KM)^2 + 5C_3 K (KM)^4} \qquad (5.5b)$$

with the initial stiffness S_c^o as

$$S_c^o = \left.\frac{dM}{d\phi_c}\right|_{M=0} = \frac{1}{C_1 K}. \qquad (5.5c)$$

The value of K can be determined by a parametric study and has a general form of

$$K = \prod_{j=1}^{m} p_j^{a_j}, \qquad (5.5d)$$

where p_j is the j-th connection size parameter, a_j is a dimensionless exponent which indicates the influence of the j-th size parameter on the connection stiffness and m is the number of size parameters.

Based on the test data, standardized M–ϕ_c functions for each of seven connection types shown in Fig. 5.5 are listed in Table 5.2 by Frye and Morris (1975). The maximum deviations of the standardized moment–rotation curves from the corresponding experimental curves are also indicated in the table.

The method of least squares is used to determine the constants of the polynomial model. The primary disadvantage of this model is that the connection stiffness (i.e. the first-derivative or the slope of the function) may be discontinuous and may be negative, which is undesirable from the physical and numerical points of view.

5.5.4. Cubic B-spline model

Jones et al. (1980) employed the B-spline curve fitting technique (Cox, 1972) and proposed a cubic B-spline model to represent the non-linear moment–rotation curve for connections. The basic idea of the cubic B-spline model is to divide the M–ϕ_c curve from experimental results into a number of segments. The data falling within each segment is then fitted by a cubic polynomial. To maintain the smoothness of the non-linear M–ϕ_c curve, the continuity of the first and second derivatives of each curve segments is enforced. If the experimental data is divided into m intervals (i.e. $m + 1$ knots), the cubic B-spline model can be expressed as

$$\phi_c- = \sum_{k=0}^{3} a_k M^k + \sum_{k=1}^{m} b_k \langle M - M_k \rangle^3, \qquad (5.6a)$$

where

$$\langle M - M_k \rangle = \begin{cases} M - M_k & \text{for } (M - M_k) \geqslant 0, \\ 0 & \text{for } (M - M_k) < 0, \end{cases} \qquad (5.6b)$$

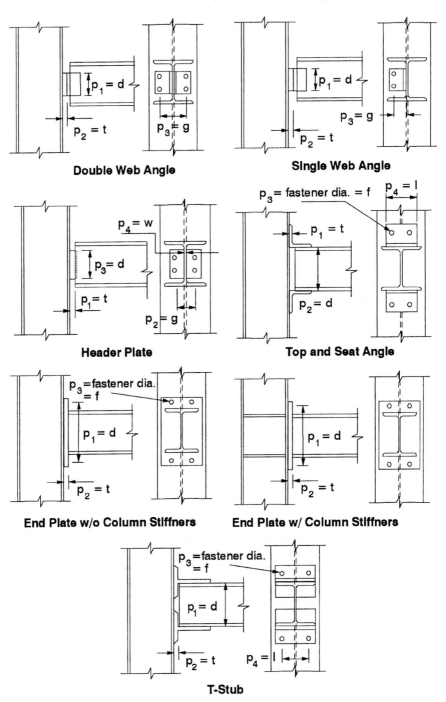

Fig. 5.5. Connection types and standardization parameters for the polynomial model used by Frye and Morris (1975).

Table 5.2
Standardized constants of the polynomial model (Frye and Morris, 1975)

Connection type	Curve-fitting coefficients	Standardized constant	Tests
Double web angle	$C_1 = 3.66 \times 10^{-4}$ $C_2 = 1.15 \times 10^{-6}$ $C_3 = 4.57 \times 10^{-8}$	$K = d^{-2.4} t^{-1.81} g^{0.15}$	Munse et al. (1959) Sommer (1969)
Single web angle	$C_1 = 4.28 \times 10^{-3}$ $C_2 = 1.45 \times 10^{-9}$ $C_3 = 1.51 \times 10^{-16}$	$K = d^{-2.4} t^{-1.81} g^{0.15}$	Lipson (1968)
Header plate	$C_1 = 5.1 \times 10^{-5}$ $C_2 = 6.2 \times 10^{-10}$ $C_3 = 2.4 \times 10^{-13}$	$K = t^{-1.6} g^{1.6} d^{-2.3} w^{0.5}$	Sommer (1969)
Top and seat angle	$C_1 = 8.46 \times 10^{-4}$ $C_2 = 1.01 \times 10^{-4}$ $C_3 = 1.24 \times 10^{-8}$	$K = t^{-0.5} d^{-1.5} f^{-1.1} l^{-0.7}$	Rathbun (1936) Hechtman and Johnston (1947) Brandes and Mains (1944)
End plate without column stiffeners	$C_1 = 1.83 \times 10^{-3}$ $C_2 = -1.04 \times 10^{-4}$ $C_3 = 6.38 \times 10^{-6}$	$K = d^{-2.4} t^{-0.4} f^{1.1}$	Ostrander (1970) Sherbourne (1948)
End plate with column stiffeners	$C_1 = 1.79 \times 10^{-3}$ $C_2 = 1.76 \times 10^{-4}$ $C_2 = 2.04 \times 10^{-4}$	$K = d^{-2.4} t^{-0.6}$	Sherbourne (1948) Johnson et al. (1960) Ostrander (1970)
T-stub	$C_1 = 2.1 \times 10^{-4}$ $C_2 = 6.2 \times 10^{-6}$ $C_3 = -7.6 \times 10^{-9}$	$K = d^{-1.5} t^{-0.5} f^{-1.1} l^{-0.7}$	Rathbun (1936) Douty (1964)

in which M_k is the lower bound moment in the k-th interval of the curve and a_k and b_k are the coefficients obtained by the least-square curve fitting procedure. The initial connection stiffness is given by

$$S_c^o = \left.\frac{dM}{d\phi_c}\right|_{M=0} = \frac{1}{a_1}. \tag{5.6c}$$

Although numerical studies have shown that the cubic B-spline model represents an excellent fit to test data, it requires a large number of sampling data in the curve-fitting process. It is, therefore, seldom employed in practical use but is usually taken as a reference for other simpler models.

5.5.5. Power model

Several power models have been developed for different types of connections. Two or three parameters are required in their functions.

A two-parameter power model (Batho and Lash, 1936; Krishnamurthy et al., 1979) has the simple form as

$$\phi_c = aM^b \tag{5.7a}$$

with the joint stiffness of

$$S_c = \frac{dM}{d\phi_c} = \frac{1}{abM^{b-1}}, \tag{5.7b}$$

where a and b are two curve-fitting parameters with the conditions $a > 0$ and $b > 1$.

Colson and Louveau (1983) introduced a three-parameter power model function (Richard, 1961; Goldberg and Richard, 1963) as

$$\phi_c = \frac{|M|}{K_i} \frac{1}{1 - |M/M_u|^n} \tag{5.8a}$$

with the tangent stiffness S_c given by

$$S_c = \frac{dM}{d\phi_c} = \frac{K_i[1 - (M/M_u)^n]^2}{1 + (n-1)(M/M_u)^n} \tag{5.8b}$$

in which K_i is the initial connection stiffness; M_u is the ultimate moment capacity of connection; and n is the shape parameter of the $M-\phi_c$ curve. The shape of the Colson–Louveau power model with different values of n is shown in Fig. 5.4(b). As illustrated in the figure, when the value of n increases, the connection is stiffer. In addition, the model has no further moment resistance when the moment approaches M_u. Thus, hardening stiffness is not considered by the model.

A similar three-parameter power model was also adopted by Kishi and Chen (1987a, 1987b) as

$$\phi_c = \frac{M}{K_i[1 - (M/M_u)^n]^{1/n}} \tag{5.9a}$$

with the corresponding tangent stiffness as

$$S_c = \frac{dM}{d\phi_c} = K_i\left[1 - \left(\frac{M}{M_u}\right)^n\right]^{(n+1)/n} \tag{5.9b}$$

in which the parameters K_i, M_u and n are the same as those defined in the previous Eq. (5.8).

Besides, King and Chen (1993) proposed a three-parameter power model, which is based on the slope of the $M-\phi_c$ curve (i.e. the connection stiffness) as

$$S_c = K_i\left(1 - (M/M_u)^n\right) \tag{5.10a}$$

with the initial stiffness as

$$S_c^o = K_i, \tag{5.10b}$$

in which K_i, M_u and n are also the same as those in Eq. (5.8).

There are some advantages of employing these power models to describe the non-linear M–ϕ_c curves of connections. Firstly, these models can always guarantee a positive first-derivative (i.e. a positive value of connection stiffness), which is particularly important to prevent the occurrence of the undesired negative connection stiffness. They also give a smooth curve without abrupt change or kink of slope. Secondly, they require only a few number of parameters in the expression so that the procedure for curve fitting and the stiffness calculation in analysis will be simpler and more convenient. Thirdly, in general, these models give a good fit for the M–ϕ_c curves to experimental data.

5.5.6. Bounding-line model

Al-Bermani et al. (1994) and Zhu et al. (1995) have proposed a bounding-line model, which requires four parameters, for the M–ϕ_c relation as shown in Fig. 5.4(c). The concept of this model is to divide the curve into three segments. The first and third segments are the linear elastic and the linear plastic portions of the M–ϕ_c curve, respectively. Between these two segments, a smooth transition curve is fitted. Since the complete M–ϕ_c curve is joined by three portions, the moment cannot be written as a single function of the connection rotation. The moment of the connection is therefore determined by accumulating the incremental moments. The form of the model is given as

$$M = \sum dM, \tag{5.11a}$$

where

$$dM = S_c \, d\phi_c. \tag{5.11b}$$

The tangent joint stiffness S_c in different segments is assumed as

$$\begin{aligned} S_c &= k_o & \text{when } |M| < m_1, \\ S_c &= k_o + \frac{|M| - m_1}{M_C - M_Y}(k_p - k_o) & \text{when } m_1 \leqslant |M| < m_2, \\ S_c &= k_p & \text{when } |M| \geqslant m_2, \end{aligned} \tag{5.11c}$$

where

$$\begin{aligned} m_1 &= M_Y + k_p \phi_c, \\ m_2 &= M_C + k_p \phi_c \end{aligned} \tag{5.11d}$$

in which k_o is the initial stiffness, k_p is the bounding stiffness, M_Y is the yielding moment and M_C is the bounding moment.

Since the bounding-line model is not represented by one function in the entire M–ϕ_c curve, it requires an additional procedure for checking which segment the current point lies. However, it requires few parameters and always gives a positive connection stiffness.

5.5.7. Exponential model

Lui and Chen (1986, 1988) proposed an exponential model called the Chen–Lui exponential model. It has the form of

$$M = M_c + \sum_{j=1}^{n} C_j \left[1 - \exp\left(\frac{-|\phi_c|}{2j\alpha}\right)\right] + R_{kf}|\phi_c|, \tag{5.12a}$$

and its tangent connection stiffness is given by

$$S_c = \left.\frac{dM}{d\phi_c}\right|_{|\phi_c|=|\phi_c|} = \sum_{j=1}^{n} \frac{C_j}{2j\alpha} \exp\left(\frac{-|\phi_c|}{2j\alpha}\right) + R_{kf}, \tag{5.12b}$$

with the initial stiffness, S_c^o, can be obtained as

$$S_c^o = \left.\frac{dM}{d\phi_c}\right|_{|\phi_c|=0} = \sum_{j=1}^{n} \frac{C_j}{2j\alpha} + R_{kf}, \tag{5.12c}$$

in which M is the moment in the connection; $|\phi_c|$ is the absolute value of the rotational deformation of the connection; M_o is the initial moment; R_{kf} is the strain-hardening stiffness of the connection; α is the scaling factor; C_j is the curve-fitting coefficient, and n is the number of terms considered. Based on the past experimental results, Lui and Chen (1988) had determined the values of the curve-fitting parameters of the model for four connection types consisting of single web angle, top and bottom seated angle, flush end plate, and extended end plate joints. The values are summarized in Table 5.3 and the property of each connection type derived from the exponential function is shown in Fig. 5.6. In general, the Chen–Lui exponential model gives a good representation of non-linear connection behaviour of which the accuracy is comparable to the cubic B-spline model. However, it requires a large number of curve-fitting parameters.

The Chen–Lui exponential function was later modified by Kishi and Chen (1986a, 1986b) to accommodate the linear components. This model is referred as the modified Chen–Lui exponential model. Under the loading condition, it is written as

$$M = M_o + \sum_{j=1}^{m} C_j \left[1 - \exp\left(\frac{-|\phi_c|}{2j\alpha}\right)\right] + \sum_{k=1}^{n} D_k(\phi_c - \phi_k)H[\phi_c - \phi_k] \tag{5.13a}$$

together with the tangent joint stiffness as

$$S_c = \left.\frac{dM}{d\phi_c}\right|_{|\phi_c|=|\phi_c|} = \sum_{j=1}^{n} \frac{C_j}{2j\alpha} \exp\left(\frac{-|\phi_c|}{2j\alpha}\right) + \sum_{k=1}^{m} D_k H[\phi_c - \phi_k] \tag{5.13b}$$

Table 5.3
Connection parameters of the Chen–Lui exponential model (Lui and Chen, 1988).

	Connection type (kip·in)			
	A Single web angle (Richard et al., 1982)	**B** Top and seated angle (Azizinamini et al., 1985)	**C** Flush-end plate (Ostrander, 1970)	**D** Extended end plate (Johnson and Walpole, 1981)
M_o	0	0	0	0
R_{kf}	0.47104×10^2	0.43169×10^3	0.96415×10^3	0.41193×10^3
α	0.51167×10^{-3}	0.31425×10^{-3}	0.31783×10^{-3}	0.67083×10^{-3}
C_1	-0.43300×10^2	-0.34515×10^3	-0.25038×10^3	-0.67824×10^3
C_2	0.12139×10^4	0.52345×10^4	0.50736×10^4	0.27084×10^4
C_3	-0.58583×10^4	-0.26762×10^5	-0.30396×10^5	-0.21389×10^5
C_4	0.12971×10^5	0.61920×10^5	0.75338×10^5	0.78563×10^5
C_5	-0.13374×10^5	-0.65114×10^5	-0.82873×10^5	-0.99740×10^5
C_6	0.52224×10^4	0.25506×10^5	0.33927×10^5	0.43042×10^5
S_c^o	0.48000×10^5	0.95219×10^5	0.11000×10^6	0.30800×10^6

and the initial stiffness as

$$S_c^o = \sum_{j=1}^{n} \frac{C_j}{2j\alpha} + D_k H\big[-|\phi_k|\big]_{k=1} \tag{5.13c}$$

in which M_o is the initial connection moment; α is the scaling factor; ϕ_k is the starting rotations of linear components; C_j and D_k are the curve-fitting constants; and $H[\phi_c]$ is the Heaviside step function defined as

$$\begin{aligned} H[\phi_c] &= 1 \quad \text{when } \phi_c \geqslant 0, \\ H[\phi_c] &= 0 \quad \text{when } \phi_c < 0. \end{aligned} \tag{5.13d}$$

This model is similar to that by Eq. (5.12), except the last term $R_{kf}|\phi_c|$ in Eq. (5.12a) is replaced by a set of Heaviside step functions $H[\phi_c]$.

Yee and Melchers (1986) proposed a simplified four-parameter exponential model as

$$M = M_p \left[1 - \exp \left| \frac{-(K_i - K_p + C\phi_c)\phi_c}{M_p} \right| \right] + K_p \phi_c, \tag{5.14a}$$

in which M_p is the plastic moment; K_i is the initial stiffness; K_p is the strain-hardening stiffness; and C is a constant controlling the slope of the curve. The tangent stiffness is given by

$$S_c = K_p + (K_i - K_p + 2C\phi_c) \exp \left| \frac{-(K_i - K_p + C\phi_c)\phi_c}{M_p} \right| \tag{5.14b}$$

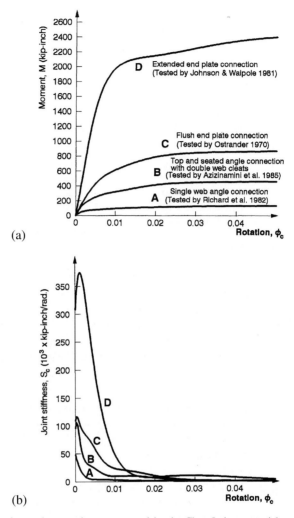

Fig. 5.6. Properties of several connections represented by the Chen–Lui exponential model (Lui and Chen, 1988). (a) Moment–rotation curves; (b) Connection stiffness–rotation curves.

and the initial stiffness is

$$S_c^o = K_i. \tag{5.14c}$$

For simplicity, Wu and Chen (1990) proposed a three-parameter exponential model in the form of

$$\frac{M}{M_u} = n \ln\left(1 + \frac{\phi_c}{n\phi_o}\right) \tag{5.15a}$$

with the tangent stiffness as

$$S_c = \frac{M_u/\phi_o}{1+\phi/(n\phi_o)}, \qquad (5.15b)$$

and the initial stiffness as

$$S_c^o = \frac{M_u}{\phi_o}, \qquad (5.15c)$$

in which the three parameters are the idealized elastic-plastic mechanism moment M_u; a reference rotation $\phi_o = M_u/K_i$, where K_i is the initial connection stiffness; and the shape parameter n which is determined empirically from test data.

5.5.8. Ramberg–Osgood model

The Ramberg–Osgood model was originally proposed for non-linear stress–strain relationships by Ramberg and Osgood (1943) and then standardized by Ang and Morris (1984). The moment–rotation curve of the model is expressed as

$$\frac{\phi_c}{\phi_o} = \frac{|KM|}{(KM)_o}\left[1+\left(\frac{|KM|}{(KM)_o}\right)^{n-1}\right] \qquad (5.16a)$$

with the tangent connection stiffness S_c as

$$S_c = \frac{(KM)_o/\phi_o}{1+n\left[\dfrac{|KM|}{(KM)_o}\right]^{n-1}} \qquad (5.16b)$$

in which $(KM)_o$ and ϕ_o are constants defining the position of the intersection point A, through which a family of Ramberg–Osgood curves passes (see Fig. 5.4(d)); n is a parameter defining the sharpness of the curve; and K is a dimensionless factor dependent on the connection type and geometry. K has the form of

$$K = \prod_{j=1}^{m} q_j^{a_j} \qquad (5.16c)$$

in which q_j is the numerical value of the j-th size parameter; a_j is the dimensionless exponent which indicates the effect of the j-th size parameter on the moment–rotation curve; and m is the number of size parameters for a particular connection type. The initial connection stiffness of the connections S_c^o can be evaluated as

$$S_c^o = \left.\frac{dM}{d\phi}\right|_{M=0} = \frac{(KM)_o}{\phi_o}. \qquad (5.16d)$$

Table 5.4
Connection parameters of the Ramberg–Osgood model (Ang and Morris, 1984)

Connection type	Dimensionless factor	Curve-fitting parameters	Tests
Single web angle	$K = d^{-2.09} t^{-1.64} g^{2.06}$	$\phi_0 = 1.03 \times 10^{-2}$ $(KM)_0 = 32.75$ $n = 3.93$	Lipson (1968)
Double web angle	$K = d^{-2.2} t^{0.08} g^{-0.28}$	$\phi_0 = 3.98 \times 10^{-3}$ $(KM)_0 = 0.63$ $n = 4.94$	Batho and Rowan (1934) Lewitt et al. (1966)
Header plate	$K = d^{-2.41} g^{-1.54} g^{2.12} w^{-0.45}$	$\phi_0 = 7.04 \times 10^{-3}$ $(KM)_0 = 186.77$ $n = 4.32$	Sommer (1969)
Top and seat angle	$K = d^{-1.06} t^{-0.54} l^{0.85} f^{-1.28}$	$\phi_0 = 5.17 \times 10^{-3}$ $(KM)_0 = 745.94$ $n = 5.61$	Hechtman and Johnston (1947)
Strap angle	$K = h^{-0.059} t^{-0.85} r^{-1.06}$	$\phi_0 = 4.58 \times 10^{-5}$ $(KM)_0 = 753.26$ $n = 5.98$	Beaulieu and Giroux (1974) Brun and Picard (1976)

The influence of the parameter, n, to the shape of the Ramberg–Osgood model is shown in Fig. 5.4(d). It can be seen that the ultimate moment is flattened with increasing value of the parameter, n.

Using the curve-fitting technique with available experimental results, Ang and Morris (1984) presented the standardized Ramberg–Osgood moment–rotation functions for five types of connections in Table 5.4.

The Ramberg–Osgood model requires only three parameters and can represent a non-linear and smooth M–ϕ_c curve very well. It is quite widely used for description of the stiffness of semi-rigid joints.

5.5.9. Richard–Abbott model

This four-parameter model was originally proposed by Richard and Abbott (1975). In the virgin loading path, the moment–rotation behaviour is written by the following expression as

$$M = \frac{(k - k_p)|\phi_c|}{\left[1 + \left|\frac{(k - k_p)|\phi_c|}{M_o}\right|^n\right]^{1/n}} + k_p|\phi_c|, \tag{5.17a}$$

and the corresponding tangent stiffness by

$$S_c = \left.\frac{dM}{d\phi_c}\right|_{|\phi_c| = |\phi_c|} = \frac{(k - k_p)}{\left[1 + \left|\frac{(k - k_p)|\phi_c|}{M_o}\right|^n\right]^{(n+1)/n}} + k_p, \tag{5.17b}$$

Chapter 5. *Connection behaviour and models* 113

in which k is the initial stiffness, k_p is the strain-hardening stiffness, M_o is a reference moment and n is a parameter defining the sharpness of the curve. The Richard–Abbott model is depicted in Fig. 5.4(e), corresponding to three values of n.

Since this model needs only four parameters to define the M–ϕ_c curve and always gives a positive stiffness, it is computational effective and is a good physical model for semi-rigid connection. It is currently one of the most popular models used.

5.6. Connection spring element

A semi-rigid joint can be modelled as a connection spring element inserted at the intersection point between a beam and a column. A typical connection spring is shown in Fig. 5.8. For most steel frame structures, the effects of the axial and shear forces in the connection deformations are small when compared with that of bending moment. For this reason, only the rotational deformation of the spring element is considered in practical analysis. For simplicity of calculation, the connection spring element is assumed dimensionless in size.

Owing to the flexibility of a semi-rigid joint, the rotational displacements on the two sides of the connection are generally unequal (see Fig. 5.7) and they are termed as the

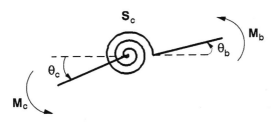

Fig. 5.7. Connection spring element.

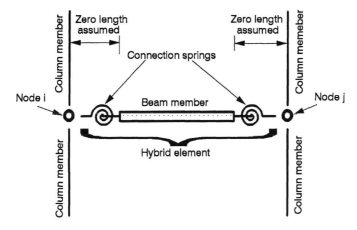

Fig. 5.8. Idealized element model.

connection and beam rotations, θ_c and θ_b, respectively. The subscripts of 'c' and 'b' refer to the connection and the beam. Consider the moment equilibrium condition at the connection, we can obtain

$$M_c + M_b = 0, \tag{5.18}$$

where

$$M_c = S_c(\theta_c - \theta_b), \tag{5.19a}$$

$$M_b = -M_c = S_c(\theta_b - \theta_c), \tag{5.19b}$$

in which M_c and M_b are the connection and beam moments at the connection spring; and S_c is the connection stiffness. Equation (5.19) can be rearranged to an incremental tangent stiffness matrix form as

$$\begin{Bmatrix} \Delta M_c \\ \Delta M_b \end{Bmatrix} = \begin{bmatrix} S_c & -S_c \\ -S_c & S_c \end{bmatrix} \begin{Bmatrix} \Delta \theta_c \\ \Delta \theta_b \end{Bmatrix}, \tag{5.20}$$

in which ΔM_c and ΔM_b are the incremental nodal moments at the junctions between the spring and the global node and between the beam and the spring; $\Delta \theta_c$ and $\Delta \theta_b$ are the incremental nodal rotations corresponding to these moments; and S_c is the instantaneous tangent connection stiffness given by

$$S_c = \frac{dM}{d\phi_c}, \tag{5.21}$$

in which M is the moment at the joint; and ϕ_c is the slip angle (i.e. the connection rotation) defined by

$$\phi_c = \theta_c - \theta_b. \tag{5.22}$$

From Eq. (5.20), the tangent element stiffness matrix of a connection spring can be derived as

$$\begin{bmatrix} S_c & -S_c \\ -S_c & S_c \end{bmatrix}. \tag{5.23}$$

This connection stiffness matrix can be combined into the conventional beam-column element to form a new element stiffness matrix with the consideration of the semi-rigid joint flexibility at the two ends of an element. The procedure is described in the following section.

Chapter 5. *Connection behaviour and models*

5.7. Modified element stiffness matrix accounting for semi-rigid joints

The connection spring and the beam-column element is combined to form a hybrid element. Figure 5.8 shows the hybrid element in the idealized finite element analysis model for a framed structure. One side of a connection spring is connected to the beam-column element and the other side is connected to a global node. In the analysis, for the column members in the figure, two connection springs each is located at the column end nodes. However, since most steel columns are designed and built as continuous columns between two successive columns, their end connection springs are usually assumed to be perfectly rigid and are therefore not shown in the figure. The deformation configuration of a hybrid element with end-springs is shown in Fig. 5.9(a). The element is initially straight in the undeformed shape, and then deforms to a bent shape by translational and rotational movements. The details of internal forces and deformations at a connection spring are illustrated in Fig. 5.9(b). The rotation of a connection is defined as the angle difference between the angles on the two sides of the connection (i.e. $\phi_c = \theta_c - \theta_b$).

In the presence of the connection springs attached at the ends of a beam-column element, the conventional stiffness matrix of the beam-column should be modified so as to

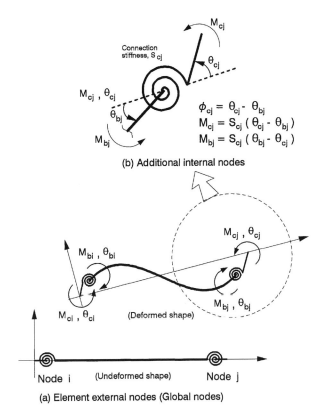

Fig. 5.9. Beam-column element with connection springs attached.

account for the semi-rigid joints in the formulation. The resultant stiffness matrix can be used in the analysis.

Considering the inner beam-column part within the connection springs at two ends of a hybrid element, the element stiffness matrix of the Hermitian cubic element is given by

$$\begin{Bmatrix} \Delta M_{bi} \\ \Delta M_{bj} \end{Bmatrix} = \begin{bmatrix} K_{ii} & K_{ij} \\ K_{ji} & K_{jj} \end{bmatrix} \begin{Bmatrix} \Delta \theta_{bi} \\ \Delta \theta_{bj} \end{Bmatrix}, \tag{5.24}$$

in which the subscripts 'i' and 'j' are referred to as end nodes i and j of the beam-column element; and K_{ij} is the element flexural stiffness of the beam-column element.

Combining the spring stiffness to the two ends of an element, we have

$$\begin{Bmatrix} \Delta M_{ci} \\ \Delta M_{bi} \\ \Delta M_{bj} \\ \Delta M_{cj} \end{Bmatrix} = \begin{bmatrix} S_{ci} & -S_{ci} & 0 & 0 \\ -S_{ci} & S_{ci}+K_{ii} & K_{ij} & 0 \\ 0 & K_{ji} & S_{cj}+K_{jj} & -S_{cj} \\ 0 & 0 & -S_{cj} & S_{cj} \end{bmatrix} \begin{Bmatrix} \Delta \theta_{ci} \\ \Delta \theta_{bi} \\ \Delta \theta_{bj} \\ \Delta \theta_{cj} \end{Bmatrix}. \tag{5.25}$$

Assuming that the loads are applied only at the global nodes and both ΔM_{bi} and ΔM_{bj} are equal to zero, we thus obtain

$$\begin{Bmatrix} \Delta \theta_{bi} \\ \Delta \theta_{bj} \end{Bmatrix} = \begin{bmatrix} S_{ci}+K_{ii} & K_{ij} \\ K_{ji} & S_{cj}+K_{jj} \end{bmatrix}^{-1} \begin{bmatrix} S_{ci} & 0 \\ 0 & S_{cj} \end{bmatrix} \begin{Bmatrix} \Delta \theta_{ci} \\ \Delta \theta_{cj} \end{Bmatrix}. \tag{5.26}$$

Eliminating the internal degrees of freedom by substituting Eq. (5.26) into (5.25), we have

$$\begin{Bmatrix} \Delta M_{ci} \\ \Delta M_{cj} \end{Bmatrix} = \left(\begin{bmatrix} S_{ci} & 0 \\ 0 & S_{cj} \end{bmatrix} - \frac{\begin{bmatrix} S_{ci} & 0 \\ 0 & S_{cj} \end{bmatrix} \begin{bmatrix} S_{cj}+K_{jj} & -K_{ij} \\ -K_{ji} & S_{ci}+K_{ii} \end{bmatrix} \begin{bmatrix} S_{ci} & 0 \\ 0 & S_{cj} \end{bmatrix}}{(S_{ci}+K_{ii})(S_{cj}+K_{jj}) - K_{ij}K_{ji}} \right)$$

$$\times \begin{Bmatrix} \Delta \theta_{ci} \\ \Delta \theta_{cj} \end{Bmatrix}. \tag{5.27}$$

Transforming the incremental moments, ΔM_{ci} and ΔM_{cj}, in Eq. (5.27) to the moments and shears about the last known axes, we have

$$\begin{Bmatrix} \Delta M_i \\ \Delta Q_i \\ \Delta M_j \\ \Delta Q_j \end{Bmatrix} = \begin{bmatrix} 1 & 0 \\ 1/L & 1/L \\ 0 & 1 \\ -1/L & -1/L \end{bmatrix} \left(\begin{bmatrix} S_{ci} & 0 \\ 0 & S_{cj} \end{bmatrix} \right.$$

$$- \frac{\begin{bmatrix} S_{ci} & 0 \\ 0 & S_{cj} \end{bmatrix} \begin{bmatrix} S_{cj}+K_{jj} & -K_{ij} \\ -K_{ji} & S_{ci}+K_{ii} \end{bmatrix}}{(S_{ci}+K_{ii})(S_{cj}+K_{jj}) - K_{ij}}$$

$$\left. \times \begin{bmatrix} 1 & 1/L & 0 & -1/L \\ 0 & 1/L & 1 & -1/L \end{bmatrix} \begin{Bmatrix} \Delta \theta_i \\ \Delta v_i \\ \Delta \theta_j \\ \Delta v_j \end{Bmatrix} \right), \tag{5.28}$$

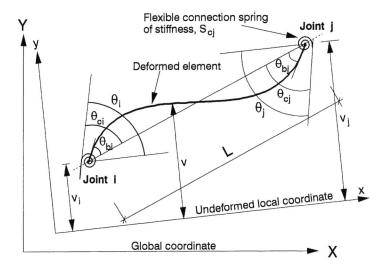

Fig. 5.10. The lateral deflections and rotations of a deformed element with flexible connection springs at the ends.

in which ΔM_i and ΔM_j are the incremental moments and ΔQ_i and ΔQ_j are the incremental shears at the two ended nodes of the hybrid element; S_{ci} and S_{cj} are the tangent spring stiffness at the ends of the element; $\Delta \theta_i$ and $\Delta \theta_j$ are the incremental rotations at the two ends of an element about the axis parallel to the last axis; Δv_i and Δv_j are the incremental lateral displacements at the two elemental nodes and projected onto the last configuration of the element; and L is the element length (see Fig. 5.10).

The axial and the torsional stiffness given by EA/L and GJ/L will be superimposed to the above bending stiffness to form a complete linear stiffness matrix of the element.

5.8. Conclusions

Connection flexibility is known to be a major source of non-linearity in structural steel frame. In many practical structures, the connections are semi-rigid that ignorance of connection flexibility may affect the accuracy of analysis. This Chapter describes the modeling of semi-rigid connections. They can be broadly classified into three types, namely, the analytical, mathematical and mixed models. Several examples of the mathematical and mixed models have been described and compared in details in this chapter. Modelling of a semi-rigid joint is carried out by the connection spring with degradable rotational stiffness. The tangent connection stiffness can be obtained as the first-derivative at the current point on the moment–rotation curve. By attaching the connection spring at the ends of the conventional beam-column element, a hybrid beam-column element is formed. Eliminating the internal degrees of freedom of the connection springs, the modified element stiffness matrix accounting for the semi-rigid joints can be formulated and used for the steel frame analysis.

References

Al-Bermani, F.G.A., Li, B., Zhu, K. and Kitipornchai, S. (1994): Cyclic and seismic response of flexibly jointed frames. *Eng. Struct.* **16**(4), 249–255.
American Institute of Steel Construction (1986): LRFD Load and Resistance Factor Design Specification for Structural Steel Buildings, AISC, Chicago.
Ang, K.M. and Morris, G.A. (1984): Analysis of three-dimensional frames with flexible beam-column connections. *Can. J. Civil Eng.* **11**, 245–254.
Arbabi, F. (1982): Drift of flexibly connected frames. *Comp. Struct.* **15**(2), 103–108.
Australian Standards (1990): AS4100-1990 Steel Structures, Australian Institute of Steel Construction, Sydney, Australia.
Azizinamini, A., Bradburn, J.H. and Radziminski, J.B. (1985): Static and Cyclic Behavior of Semi-Rigid Steel Beam-Column Connections, Technical Report, Dept. of Civil Engineering., Univ. of South Carolina, Columbia, SC.
Bailey, J.R. (1970): Strength and Rigidity of Bolted Beam-to-Column Connections, Conference on Joints in Structures, Univ. of Sheffield.
Baker, J.F. (1934): A Note on the Effective Length of a Pillar Forming Part of a Continuous Member in a Building Frame, 2nd Report, Steel Structures Research Committee, Dept. of Scientific and Industrial Research, HMSO, London, pp. 13–34.
Batho, C. (1931): Investigation on Beam and Stanchion Connections, 1st Report, Steel Structures Research Committee, Dept. of Scientific and Industrial Research, His Majesty's Stationary Office, London, Vols. 1–2, pp. 61–137.
Batho, C. and Lash, S.D. (1936): Further Investigations on Beam and Stanchion Connections Encased in Concrete, Together with Lab. Investigation on a Full Scale Steel Frame, Final Report, Steel Structures Research Committee, Dept. of Scientific and Industrial Research, HMSO, London, pp. 92.
Batho, C. and Rowan, H.C. (1934): Investigation on Beam and Stanchion Connections, 2nd Report, Steel Structures Research Committee, Dept. of Scientific and Industrial Research, His Majesty's Stationary Office, London, Vols. 1–2, pp. 61–137.
Beaulieu, D. and Giroux, Y.M. (1974): Etude experimentale d'un joint rigide entre un poteau tubulaire et des poutres en double-té. Rapport GCT-74-06-02. Département de Génie Civil, Université Laval, Québec, Canada.
Brun, P. and Picard, A. (1976): Etude d'un assemblage imparfaitement rigide et des effects de son utilisation dans un multi-étage. Rapport GCT-76-03, Département de Génie Civil, Université Laval, Québec, Canada.
Bell, W.G., Chesson, E.J. and Munse, W.H. (1958): Static Tests of Standard Riveted and Bolted Beamd-to-Column Connections, Univ. of Illinois, Engineering Experiment Station, Urban, IL.
Bjorhovde, R. (1984): Effect of end-restraint on column strength – practical applications. *AISC Eng. J.* **20**(1).
Brandes, J.L. and Mains, R.M. (1944): Report of tests of top plate and seat building connections. *Br. Weld. J.* **23**(3).
British Standard Institution (1990): BS5950: Part I: Structural Use of Steelwork in Building, BSI, London, England.
Chan, S.L. (1994): Vibration and modal analysis of steel frames with semi-rigid connections. *Eng. Struct.* **16**(1), 25–31.
Chen, W.F. and Kishi, N. (1989): Semirigid steel beam-to-column connections: data base and modeling. *J. Struct. Div. ASCE* **115**(1), 105–119.
Colson, A. and Louveau, J.M. (1983): Connections Incidence on the Inelastic Behavior of Steel Structural, Euromech Colloquium **174**.
Cox, M.G. (1972): The numerical evaluation of B-splines. *J. Inst. Math. Applic.* **10**, 134–139.
Davison, J.B., Kirby, P.A. and Nethercot, D.A. (1987): Rotational stiffness characteristics of steel beam-to-column connections. *J. Construct. Steel Res.* **8**, 17–54.
Douty, R.T. (1964): Strength Characteristics of High Strength Bolted Connections with Particular Application to the Plastic Design of Steel Structures, Ph.D. Thesis, Cornell Univ., Ithaca, NY.
Eurocode 3 (1992): Eurocode 3, Design of Steel Structures: Part 1.1, General Rules and Rules for Buildings, DD ENV 1993-1-1.

Frye, M.J. and Morris, G.A. (1975): Analysis of flexibly connected steel frames. *Can. J. Civil Eng.* **2**(3), 280–291.

Gao, L. and Haldar, A. (1995): Nonlinear seismic analysis of space structures with partially restrained connections. *Microcomput. Civil Eng.* **10**, 27–37.

Gerstle, K.H. (1988): Effect of connections on frames. *J. Construct. Steel Res.* **10**, 241–267.

Goldberg, J.E. and Richard, R.M. (1963): Analysis of nonlinear structures. *J. Struct. Div. ASCE* **89**(ST4).

Goverdhan, A.V. (1984): A Collection of Experimental Moment–Rotation Curves and Evaluation of Prediction Equations for Semi-Rigid Connections, Thesis for Master of Science, Vanderbilt Univ., Nashville, TN.

Hechtman, R.A. and Johnston, B.G. (1947): Riveted Semi-Rigid Beam-to-Column Building Connections, Committee of Steel Structures Research, AISC, Progress Report No. 1, November.

Johnson, L.G., Cannon, J.C. and Spooner, L.A. (1960): High tensile preloaded bolted joints. *Br. Weld. J.* **7**(9).

Johnson, N.D. and Walpole, W.R. (1981): Bolted End-Plate Beam-to-Column Connections Under Earthquake Type Loading, Research Report 81-7, Dept. of Civil Engineering, Univ. of Canterbury, Christchurch, New Zealand.

Jones, S.W., Kirby, P.A. and Nethercot, D.A. (1980): Effect of semi-rigid connections on steel column strength. *J. Construct. Steel Res.* **1**, 38–46.

Kawashima, S. and Fujimoto, T. (1984): Vibration analysis of frames with semi-rigid connections. *Comput. Struct.* **19**, 85–92.

Kennedy, D.J.L. and Hafez, M.A. (1986): A study of end plate connections for steel beams. *Can. J. Civil Eng.*

King, W.S. and Chen, W.F. (1993): A LRFD-Based Analysis Method for Semi-Rigid Frame Design, Structural Engineering Report No. CE-STR-93-15, School of Civil Engineering, Purdue Univ., West Lafayette, IN.

Kishi, N. and Chen, W.F. (1986a): Data Base of Steel Beam-to-Column Connections, Structural Engineering Report No. CE-STR-86-26, School of Civil Engineering, Purdue Univ., West Lafayette, IN.

Kishi, N. and Chen, W.F. (1986b): Steel Connection Data Bank Program, Structural Engineering Report No. CE-STR-86-18, School of Civil Engineering, Purdue Univ., West Lafayette, IN.

Kishi, N. and Chen, W.F. (1987a): Moment–Rotation Relation of Top-and Seat-Angle Connections, Structural Engineering Report No. CE-STR-87-4, School of Civil Engineering, Purdue Univ., West Lafayette, IN.

Kishi, N. and Chen, W.F. (1987b): Moment–Rotation of Semi-Rigid Connections, Structural Engineering Report No. CE-STR-87-29, School of Civil Engineering, Purdue Univ., West Lafayette, IN.

Krishnamurthy, N., Huang, H.T., Jeffrey, P.K. and Avery, L.K. (1979): Analytical M–ϕ curves for end-plate connections. *J. Struct. Div. ASCE* **105**(ST1), 133–145.

Lewitt, C.W., Chesson, E.J. and Munse, W.H. (1966): Restraint Characteristics of Flexible Riveted and Bolted Beam-to-Column Connections, Structural Research Series No. 296, Dept. of Civil Engineering, Univ. of Illinois, Urbana, IL.

Lightfoot, F. and LeMessurier, A.P. (1974): Elastic analysis of frameworks with elastic connections. *J. Struct. Div. ASCE* **100**(ST6), 1297–1309.

Lionberger, S.R. (1967): Statics and Dynamics of Building Frames with Non-Rigid Connections, Ph.D. Thesis, Stanford Univ., Stanford, CA.

Lionberger, S.R. and Weaver, J.W. (1969): Dynamic response of frames with non-rigid connections. *J. Eng. Mech. Div. ASCE* **95**(EM1), 95–114.

Lipson, S.L. (1968): Single-Angle and Single Plate Beam Framing Connections, First Canadian Structural Engineering Conference.

Lipson, S.L. (1977): Single-angle welded-bolted connections. *J. Struct. Div. ASCE* **103**(ST3), 559–571.

Lipson, S.L. and Antonio, M. (1980): Single-angle welded-bolted beam connections. *Can. J. Civil Eng.* **7**, 315–324.

Lui, E.M. and Chen, W.F. (1983): Strength of H-Columns with Small End Restraints. *J. Inst. Struct. Eng. (London)* **61B**(1) 17–26.

Lui, E.M. and Chen, W.F. (1986): Analysis and behaviour of flexibly-jointed frames. *Eng. Struct.* **8**, 107–118.

Lui, E.M. and Chen, W.F. (1988): Behavior of braced and unbraced semi-rigid frames. *Int. J. Solids Struct.* **24**(9), 893–913.

Maxwell, S.M., Jenkins, W.M. and Howlett, J.N. (1981): A Theoretical Approach to the Analysis of Connection Behaviour, Proc. of Conference on Joints in Steelwork, Teeside Polytechnic, England.

Melchers, R.E. and Kaur, D. (1982): Behaviour of Frames with Flexible Joints, Proc. of the 8th Australian Conference on Mechanics of Structural Materials, Newcastle, Australia, pp. 271–275.

Moncarz, P.D. and Gerstle, K.H. (1981): Steel Frames with Nonlinear Connections. *J. Struct. Div. ASCE* **107**(ST8), 1427–1441.

Monforton, A.R. and Wu, T.S. (1963): Matrix analysis of semi-rigid connected frames. *J. Struct. Div. ASCE* **89**(ST6), 13–42.

Moore, D.B., Nethercot, D.A. and Kirby, P.A. (1993a): Testing steel frames at full scale. *Struct. Engineer* **71**, 418–427.

Moore, D.B., Nethercot, D.A. and Kirby, P.A. (1993b): Testing steel frames at full scale: Appraisal of results and implications for design. *Struct. Engineer* **71**, 428–435.

Munse, W.H., Bell, W.G. and Chesson, E.J. (1959): Behaviour of Beam-to-Column Connections, Dept. of Civil Engineering, Univ. of Illinois, Urbana, IL.

Nader, M.N. and Astaneh, A. (1991): Dynamic behaviour of flexible, semirigid and rigid steel frames. *J. Construct. Steel Res.* **18**, 179–192.

Nethercot, D.A. (1985): Steel Beam-to-Column Connections – A Review of Test Data, Construction Industry Research and Information Association, London, England.

Ostrander, J.R. (1970): An Experimental Investigation of End-Plate Connections, Master's Thesis, Univ. of Saskatchewan, Saskatoon, SK, Canada.

Popov, E.P. (1983): Seismic Moment Connections for Moment-Resisting Steel Frames, Report No. UCB/EERC-83/02, Earthquake Engineering Research Center, Univ. of California, Berkeley, CA.

Popov, E.P. and Stephen, R.M. (1970): Cyclic Loading of Full-Size Steel Connections, Report No. UCB/EERC-70/03, Earthquake Engineering Research Centre, Univ. of California, Berkeley. (Republished as Bulletin No. 21, American Iron and Steel Institute, 1972).

Ramberg, W. and Osgood, W.R. (1943): Description of Stress–Strain Curves by Three Parameters, Technical Report No. 902, National Advisory Committee for Aeronautics, Washington, DC.

Rathbun, J.C. (1936): Elastic properties of riveted connections. *ASCE Trans.* **101**(1933), 524–563.

Razzaq, Z. (1983): End restraint effect on steel column strength. *J. Struct. Div. ASCE* **109**(ST2), 314–334.

Richard, R.M. (1961): A Study of Structural Systems Having Conservative Nonlinearity, Ph.D. Thesis, Purdue Univ., West Lafayette, IN.

Richard, R.M., Kreigh, J.D. and Hormby, D.E. (1982): Design of single plate framing connections with A307 bolts. *Eng. J. AISC* **19**(4), 209–213.

Richard, R.M. and Abbott, B.J. (1975): Versatile elastic–plastic stress–strain formula. *J. Eng. Mech. Div. ASCE* **101**(4), 511–515.

Romstad, K.M. and Subramanian, C.V. (1970): Analysis of frames with partial connection rigidity. *J. Struct. Div. ASCE* **96**(ST11), 2283–2300.

Sherbourne, A.N. (1948): Bolted beam-to-column connections. *Struct. Eng.* **27**.

Sherbourne, A.N. (1961): Bolted beam-to-column connections. *Struct. Engineer* **39**(6), 203–210.

Shi, G. and Atluri, S.N. (1989): Static and dynamic analysis of space frames with nonlinear flexible connections. *Int. J. Num. Methods Eng.* **28**, 2635–2650.

Shi, Y.J., Chan, S.L. and Wong, Y.L. (1996): Modelling for moment–rotation characteristics for end-plate connections. *J. Struct. Eng. ASCE* **112**(11), 1300–1306.

Sivakumaran, K.S. (1988): Seismic response of multi-storey steel buildings with flexible connections. *Eng. Struct.* **10**, 239–248.

Sommer, W.H. (1969): Behaviour of Welded Header Plate Connections, Master's Thesis, Univ. of Toronto, ON, Canada.

Stelmack, T.W., Marley, M.J. and Gerstle, K.H. (1986): Analysis and tests of flexibly connected steel frames. *J. Struct. Eng. ASCE* **112**(7), 1573–1588.

Sugimoto, H. and Chen, W.F. (1982): Small end restraint effects on the strength of H-columns. *J. Struct. Div. ASCE* **108**(ST3), 661–681.

Tarpy, T.S. and Cardinal, J.W. (1981): Behaviour of Semi-Rigid Beam to Column End Plate Connections, Proc. of Conference on Joints in Structural Steelwork, Pentach Press, England.

Thompson, L.E., McKee, R.J. and Visintainer, D.A. (1970): An Investigation of Rotation Characteristics of Web Shear Framed Connections Using A-36 and A-441 Steels, Dept. of Civil Engineers, Univ. of Missouri-Rolla, Rolla, MO.

Vinnakota, S. (1983): Planar strength of restrained beam-columns. *J. Struct. Div. ASCE* **108**(11), 2496–2515.

Wilson, W.M. and Moore, H.F. (1917): Tests to Determine the Rigidity of Riveted Joints in Steel Structures, Bulletin No. 104, Engineering Experiment Station, Univ. of Illinois, Urbana, IL.

Wu, F.H. and Chen, W.F. (1990): A design model for semi-rigid connections. *Eng. Struct.* **12**(April), 88–97.

Yee, Y.L. and Melchers, R.E. (1986): Moment–rotation curves for bolted connections. *J. Struct. Eng. ASCE* **112**(ST3), 615–635.

Young, C.R. and Jackson, K.B. (1934): The relative rigidity of welded and riveted connections. *Can. J. Res.* **11**(1–2), 62–134.

Youssef-Agha, W., Aktan, H.M. and Olowokere, O.D. (1989): Seismic response of low-rise steel frames. *J. Struct. Div. ASCE* **115**(3), 594–607.

Yu, C.H. and Shanmugam, N.E. (1986): Stability of frames with semi-rigid joints. *Comput. Struct.* **23**(5), 639–648.

Zhu, K., Al-Bermani, F.G.A., Kitipornchai, S. and Li, B. (1995): Dynamic response of flexibly jointed frames. *Eng. Struct.* **17**(8), 575–580.

Zoetemeijer, P. and Kolstein, M.H. (1975): Flush End Plate Connections, Stevin Laboratory Report No. 6-75-20, Delft Univ. of Technology, Delft, The Netherlands.

Chapter 6

NON-LINEAR STATIC ANALYSIS ALLOWING FOR PLASTIC HINGES AND SEMI-RIGID JOINTS

6.1. Abstract

A simple, accurate and efficient numerical procedure for collapse analysis of inelastic steel frames with semi-rigid connections is presented in this Chapter. Non-linearities due to geometrical change or the geometrical second-order effect, connection flexibility and material yielding are considered. The refined-plastic hinge model is proposed for smooth transition from the elastic to the plastic state. The plastic hinge is modelled by a section spring of degradable stiffness accounting for the gradual plastification of a cross-section at the two ends of an element. To determine a more rigorous full-yield strength surface of I- or H-sections under moment and axial force, the "section assemblage approach" based on the simple concept of the central cored web taking axial force, and the remaining area taking moments is proposed. The formulation of cross-sectional strength is based solely on the fundamental sectional dimensions. The approach is more consistent and rational than other existing methods including the bi-linear equation in AISC-LRFD (1986) and the artificially prescribed full-yield surfaces covering all section sizes. As described in Chapter 5, the deformation behaviour of beam-column connections is modelled by a connection spring. The section and connection springs can be combined to form a resultant spring termed as the springs-in-series. This hybrid spring is capable of modelling the combined effect of material yielding and connection flexibility without numerical divergence and ill-conditioning. The spring can be easily incorporated into the element stiffness matrix via the static condensation procedure so that no additional degree-of-freedoms will be created in the global stiffness equation which leads to much convenience and simplicity. The proposed refined-plastic hinge method with the springs-in-series model is incorporated in the computer program GMNAF and verified against the European calibration frames. It was found that the program is accurate and efficient in tracing the non-linear load–deformation curve and in predicting the load-carrying capacity of the structure.

6.2. Introduction

In conventional design, the first-order linear analysis for elastic structures with pinned or rigid connections is commonly employed. Linear analysis results, however, are only valid for structures in the small deformation and elastic range. When the geometrical change is significant or the material yields, the linear assumptions will become invalid.

Non-linear analysis of steel structures therefore becomes an essential part of structural design, especially for slender structures at ultimate limit state.

Non-linearity due to beam-to-column connection deformation, member material yielding, and the second-order effects of geometric changes govern the load-carrying capacity of framed structures. These non-linear effects are seldom considered in analysis but are allowed for in the conventional design procedure using the design code of practice. The availability of fast-speed and low-cost personal computers has enabled analysis methods requiring extensive computational time widely used in practice. Naturally, the more advanced, accurate and sophisticated analysis techniques are demanded. Recently, the Australian Limit States Standard for Structural Steelwork (AS4100, 1990) included an Appendix offering a rough guideline for the use of advanced analysis, which can be defined as any analysis method capturing the important features for the behaviour of practical steel structures where only section capacity check is needed and individual member capacity check is not necessary. An important advantage of the method is that the assumption of effective length for a particular column is not required and this leads to much convenience and improvement in accuracy since this assessment of effective length is difficult to determine in many cases.

Material yielding and instability of structural members are major factors controlling the ultimate load of a structure. For framed structures, plastic analysis can be broadly classified into two main types; the distributed plasticity and lumped plasticity approaches. The distributed plasticity method, also referred to as the plastic zone method, models the spreading of plasticity within the whole volume of the structure. The procedure is to discretize each member into many elements longitudinally and transversely. Monitoring of stress and strain is then carried out for all fibres in the analysis and residual stresses can be explicitly included by assigning a stress for each sub-area before loading. This stress can be varied along the side of the section and across the section thickness (Chan, 1990). On the other hand, the lumped plasticity approach, also referred to as the plastic hinge method, simulates the spreading of cross-section plastification at the two ends of elements through the use of spring elements in the computer analysis.

Generally speaking, the plastic zone approach is more accurate than the plastic hinge method because the fundamental stress–strain relationship is explicitly and directly used for moments and forces computation. Owing to the huge amount of calculation work required, the plastic zone approach is only suitable for simple structures. In contrast to this approach, an equivalent force–deformation relationship derived from the stress–strain relationship is adopted in the plastic hinge approach to control the cross-sectional plastification at element ends. Since no complicated integration procedure is needed to determine the resisting forces from the fibre stress distribution of a cross-section, the plastic hinge approach is more efficient and thus preferred in practical engineering design. Furthermore, the moment and force interactive equation on various design codes can be directly incorporated into the analysis program such that the method can exercise the design code requirements directly.

This Chapter is confined to the study of second-order inelastic analysis for semi-rigid steel frames using the plastic hinge approach. A literature review on the static plastic analysis methods is given in Section 6.3. Some commonly used inelastic models are described in Section 6.4. A refined-plastic hinge method based on the plastic hinge approach is proposed and formulated in Section 6.5. Gradual material yielding and member residual stress are also considered in the method. The concept of section assemblage used to determine the full-yield strength moment–axial force surface of a cross-section is introduced. Using the springs-in-series model, the numerical procedure for non-linear analysis allowing for the combined effects of connection flexibility and material yielding is formulated in Section 6.6. The procedure has been coded into the developed computer static analysis program GMNAF. Through the numerical examples illustrated in Section 6.7, this new program is verified against a number of well-known benchmark and calibration frames. Finally, the behaviour and load-carrying capacity of inelastic steel frames with semi-rigid joints are investigated and discussed.

6.3. Literature review

A literature review on the research work for the joint component of a semi-rigid frame is given in Chapter 5. Previous work on static inelastic analysis of steel structures is reported in this Section.

Since the early twentieth century, a series of tests to determine the strength of beam-column members under various combined axial force and end moment have been conducted (Johnston and Cheney, 1942; Massonnet and Campus, 1956; Ketter et al., 1952; Ketter et al., 1955; Mason et al., 1958). Based on the test results, Galambos and Ketter (1961) proposed a set of interaction curves for determining the in-plane strength of wide flange beam columns. The influence of residual stresses on cross-sections was considered in the interaction curves. For ease of design, the interaction curves were fitted into approximate equations and represented graphically on the design charts. They could be used directly and conveniently for member capacity check by designers. Recently, Duan and Chen (1989a) proposed a unified design interaction equation for steel beam-columns subjected to compression combined with bending moments either about the major axis, the minor axis or with biaxial bending moments. These equations are convenient for manual checking of member capacity.

Since the results computed by the plastic zone approach are accurate, many researchers (Chu and Pabarcius, 1964; Alvarez and Birnstiel, 1969; El-Zanaty et al., 1980; Yang and Saigal, 1984; White, 1988; Hsiao et al., 1988; Karamanlidis, 1988; Meek and Loganathan, 1990; Kitipornchai et al., 1990; Kitipornchai and Chan, 1990; and Izzuddin and Smith, 1996a, 1996b) employed this approach to study the elasto-plastic structural responses. Their results are usually considered as the exact or benchmark solutions that are used for verifying the accuracy and versatility of developed computer programs. Toma and Chen (1992) studied three European calibration frames as benchmark examples for second-order inelastic analysis of frames. Toma and Chen (1992) employed the plastic zone method to study the inelastic responses of the frames. These calibration frames include a portal frame, a gable frame and a six-storey frame and they are sometimes referred to as the European calibration frames. Recently, Toma et al. (1995) further selected some useful calibration frames as benchmark examples for verification studies. The necessity of calibration frames for program quality assurance was emphasized by Task Group 28 (Computer Applications) of the Structural Stability Research Council (SSRC) at the 1990

Annual Meeting (Basu, 1990). In Australia, Bridge et al. (1991) suggested the need for establishing a suitable set of benchmark problems to validate developed computer programs. In Japan, Toma and Chen (1994) provided detailed information for four test results of frames subjected to cyclic loading up to inelastic deformation ranges. These four frames were a full-scaled portal and a gable frame, and a one-quarter-scaled one-storey and a two-storey frame. The test results provide model solutions for verifying existing inelastic analysis methods.

As pointed out by Ziemian (1993), the computational expense required for the plastic zone approach could be one hundred times higher than that for the plastic hinge method. For this reason, the plastic hinge method is an alternative approach offering the advantage of cost-effectiveness and accuracy meeting the requirement of practicing engineers. In the past two decades, research work on developing and applying the second-order plastic-hinge-based inelastic analysis of steel structures has been extensive. Material yielding is commonly assumed to take place only at the zero-length ends of elements. Nedergaard and Pedersen (1985) presented a structural analysis procedure for space frames with material and geometrical non-linearities. Material non-linearity was taken into account through the plastic hinge method including the effect of gradual plastification for thin-walled circular tube sections. A linear conical shape initial yield and a spherical shape fully plastic yield surface functions were proposed for the interaction of axial force and the two biaxial bending moments. For a force point between the surfaces of full plastic and first yield, the corresponding degree of plastification at the two beam ends was determined by a plastic reduction matrix to modify the element stiffness. Chen and Powell (1982, 1986) adopted a generalized plastic hinge concept for three-dimensional beam-column elements. They assumed that the plastic hinge was lumped to the two ends of an element. In their inelastic model, four sub-hinge elements were used to represent the complete plastic hinge. Each element accounts for an action–deformation pair, namely, the axial force and corresponding axial deformation, torque and corresponding twist-rotation, and two bending moments about the two principal axes and the corresponding rotations. Their relationship is rigid-plastic and may have up to three linear segments. The final plastic hinge is represented by a series model composed of these sub-hinges. Each sub-hinge has its first yield and full yield surfaces. The overall element behaviour of a complete plastic hinge is, thus, multi-linearly inelastic. The approach using the concept of sub-hinge elements is a generalized method to consider varying degree of plastic deformations at a plastic hinge.

Recently, King et al. (1992) presented two simplified methods for inelastic analysis of steel frames with geometric non-linearity. These are the modified plastic-hinge and the beam-column strength approaches. A linear initial-yield and a fixed full-yield surface of compact I-shaped cross-sections with residual stress were used to account for material non-linearity. To simulate the gradual plastification behaviour, the element stiffness was degraded by a set of reduction parameters in the stiffness formulation. Liew et al. (1992, 1993) proposed a second order refined-plastic hinge analysis method for steel frame design. This method accounts for inelastic stiffness degradation. It was based on a tangent modulus concept in which the modulus of elasticity was reduced continuously from the elastic stage to the plastic stage. The method was evaluated from the column equations of the specifications (AISC-LRFD, 1986; Galambos, 1988). Three types of reduction functions for the modulus of elasticity were suggested. White (1993) compared

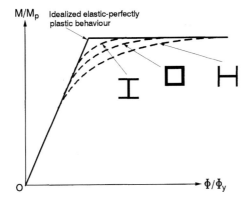

Fig. 6.1. Diagrammatic moment–curvature relationship.

several plastic-hinge methods for second-order elasto-plastic analysis of steel frames. Al-Mashary and Chen (1991) and Yau and Chan (1994) presented simplified second-order inelastic analysis methods of steel frames, in which the gradual cross-section plastification at member ends was modelled by a zero-length pseudo-spring with variable rotational stiffness. A fixed full-yield surface is used to determine moment strength capacity. Chen and Chan (1995) formulated an extended pseudo-end-spring model and an efficient method for elasto-plastic large-defection analysis of steel frames under uniformly distributed load by considering an element with plastic hinges at mid-span and two ends. Kato et al. (1998) used the plastic hinge method to study the collapse load of semi-rigid reticulated domes.

In order to further reduce computational effort, several simplified elasto-plastic analysis methods are proposed. Guralnick and He (1992) presented an efficient energy method for the collapse analysis of elastic–perfectly plastic frames. However, the approach is designed for structures under symmetric loading. Wong and Tin-Loi (1990) presented a simple procedure for tracing the load–deformation curve of elastic–perfectly plastic frames. The basis of this method was a direct combination of two separate formulations for large deflection elastic analysis and for small deflection plastic analysis. This method is simple but it may contain kinks in the load–deformation curve.

6.4. Existing inelastic analysis methods

In the inelastic range, material yielding limits the moment capacity of a cross-section. The cross-section cannot resist further moment when the capacity is reached. Material yielding affects significantly the ultimate load capacity for most structures. Figure 6.1 shows the schematic representation of a normalized moment–curvature relationship for several types of cross-sections. For simplicity, the strain-hardening effect of yielding is not illustrated in the diagram and is not considered in analysis. From the Figure, the moment–curvature curves converge to the fully plastic moment with different rates and curvatures. These variations are mainly due to the shape factor of the cross-sections, which is defined

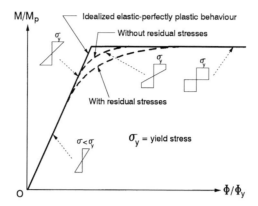

Fig. 6.2. Moment–curvature relationship for I-section with and without residual stresses.

as the ratio of plastic section modulus to elastic section modulus. For a section with a shape factor value close to unity, the moment versus curvature curve will be close to the idealized bi-linear elastic–perfectly plastic moment–curvature relation. As moment at the elastic limiting moment (i.e. $M = M_y$), the section begins to yield at its outermost fibre. The section continues to yield until the full plastic moment is developed (i.e. $M = M_p = Z_p \sigma_y$, where Z_p is the plastic modulus and σ_y is the yield stress of the material) and the whole section is completely yielded at this stage. Since residual stresses in hot-rolled sections are unavoidable during the cooling process after welding and fabrication, the presence of residual stresses in a member will result in an early yielding of the section, as graphically illustrated in Fig. 6.2.

Second-order plastic analysis of steel structures can be broadly classified into two main approaches, which are the lumped plasticity (or the plastic hinge) and the distributed plasticity (or the plastic zone) approaches. In the former method, progressive cross-section yielding at the ends of members is assumed. In the latter approach, spread of yielding in the finely meshed fibres for the whole volume of a structure is considered. Several commonly used plasticity methods are described below.

6.4.1. Plastic hinge approach

The basis of this approach is cross-section plastification. Material yielding is accounted for by zero-length plastic hinges at the end(s) of each element. Plasticity is assumed to be lumped only at the two ends of an element, while the portion within the element is assumed to remain elastic throughout the analysis. Several methods of modelling the cross-section plastification at the ends of an element are summarised as follows.

6.4.1.1. Elastic–plastic hinge method

In the elastic–plastic hinge method, the beam–column element is assumed to remain elastic before the bending moment reaches the cross-section plastic strength at the critical end(s) of the element. The cross-section is assumed to be either ideally elastic or fully plastic (i.e. elastic–perfectly plastic or elastic–plastic). When the moment reaches the

Chapter 6. Non-linear static analysis allowing for plastic hinges and semi-rigid joints

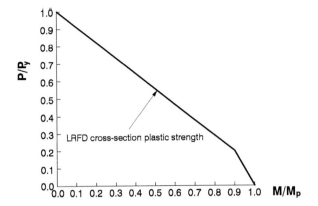

Fig. 6.3. AISC-LRFD bilinear plastic strength curve (AISC-LRFD, 1986).

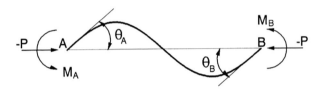

Fig. 6.4. Beam–column element.

yield criteria at a cross-section, a fully plastic hinge is assumed to have been developed. In the design of steel beams and columns, the AISC-LRFD bi-linear interaction equation (AISC-LRFD, 1986) for the cross-section plastic strength shown in Fig. 6.3 can be used and is given by

$$\frac{P}{P_y} + \frac{8}{9}\frac{M}{M_p} = 1.0 \quad \text{for } \frac{P}{P_y} \geqslant 0.2,$$
$$\frac{P}{2P_y} + \frac{M}{M_p} = 1.0 \quad \text{for } \frac{P}{P_y} < 0.2, \tag{6.1}$$

in which $P_y = \sigma_y A$ is the squash load of the cross-section; $M_p = \sigma_y Z_p$ is the plastic moment capacity for member under pure bending action; σ_y is the yield stress of material; A is the cross-sectional area; Z_p is the plastic modulus; and P and M are the axial force and bending moment at the cross-section being considered. The equation provides a reasonable lower-bound fit to most cross sections bent about their principal axes. Although, the elastic–plastic hinge method is simple, it does not account for partial yielding.

Liew et al. (1993) adopted the cross-section plastic strength equation (6.1) in their elastic–plastic analysis. The internal forces of an element are shown in Fig. 6.4. Within the elastic range, the incremental force–displacement relationship of an element has the familiar form as

$$\begin{pmatrix} \Delta M_A \\ \Delta M_B \\ \Delta P \end{pmatrix} = \frac{EI}{L} \begin{bmatrix} S_1 & S_2 & 0 \\ S_2 & S_1 & 0 \\ 0 & 0 & A/I \end{bmatrix} \begin{pmatrix} \Delta \theta_A \\ \Delta \theta_B \\ \Delta e \end{pmatrix}, \qquad (6.2)$$

in which ΔM_A, ΔM_B, $\Delta \theta_A$, $\Delta \theta_B$ are the incremental nodal moments and the corresponding joint rotations at element ends A and B, respectively; ΔP and Δe are the incremental axial force and displacement; S_1 and S_2 are the stability functions for stiffness under the action of the axial force (Chen and Lui, 1991), E is the elastic modulus, I is the moment of inertia of the cross-section; and L is the length of member.

To account for the formation of plastic hinges, the beam–column stiffness must be modified. If a plastic hinge is formed at element end A, the incremental force–displacement relationships is written as

$$\begin{pmatrix} \Delta \overline{M}_{pcA} \\ \Delta M_B \\ \Delta P \end{pmatrix} = \frac{EI}{L} \begin{bmatrix} S_1 & S_2 & 0 \\ S_2 & S_1 & 0 \\ 0 & 0 & A/I \end{bmatrix} \begin{pmatrix} \Delta \theta_A \\ \Delta \theta_B \\ \Delta e \end{pmatrix}, \qquad (6.3)$$

in which $\Delta \overline{M}_{pcA}$ is the incremental change in plastic moment capacity at end A as P changes. From the first row of Eq. (6.3), $\Delta \theta_A$ can be written as

$$\Delta \theta_A = \frac{\Delta \overline{M}_{pcA}}{S_1} \left(\frac{L}{EI} \right) - \frac{S_2 \Delta \theta_B}{S_1}. \qquad (6.4)$$

Substituting Eq. (6.4) into (6.3), the modified stiffness matrix is further expressed as

$$\begin{pmatrix} \Delta M_A \\ \Delta M_B \\ \Delta P \end{pmatrix} = \frac{EI}{L} \begin{bmatrix} 0 & 0 & 0 \\ 0 & (S_1^2 - S_2^2)/S_1 & 0 \\ 0 & 0 & A/I \end{bmatrix} \begin{pmatrix} \Delta \theta_A \\ \Delta \theta_B \\ \Delta e \end{pmatrix} + \begin{pmatrix} 1 \\ S_2/S_1 \\ 0 \end{pmatrix} \Delta \overline{M}_{pcA}. \qquad (6.5)$$

Equation (6.5) represents the modified element stiffness accounting for the presence of a plastic hinge formed at the end A of an element.

If plastic hinges are formed at both ends (A and B) of the element, Eq. (6.5) is rewritten by the similar procedure as

$$\begin{pmatrix} \Delta M_A \\ \Delta M_B \\ \Delta P \end{pmatrix} = \frac{EI}{L} \begin{bmatrix} 0 & 0 & 0 \\ 0 & 0 & 0 \\ 0 & 0 & A/I \end{bmatrix} \begin{pmatrix} \Delta \theta_A \\ \Delta \theta_B \\ \Delta e \end{pmatrix} + \begin{pmatrix} \Delta \overline{M}_{pcA} \\ \Delta \overline{M}_{pcB} \\ 0 \end{pmatrix}. \qquad (6.6)$$

Although the elastic–plastic hinge method is simple, it cannot account for partial yielding of material and does not give reasonable results. To overcome this defect, the following methods to account for the material gradual yielding can be used.

6.4.1.2. Column tangent modulus method

This method models the plastic strength of heavily axial loaded column members. The concept of the column tangent modulus method is to reduce the modulus of elasticity in

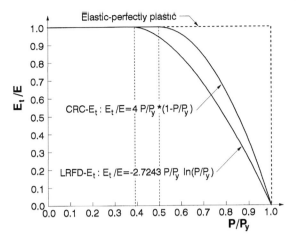

Fig. 6.5. Column equations: inelastic stiffness reductions for axial force effect (Liew et al., 1993).

the element stiffness calculation. The tangent modulus may be evaluated from the column equations of the specifications used by the designers. This method has been employed by Liew et al. (1993).

In the AISC-LRFD, the tangent modulus E_t is directly related to the inelastic stiffness reduction factors for calculation of column strength in the inelastic range (AISC-LRFD, 1986). Liew et al. (1993) rewrote the expression in terms of the ratio of the tangent modulus to the elastic modulus E_t/E as

$$\frac{E_t}{E} = 1.0 \quad \text{for } P \leqslant 0.39\, P_y, \tag{6.7a}$$

$$\frac{E_t}{E} = -2.7243 \frac{P}{P_y} \ln\left(\frac{P}{P_y}\right) \quad \text{for } P > 0.39\, P_y, \tag{6.7b}$$

in which E_t and E are the tangent and elastic moduli and P and P_y are the axial force and the squash axial load capacity, respectively. This LRFD-E_t model is derived from the LRFD column strength formula and includes the effects of member initial out-of-straightness and residual stress in member effective stiffness. The plot of the LRFD column equation is shown in Fig. 6.5. The tangent elasticity E_t remains E before the axial force P reaches $0.39 P_y$. The tangent elasticity E_t then degrades gradually from E to zero as the axial force increases from $0.39 P_y$ to P_y. Based on the LRFD-E_t model in Eqs. (6.7a) and (6.7b), the compressive axial load–deformation relationship is derived as

$$e = \frac{PL}{EA} \quad \text{for } P \leqslant 0.39\, P_y, \tag{6.8a}$$

$$e = \frac{0.39\, P_y L}{EA} + \int_{0.39 P_y}^{P} \frac{L}{AE_t}\, dP \quad \text{for } P > 0.39\, P_y. \tag{6.8b}$$

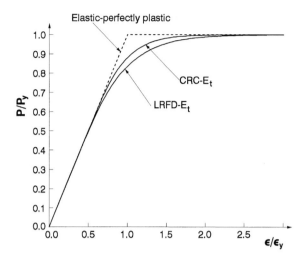

Fig. 6.6. Column equations: normalized axial force–displacement relationship (Liew et al., 1993).

By substituting Eq. (6.7) for E_t into Eq. (6.8) and then taking integration for Eq. (6.8b), the LRFD-E_t normalized axial force–displacement relationship is obtained as

$$\frac{P}{P_y} = \frac{\varepsilon}{\varepsilon_y} \quad \text{for } P \leqslant 0.39 P_y, \tag{6.9a}$$

$$\frac{P}{P_y} = \exp\left\{-0.9416 \exp\left[2.7243\left(0.39 - \frac{\varepsilon}{\varepsilon_y}\right)\right]\right\} \quad \text{for } P > 0.39 P_y, \tag{6.9b}$$

where $\varepsilon/\varepsilon_y = Ee/(\sigma_y L)$ is the normalized axial strain. The relationship in Eqs. (6.9a) and (6.9b) is depicted in Fig. 6.6.

The Column Research Council (CRC) has adopted column equations based on the tangent modulus concept as

$$\frac{E_t}{E} = 1.0 \quad \text{for } P \leqslant 0.5 P_y, \tag{6.10a}$$

$$\frac{E_t}{E} = \frac{4P}{P_y}\left(1 - \frac{P}{P_y}\right) \quad \text{for } P > 0.5 P_y. \tag{6.10b}$$

Before axial force reaches 0.5 P_y, the effective tangent modulus remains the elastic modulus (i.e. $E_t = E$). For further increases in axial force, the tangent modulus stiffness reduces continuously. Equations (6.10) is plotted in Fig. 6.5. Based on the CRC-E_t model in Eqs. (6.10a) and (6.10b), the normalized total axial load–displacement is derived as

$$e = \frac{PL}{EA} \quad \text{for } P \leqslant 0.5 P_y, \tag{6.11a}$$

$$e = \frac{0.5 P_y L}{EA} + \int_{0.5 P_y}^{P} \frac{L}{AE_t} dP \quad \text{for } P > 0.5 P_y. \tag{6.11b}$$

Similarly, substituting Eqs. (6.10a) and (6.10b) into (6.11a) and (6.11b) and then integrating Eqs. (6.11a) and (6.11b), the CRC-E_t normalized axial force–displacement relationship can be written as

$$\frac{P}{P_y} = \frac{\varepsilon}{\varepsilon_y} \quad \text{for } P \leqslant 0.5 P_y, \tag{6.12a}$$

$$\frac{P}{P_y} = \frac{1}{1 + \exp(2 - 4\varepsilon/\varepsilon_y)} \quad \text{for } P > 0.5 P_y. \tag{6.12b}$$

Equations (6.12a) and (6.12b) is plotted in Fig. 6.6.

6.4.1.3. Beam–column stiffness degradation method

In the beam–column stiffness degradation method, the element stiffness is reduced gradually from the purely elastic to perfectly plastic states. Liew et al. (1992, 1993) proposed the reduced element stiffness considering the partial yielding at both ends of an element as

$$\begin{pmatrix} \Delta M_A \\ \Delta M_B \end{pmatrix} = \frac{E_t I}{L} \begin{bmatrix} \phi_A \left[S_1 - \frac{S_2^2}{S_1}(1 - \phi_B) \right] & \phi_A \phi_B S_2 \\ \phi_A \phi_B S_2 & \phi_B \left[S_1 - \frac{S_2^2}{S_1}(1 - \phi_A) \right] \end{bmatrix} \begin{pmatrix} \Delta \theta_A \\ \Delta \theta_B \end{pmatrix}, \tag{6.13}$$

in which S_1 and S_2 are the traditional beam-column stability functions (Livesley and Chandler, 1956); E_t is the tangent modulus which is based on the LRFD-E_t or CRC-E_t equation (i.e. Eq. (6.7) or (6.10)); and ϕ_A and ϕ_B are scalar parameters that allow for progressive reduction of inelastic element stiffness associated with the effect of cross-section plastification at the element ends A and B, respectively. It is noted that when $\phi = 1$, the cross-section remains elastic. When $0 < \phi < 1$, the cross-section is under partial plastification. When $\phi = 0$, a fully plastic hinge is formed at the section. The parameter ϕ is a prescribed function given by

$$\phi = h(\alpha) = h(M, P), \tag{6.14}$$

in which α is a yielding index measuring the magnitude of axial force and moment at the element end. To account for partial yielding, Liew et al. (1992, 1993) employed the LRFD bilinear functions (AISC-LRFD, 1986) given by Eq. (6.1) to define α as

$$\alpha = \frac{P}{P_y} + \frac{8}{9} \frac{M}{M_p} \quad \text{for } \frac{P}{P_y} \geqslant 0.2, \tag{6.15a}$$

$$\alpha = \frac{P}{2P_y} + \frac{M}{M_p} \quad \text{for } \frac{P}{P_y} < 0.2. \tag{6.15b}$$

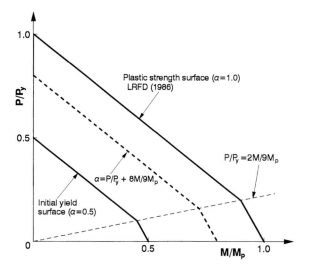

Fig. 6.7. Initial yield and plastic strength surfaces employed by Liew et al. (1993).

Liew et al. (1992, 1993) assumed that the initial yield function (e.g. for $\alpha = 0.5$ or 0.7) is similar to the full yield function shape (i.e. $\alpha = 1$), shown in Fig. 6.7. To account for partial yielding, they adopted three reduction schemes as

(A) Parabolic stiffness function with initial yield surface of $\alpha = 0.5$ as

$$\phi = 4\alpha(1-\alpha) \quad \text{for } \alpha > 0.5; \tag{6.16a}$$

(B) Parabolic stiffness function with initial yield surface of $\alpha = 0.7$ as

$$\phi = \frac{-\alpha^2 + 1.4\alpha - 0.4}{0.09} \quad \text{for } \alpha > 0.7; \tag{6.16b}$$

(C) Linear stiffness function with initial yield surface of $\alpha = 0.7$ as

$$\phi = \frac{1-\alpha}{0.3} \quad \text{for } \alpha > 0.7. \tag{6.16c}$$

King et al. (1992) proposed a similar beam-column stiffness degradation method with the use of elastic modulus E is place of the tangent modulus E_t. The reduced stiffness equation accounting for the plasticity effect at the ends A and B of an element is expressed as

$$\begin{pmatrix} \Delta M_A \\ \Delta M_B \end{pmatrix} = \begin{bmatrix} \left(K_{AA} - K_{AB} \dfrac{K_{BA}}{K_{BB}} \rho_B\right)(1-\rho_A) & K_{AB}(1-\rho_A)(1-\rho_B) \\ K_{BA}(1-\rho_B)(1-\rho_A) & \left(K_{BB} - K_{BA} \dfrac{K_{AB}}{K_{AA}} \rho_A\right)(1-\rho_B) \end{bmatrix} \begin{pmatrix} \Delta\theta_A \\ \Delta\theta_B \end{pmatrix}, \tag{6.17}$$

in which ρ_A and ρ_B are scalar parameters accounting for the plasticity effect at the ends of an element; and K_{ij} are the conventional element stiffness coefficients which can be expressed in terms of axial force (Goto and Chen, 1987) as

$$K_{AA} = K_{BB} = \frac{4EI}{L} + \frac{2PL}{15} + \frac{44P^2L^3}{25,000\,EI},$$
$$K_{AB} = K_{BA} = \frac{2EI}{L} - \frac{PL}{30} - \frac{26P^2L^3}{25,000\,EI}. \quad (6.18)$$

If the effect of residual stresses is ignored, an initial yield surface may be assumed as

$$\frac{P}{P_y} + \frac{fM}{M_p} = 1.0. \quad (6.19)$$

Partial yielding is considered when the interaction of axial force and moment lies outside the initial yield surface function. The cross-section i ($i = A$ or B) is purely elastic for $\rho_i = 0$, perfectly plastic for $\rho_i = 1$ and partially yielded for $0 < \rho_i < 1$. To cater for the effect of residual stresses, King et al. (1992) proposed an initial yield surface as

$$\frac{P}{0.8\,P_y} + \frac{fM}{0.9\,M_p} = 1.0, \quad (6.20)$$

in which f is the shape factor of the cross-section; $P_y = A\sigma_y$ is the squash load of the element; $M_p = Z_p\sigma_y$ is the plastic moment capacity in the absence of axial load; and P and M are the axial force and bending moment. King et al. (1992) adopted the full yield surface (Duan and Chen, 1989b) as

$$\left(\frac{P}{P_y}\right)^{1.3} + \frac{M}{M_p} = 1.0, \quad (6.21)$$

and defined the parameter ρ_i as

$$\rho_i = 0 \quad \text{for } M_i < M_{yc},$$
$$0 \leqslant \rho_i = \frac{M_i - M_{yc}}{M_{pc} - M_{yc}} \leqslant 1 \quad \text{for } M_{yc} \leqslant M_i \leqslant M_{pc}, \quad (6.22)$$

in which M_i is the moment at node i ($i = A$ or B); M_{yc} and M_{pc} are the initial and full yield moments for the current axial force P on the interaction of axial force and moment. In their approach, ρ_i is equal to zero and the stiffness is ideally elastic when M_i is less than M_{yc}. However, when M_i is equal to or greater than M_{yc}, ρ_i increases linearly from zero to unity and the stiffness reduces to zero as M_i approaches M_{pc}. The initial and the full yield surfaces are depicted in Fig. 6.8.

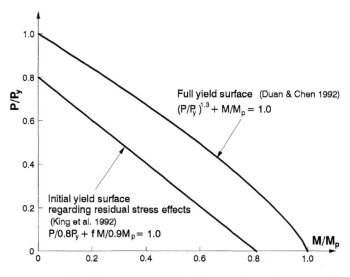

Fig. 6.8. Cross-section strength surface and initial yield surface regarding residual stress effects (King et al., 1992).

6.4.1.4. Beam–column strength degradation method

Instead of considering the plasticity effect only at the ends of an element by the cross-section strength degradation method, King et al. (1992) proposed the use of overall member strength to simulate the inelastic behaviour of beam-column elements. Similar to the cross-section strength method, a plasticity parameter is employed to model the gradual reduction of the element stiffness from the ideally elastic to fully plastic state. Only one plasticity parameter is used to represent the overall member strength. The inelastic element stiffness relationships are given by

$$\begin{pmatrix} \Delta M_A \\ \Delta M_B \end{pmatrix} = \begin{bmatrix} \left(K_{AA} - K_{AB}\dfrac{K_{BA}}{K_{BB}}\rho\right)(1-\rho) & K_{AB}(1-\rho)(1-\rho) \\ K_{BA}(1-\rho)(1-\rho) & \left(K_{BB} - K_{BA}\dfrac{K_{AB}}{K_{AA}}\rho\right)(1-\rho) \end{bmatrix} \begin{pmatrix} \Delta\theta_A \\ \Delta\theta_B \end{pmatrix}, \quad (6.23)$$

in which the plasticity parameter ρ is defined as

$$\begin{aligned} \rho &= 0 & \text{for } \gamma \leqslant 0.7, \\ \rho &= \frac{\gamma - 0.7}{0.3} \leqslant 1 & \text{for } \gamma > 0.7, \end{aligned} \quad (6.24)$$

where γ is the element strength ratio which can be evaluated from the beam-column strength equation about the strong axis (i.e. the x–x axis) as

$$\gamma = \left(\frac{P}{P_y}\right)^{1.3 + 0.002(KL/r)_x} + \frac{M}{M_p}, \quad (6.25)$$

in which P is the axial force, P_y is the column strength of member, M is the maximum second-order elastic moment within the length of the beam-column, M_p is the plastic moment capacity of the section and $(KL/r)_x$ is the effective slender ratio about the major axis.

6.4.1.5. End-spring method

The concept of this method is that the cross-section strength of an element is assumed to be represented by the stiffness of a zero-length pseudo-spring element which is attached to one end of each element. The spring stiffness reduces gradually as partial yielding takes place in the cross-section located at the end of an element. However, the portion within the beam-column element is assumed to remain elastic throughout the analysis. The stiffness of the spring element is defined by the degree of section plastification such that when the element is perfectly elastic, it remains infinite until the section material starts to yield. When the member end reaches the fully plastic moment capacity, the tangential spring stiffness will vanish to simulate the formation of a full plastic hinge at the cross-section. Between these two extreme cases, Al-Mashary and Chen (1991) suggested the following expression for the spring stiffness s as

$$s = \frac{EI}{L} \frac{2}{\frac{1-\lambda_i}{1-\lambda} - 1}, \tag{6.26}$$

in which λ is the current plasticity index defined in the following Eqs. (6.27) and (6.28); and λ_i is the plasticity index monitoring the initial yield at the cross-section. Al-Mashary and Chen (1991) adopted the full plastic criterion for a section recommended by Duan and Chen (1989b) as

$$\lambda = \left(\frac{P}{P_y}\right)^{1.3} + \frac{M}{M_p} = 1.0, \tag{6.27}$$

and also adopted the full plastic criterion about the strong-axis suggested by Orbison et al. (1982) as

$$\lambda = 1.15\, p^2 + m^2 + 3.67\, p^2 m^2 = 1.0, \tag{6.28}$$

in which $p = P/P_y$ and $m = M/M_p$. Both Eqs. (6.27) and (6.28) are for the fully plastic state (i.e. $\lambda = 1.0$). For an initially yielded state, the plasticity index can be assigned to a scaled value depending on the shape of the cross-section and the distribution of residual stresses.

6.4.2. Plastic zone approach

In the traditional plastic zone approach, all the beam and column members of a structure are discretized into a number of elements, and each cross-section is also further divided into a number of small fibres, as shown in Fig. 6.9. Physical attributes such as initial

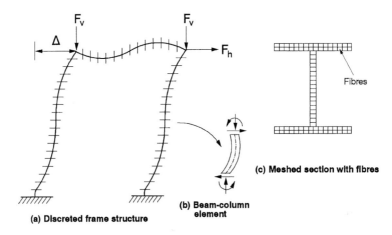

Fig. 6.9. Discretization of frame by plastic zone theory.

imperfections and residual stresses can be taken into account. The analysis results, therefore, are generally regarded as the exact solutions. However, it requires huge computer memory to store all the data from the whole volume of the structure. In addition, the analysis requires heavy and intensive calculation effort. Ziemian (1993) indicated that the computational time required for the plastic zone approach can be a hundred times greater than that for plastic hinge based calculations. Consequently, the plastic zone approach is suitable only for simple structures and it is usually used to provide the benchmark solutions for inelastic calibration problems to verify the validity of developed plastic hinge based methods. It is expected that the method cannot gain widespread acceptance by the practicing engineer in routine design, at least in the near future.

6.4.2.1. Traditional plastic zone method
In the traditional plastic zone method, each member of a structure is finely divided into many elements and every cross-section at a node is further divided into a mesh. Once the axial force, moment, axial strain and curvature are obtained, the stress in each fibre can be determined by the fundamental stress–strain relationships of the material. Initial geometrical imperfections and residual stresses can, therefore, be easily included in the calculation of stresses in fibres as geometry and state of stress before the structure is loaded. The stress–strain relationship is usually assumed to be elastic–perfectly plastic with or without strain hardening effect, as shown in Fig. 6.10(a). The stress–strain σ–ε relationship accounting for the strain hardening effect can be written as

$$\begin{aligned}
\sigma &= E\varepsilon & \text{for } \varepsilon < \varepsilon_y, \\
\sigma &= E\varepsilon_y = \sigma_y & \text{for } \varepsilon_y \leqslant \varepsilon < \varepsilon_{st}, \\
\sigma &= E\varepsilon_y + E_{st}(\varepsilon - \varepsilon_{st}) & \text{for } \varepsilon_{st} \leqslant \varepsilon,
\end{aligned} \qquad (6.29)$$

in which σ_y and ε_y are the yield strain and corresponding yield stress; ε_{st} is the strain at which the strain-hardening effect starts; E is elastic Young's modulus; and E_{st} is the strain-hardening modulus. A typical stress–strain relationship is shown in Fig. 6.10(b).

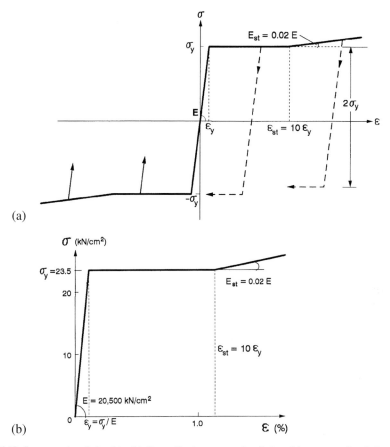

Fig. 6.10. Stress–strain relationship. (a) Generalized stress–strain relation; (b) stress–strain relationship.

6.4.2.2. Simplified plastic zone method

To take advantage of the accuracy by the plastic zone approach as well as the efficient computation by the plastic hinge approach, a simplified plastic zone method was suggested by Li and Lui (1995). The concept of this method is that the cross-section plastification is accounted for by the use of an effective cross-sectional stiffness derived from a predetermined set of cross-sectional moment–curvature–thrust (M–Φ–P) relationships, while member plastification is accounted by the use of an effective member stiffness. The fine meshing of cross-sections, therefore, is not required and the number of discretized elements along the member length is greatly reduced. The cross-sectional behaviour under the combined action of axial force and bending moment is represented quantitatively by a set of M–Φ–P relationships. Li and Lui (1995) adopted the M–Φ–P equations for I-shaped sections developed by Liapunov (1974) as

$$\Phi = \frac{M}{BEI_o} \qquad (6.30)$$

where

$$B = \left[1 - \left|\frac{M}{M_{pc}}\right|^{n_w}\right]^{1/n_w}, \tag{6.31}$$

$$n_w = Rn, \tag{6.32}$$

in which n is a factor to account for the effect of axial force; and R takes effect of residual stresses into consideration and is defined as

$$\begin{aligned} R &= 1 + \left(2.08 \left|\frac{P}{P_y}\right| - 0.665\right)\frac{\sigma_{rc}}{\sigma_y} & \text{for } P > 0 \text{ (in tension)}, \\ R &= 1 - 0.665 \frac{\sigma_{rc}}{\sigma_y} & \text{for } P < 0 \text{ (in compression)}, \end{aligned} \tag{6.33}$$

in which σ_{rc} is the maximum compressive residual stress in the cross-section; σ_y is the yield stress of the material; $P_y = A\sigma_y$ is the squash load; A is the cross-sectional area; and EI_o is the flexural constant of the member.

The expressions of n for major axis bending is given by

$$n = 10.6 \left|\frac{P}{P_y}\right|^2 - 14.75 \left|\frac{P}{P_y}\right| + 7. \tag{6.34}$$

The term M_{pc} in Eq. (6.31) is the reduced plastic moment capacity of a cross-section in the presence of axial force P and, for I-shaped sections, is given by

$$\begin{aligned} M_{pc} &= M_{px} & \text{if } 0 < |P/P_y| \leqslant 0.15, \\ M_{pc} &= 1.18(1 - |P/P_y|)M_{px} & \text{if } 0.15 < |P/P_y| \leqslant 1, \end{aligned} \tag{6.35}$$

where

$$M_p = Z_p\sigma_y, \tag{6.36}$$

in which M_p is the plastic moments of the cross-section about the major axis and Z_p is the corresponding plastic modulus. Equation (6.35) for an I-shaped cross-section strength is plotted in Fig. 6.11.

Based on Eq. (6.30), a series of non-dimensional M–Φ–P curves can be determined. The effective stiffness S of the cross-section is defined as the tangent slope of the curves and can be determined as

$$S = \frac{dM}{d\phi} = S_o\left[1 - \left|\frac{M}{M_{pc}}\right|^{n_w}\right]^{(n_w+1)/n_w} \tag{6.37}$$

in which $S_o = EI_o$ is the elastic cross-sectional stiffness. It is noted that the stiffness S will be equal to S_o when the moment M is zero, while S will be zero as the moment M

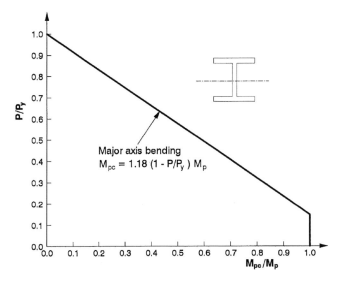

Fig. 6.11. Reduced plastic moment capacity of I-shaped cross-section under axial load (Li and Lui, 1995).

reaches the reduced plastic moment M_{pc}. As the moment gradually increases from zero to M_{pc}, the effective stiffness will be progressively degraded from S_o to zero.

The plastification in a cross-section can be reflected in the value of the effective stiffness S. The member plastification can be simplified as the summation of the effective cross-sectional stiffness at various locations along a member. The strength of the member can be related to the effective member stiffness S_{eff} which may be simply determined by

$$S_{\text{eff}} = \frac{1}{L} \int_0^L S(x)\,dx, \qquad (6.38)$$

in which L is the member length; and $S(x)$ is the cross-sectional stiffness of a cross-section located at a distance x from one end of the member. The integration of $S(x)$ can be solved numerically using the trapezoidal rule and, therefore, the effective member stiffness S_{eff} can be written as

$$S_{\text{eff}} \approx \frac{1}{8}(S_1 + S_5) + \frac{1}{4}(S_2 + S_3 + S_4), \qquad (6.39)$$

in which the subscripts 1, 2, 3, 4 and 5 are the control points at $x = 0$, $L/4$, $L/2$, $3L/4$ and L, respectively.

6.5. Proposed refined-plastic hinge method

For elasto-plastic analysis, it is important to assume a function to monitor the yielding of a section. Most analyses assume a simplified linear interaction equation to control the formation of a plastic hinge. This linear interactive function is correct for first yield analysis as it is based on the simple combined stress equation for elastic analysis. However, the interactive equation can only provide an approximation when used in plastic analysis. Researchers (Chen and Lui, 1991) adopted various equations to model the ultimate section capacity (AISC-LRFD, 1986) as follows:

$$\frac{P}{P_y} + \frac{8}{9}\frac{M}{M_p} = 1.0 \quad \text{for} \quad \frac{P}{P_y} \geqslant 0.2,$$
$$\frac{P}{2P_y} + \frac{M}{M_p} = 1.0 \quad \text{for} \quad \frac{P}{P_y} < 0.2.$$
(6.40)

This empirical formula is fitted as a lower bound capacity for a general section and, consequently, it becomes uneconomical for many other sections. Furthermore, these two discrete functions are discontinuous and lead to an abrupt change of yield function at $P = 0.2\,P_y$. It, therefore, represents an inconsistency between high and low axial load and contains a discontinuity or kink in the axial force–moment curve.

Strictly speaking, a more accurate plastic hinge function can be derived by first dividing a section into a large number of finite sub-areas. The axial and residual stress can first be used as the initial stress for the sub-areas in the section. The curvature is then gradually increased and the final strain is calculated. A set of moment versus curvature (M–Φ–P) curves can then be generated and mathematical functions can be employed to curve fit the relationship and stored in a data bank for elasto-plastic analysis. The described rigorous approach is accurate, but may become too complicated for general use. Furthermore, checking by the manual approach is difficult and cannot provide an economical evaluation as in Eq. (6.40). In addition, its use for practical design may be complicated as these moment–curvature relationships are not generally available for routine uses.

6.5.1. The full and initial yield surfaces by section assemblage method

An alternative approach which is called the section assemblage concept has been developed by Chan and Chui (1997). It is derived on the basis of plate assemblage in section. In this method, it is assumed that the web takes the axial load and the remaining unyielded part resists the moment. This simplification actually follows the rigorous formula in the reference (BS5950, 1990), and produces the same sets of rigorous moment capacity of section under axial load published by the Steel Constructional Institute (SCI, 1987). The developed software can be used for routine design as it is consistent with the current practice and can be used as an economy indicator in Eq. (6.40). Most importantly, its yield functions for both first yield and fully-plastic yield surfaces can be obtained through a data bank for sectional properties containing the dimensions of flange and web thickness. In addition, the function can be represented by a simple rule that is applicable to all sections. Empirical parameters of yield function such as Eq. (6.40) are not required in this approach.

In the new computer program GMNASF-S for analysis of frames made up of I-beams and H-columns, the section parameters such as the depth, the breadth, the flange and the web thickness are available for section capacity computation. The section is idealised as

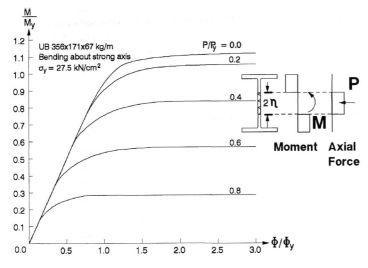

Fig. 6.12. Moment–curvature–thrust curves traced by the present theory.

an assemblage of three rectangular strips or four rectangular strips in the case of box sections. For annealed sections with no residual stress, the plastic zone for axial load resistance will be calculated as (see also Fig. 6.12)

$$\eta = \frac{P}{2\sigma_y t} \quad \text{for } \eta \leqslant \frac{d}{2},$$
$$\eta = \frac{(P - \sigma_y t d)}{2B\sigma_y} + \frac{d}{2} \quad \text{for } \frac{d}{2} < \eta \leqslant \frac{d}{2} + T, \quad (6.41)$$

in which P is the axial load; η is the half-depth of yielded area for axial load; σ_y is the yield stress; B is the breadth; t is the web thickness; T is the flange thickness; and d is the depth of web. Upon determination of the plastic zone in the section, the section moment resistance by the remaining unyielded zones can be computed as

$$M_{\text{pr}} = \left[BT(D - T) + \left(\left(\frac{d}{2}\right)^2 - \eta^2 \right) t \right] \sigma_y \quad \text{for } \eta \leqslant \frac{d}{2},$$
$$M_{\text{pr}} = \left[\left(\frac{D}{2}\right)^2 - \eta^2 \right] B\sigma_y \quad \text{for } \frac{d}{2} < \eta \leqslant \frac{d}{2} + T, \quad (6.42)$$

in which M_{pr} is the reduced plastic moment capacity of section in the presence of axial force; and D is the total depth of the section.

The section assemblage approach only uses the basic sectional parameters, the depth, breadth, web and flange thickness in the assessment of ultimate section strength by a simple concept of the central cored web taking axial force and the remaining web and flange taking moments. This idea can be applied to many typical sections including rectangular hollows, channel and I-sections. The formulation of section strength is based solely on the fundamental sectional properties. Equations (6.41) and (6.42) for I-shaped sections can produce the same values as the capacity of section under axial load published by the Steel Constructional Institute (SCI, 1987). Since the section strength equations are expressed in the interaction relationship between axial force and moment, they are more friendly to practicing engineers for section capacity check. This contrasts to other existing methods employing bi-linear full-yield surfaces or artificially prescribed shapes of yield surfaces based on curve-fitting. In these existing methods, their application to a new sectional type requires another mathematical curve-fitting procedure for generation of a new set of mathematical parameters, which is both tedious and does not carry a physical significance. This section assemblage approach is considered to be simpler, more consistent and rational than other existing methods.

When the moment at a section spring exceeds the elastic moment (i.e. $M_e = Z_e \sigma_y$), the depth of yielded zone is calculated using Eq. (6.41) by the two conditions of neutral axis. With the values of plastic moment, the normalized moment versus curvature and axial force $(M-P-\Phi)$ curves can be plotted graphically as in Fig. 6.12, together with the first yield and plastic moment envelopes shown in Fig. 6.13. When the moment–axial force coordinate lies inside the first yield envelope, the section remains elastic and no modification on the section spring is required. For any $M-P$ coordinate lying outside the full yield surface, the section spring stiffness will be reduced to zero and the moment resistance will be set to the plastic moment, enabling any unbalanced moment to be redistributed to other parts of the structure before a mechanism is formed. These two conditions are applicable for elastic–plastic hinge analysis if we define the first yield and the yield envelopes to be coincident. For elasto-plastic hinge analysis having a gradual degradation of section

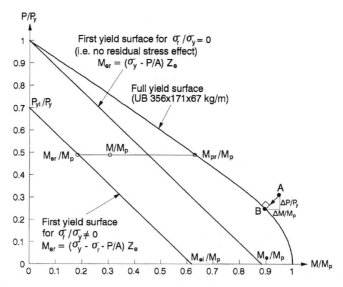

Fig. 6.13. First yield and full yield surfaces of section UB356 × 171 × 67 kg/m determined by the present theory.

stiffness and capacity in the equilibrium check, the section spring is required to reduce gradually. This condition occurs when the M–P coordinate lies between the first yield and the full yield surfaces.

For a section to initiate yielding, the first yield surface must be defined. In the presence of axial force and residual stress, the proposed first yield surface can be simply written as

$$M_{\text{er}} = \left(\sigma_{\text{y}} - \sigma_{\text{r}} - \frac{P}{A}\right) Z_{\text{e}}, \tag{6.43}$$

in which M_{er} is the reduced first yield moment, P is the axial force, σ_{r} is the maximum residual stress, A is the cross-section area and Z_{e} is the elastic modulus. It should be noted that this surface is a straight line because the maximum stress before reaching the yield stress is a linear function of strain induced by the moment and axial force.

6.5.2. Refined-plastic hinge method based on the section assemblage concept

Unlike other methods possessing an abrupt change from ideally elastic to perfectly plastic states, a refined-plastic hinge method for providing a smooth transition between these conditions (i.e. the partial yielding process) is developed (Chan and Chui, 1997). This method refines the transition of material yielding and can trace a more accurate load–deformation curve of structure. To deal with the type of elasto-plastic hinge analysis, the location of the point relative to the first and full yield surfaces is determined (see Fig. 6.13). The residual stresses adopted are based on recommendations by the European Convention for Construction Steelwork (ECCS, 1983). The maximum magnitude of residual stresses is assumed to be dependent on the depth/breadth ratio and the distribution of residual stress is shown in Fig. 6.14. This effect of residual stresses will then be included in the determination of first yield moment. Initial geometric imperfections suggested by ECCS and as shown in Fig. 6.15, are included in the analysis.

In the present refined-plastic hinge method, a pseudo-spring for modelling the decreasing rotational stiffness properties is proposed to simulate the gradual cross-sectional plas-

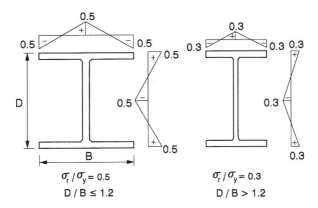

Fig. 6.14. ECCS residual stress distribution for hot-rolled I-sections (ECCS, 1983).

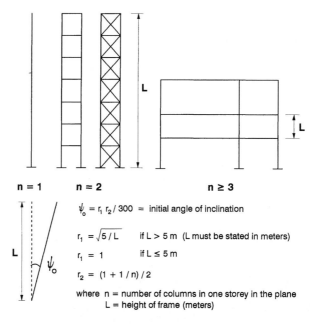

Fig. 6.15. Definition of representative column height for determining the initial out-of-plumb by ECCS (1983).

tification at an element end. With a given axial load, the elastic and plastic moments at this axial load level are then computed and used for determination of the section spring stiffness, S_s, as follows.

$$S_s = \frac{6EI}{L} \frac{|M_{pr} - M|}{|M - M_{er}|} \quad \text{for } M_{er} < M < M_{pr}, \tag{6.44}$$

in which EI is the flexural constant, L is the member length and M_{er} and M_{pr} are the reduced first yield and plastic moments in the presence of axial force and residual stress. From Eq. (6.44), the section stiffness varies from infinity to zero, representing the two extreme values of moment equal to the elastic and plastic moments. In computer analysis, however, these two extreme values are assigned as 10^{-10} and 10^{+10} of the member stiffness, respectively. Between these two values, the stiffness of section spring in Eq. (6.44) is assigned a finite value representing the degree of cross-sectional plastification at the element end.

It can be seen from Eq. (6.44) that the stiffness of a section spring is controlled by the magnitude of the existing moment, the plastic moment (M_{pr}) and the first yield moment (M_{er}) in the presence of axial load, which are computed from Eqs. (6.41) to (6.43). When the moment is less than the first yield moment, the stiffness of the section spring will be infinite, indicating that the moment can be completely transferred from one side to another of the section spring element. When the moment is equal to the plastic moment, the spring stiffness will be zero, implying that no further moment can be allowed to transfer across the spring. In case of a force point outside the full yield surface (i.e. $M > M_{pr}$), it should be moved back onto the surface to avoid the violation of plastic state and this correction procedure will be further elaborated in Section 6.5.6. The parameter $6EI/L$ is a function of element stiffness affecting the sharpness of the transition curve from the first yield to the fully plastic moments. The use of the constant 6 was found to produce a curve basically in agreement with the more rigorous moment–curvature curve for a general section. However, it was further noted that the number does not change the structural behaviour significantly provided that it does not deviate too much from the constant 6.

The proposed refined-plastic hinge method is computationally more efficient than the plastic zone method because the time-extensive integration procedure for determining the element forces from the fibre stress distribution of finely meshed section is not required. In addition, the proposed method does not require huge computer memory to store data as in the plastic zone approach. In contrast to the elastic–plastic method in which abrupt change of yielding from ideally elastic to perfectly plastic condition is assumed (Harstead et al., 1968; Alvarez and Birnstiel, 1969; Cohn and Abdel-Rohman, 1976; Kassimali, 1983; Wong and Tin-Loi, 1990; Freitas and Ribeiro, 1992; Guralnick and He, 1992), the proposed method can simulate partial yielding by providing a smooth transition path between these conditions. In some methods using the beam-column stiffness degradation approach and the stability functions, divergence occurs when the axial force of member is close to zero. In addition, the bi-linear function of full-yield surface was generally too conservative and has a discontinuity in its slope which is undesirable in computer analysis. On the other hand, the prescribed full-yield functions used by many researchers may not be general and applicable to most sections. The section assemblage concept adopted in this Chapter is simple, consistent and more rational than these existing methods. More interestingly, it does not require any pre-requisite curve-fitting procedure to obtain the yield function for a section and only the basic sectional dimensions are required.

6.5.3. The elastic–plastic hinge method based on the section assemblage method

In the present elastic–plastic hinge method, the material is assumed to be either perfectly elastic or fully plastic. This is the simplest inelastic analysis method because the computational effort accounting for partial yielding is not needed. The purpose of employing this method here is for comparison with the proposed refined-plastic hinge method. Since no partial yielding is considered in this method, the stiffness of the section spring model does not degrade and the load–deformation curve has generally one or more kinks occurring when plastic hinges are formed. The predicted ultimate load for this method is, therefore, generally higher than that for the refined-plastic hinge method. Based on the proposed full-yield strength surface determined by the section assemblage concept in Eqs. (6.41) and (6.42), the section is assumed to behave elastically when the current force-point is within this surface; and the section is assumed to form a fully plastic hinge when the force-point is on the surface. In computer analysis, the section spring is assigned a very large value of $10^{+10} EI/L$ for the elastic case (i.e. $M < M_{\text{pr}}$) and a very small value of $10^{-10} EI/L$ for the fully plastic case (i.e. $M = M_{\text{pr}}$). Thus, we have

$$\begin{aligned} S_s &= 10^{+10} EI/L \quad \text{for } M < M_{\text{pr}}, \\ S_s &= 10^{-10} EI/L \quad \text{for } M = M_{\text{pr}}. \end{aligned} \quad (6.45)$$

148   Non-linear static and cyclic analysis of steel frames with semi-rigid connections

For a force point outside the full yield surface, the point will be moved back in a normal direction to the yield surface so that the equilibrium of the member forces is maintained. In other words, moment at a section is always kept below the plastic moment capacity of a section.

6.5.4. Element stiffness formulation accounting for plasticity effect

With the values of section spring stiffness, the element stiffness matrix can be formulated and used in the second-order inelastic analysis in the following section.

The nodal rotations of a deformed beam-column element with two section springs each attached to the element end node are shown in Fig. 6.16. Owing to partial yielding of the cross-section, the rotational rigidity of a section spring softens, and rotations on the two sides of the spring are generally unequal. Considering the moment equilibrium condition at a section spring, the incremental equation can be written as

$$\begin{bmatrix} \Delta M_s \\ \Delta M_b \end{bmatrix} = \begin{bmatrix} S_s & -S_s \\ -S_s & S_s \end{bmatrix} \begin{bmatrix} \Delta \theta_s \\ \Delta \theta_b \end{bmatrix}, \tag{6.46}$$

in which ΔM_s and ΔM_b are the incremental nodal moments at the junctions between the section spring and the global node and between the beam and the section spring; $\Delta \theta_s$ and $\Delta \theta_b$ are the incremental nodal rotations corresponding to these moments. The subscripts 'b' and 's' refer to inside and outside nodes relative to the element. Combining the section spring stiffness to the ends of an element and expressing in an incremental form, we have

$$\begin{pmatrix} \Delta M_{s1} \\ \Delta M_{b1} \\ \Delta M_{b2} \\ \Delta M_{s2} \end{pmatrix} = \begin{bmatrix} S_{s1} & -S_{s1} & 0 & 0 \\ -S_{s1} & K_{11}+S_{s1} & K_{12} & 0 \\ 0 & K_{21} & K_{22}+S_{s2} & -S_{s2} \\ 0 & 0 & -S_{s2} & S_{s2} \end{bmatrix} \begin{pmatrix} \Delta \theta_{s1} \\ \Delta \theta_{b1} \\ \Delta \theta_{b2} \\ \Delta \theta_{s2} \end{pmatrix}, \tag{6.47}$$

in which the subscripts '1' and '2' refer to nodes 1 and 2, respectively. Assuming the loads are applied only at the global nodes, both ΔM_{b1} and ΔM_{b2} are equal to zero and

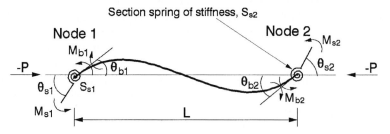

Fig. 6.16. Internal forces of an element with end-section springs accounting for cross-section plastification employed by this book.

Chapter 6. Non-linear static analysis allowing for plastic hinges and semi-rigid joints

we obtain from the second and the third rows in Eq. (6.47)

$$\begin{pmatrix} \Delta\theta_{b1} \\ \Delta\theta_{b2} \end{pmatrix} = \begin{bmatrix} K_{11}+S_{s1} & K_{12} \\ K_{21} & K_{22}+S_{s2} \end{bmatrix}^{-1} \begin{bmatrix} S_{s1} & 0 \\ 0 & S_{s2} \end{bmatrix} \begin{pmatrix} \Delta\theta_{s1} \\ \Delta\theta_{s2} \end{pmatrix}. \tag{6.48}$$

Eliminating the internal degrees of freedom by substituting Eq. (6.48) into (6.47), the final incremental stiffness relationships for the element can be formulated as

$$\begin{pmatrix} \Delta P \\ \Delta M_{s1} \\ \Delta M_{s2} \end{pmatrix} = \begin{bmatrix} EA/L & 0 & 0 \\ 0 & S_{s1}-S_{s1}^2(K_{22}+S_{s2})/\beta_s & S_{s1}S_{s2}K_{12}/\beta_s \\ 0 & S_{s1}S_{s2}K_{21}/\beta_s & S_{s2}-S_{s2}^2(K_{11}+S_{s1})/\beta_s \end{bmatrix} \begin{pmatrix} \Delta L \\ \Delta\theta_{s1} \\ \Delta\theta_{s2} \end{pmatrix}$$

$$= \begin{bmatrix} EA/L & 0 & 0 \\ 0 & k'_{22} & k'_{23} \\ 0 & k'_{32} & k'_{33} \end{bmatrix} \begin{pmatrix} \Delta L \\ \Delta\theta_{s1} \\ \Delta\theta_{s2} \end{pmatrix}, \tag{6.49}$$

in which ΔP is the axial force increment, ΔL is the axial deformation increment, and

$$\beta_s = \begin{vmatrix} K_{11}+S_{s1} & K_{12} \\ K_{21} & K_{22}+S_{s2} \end{vmatrix} = (K_{11}+S_{s1})(K_{22}+S_{s2}) - K_{12}K_{21} > 0,$$

$$K_{11} = K_{22} = \frac{4EI}{L} + \frac{2PL}{15}, \tag{6.50}$$

$$K_{12} = K_{21} = \frac{2EI}{L} - \frac{PL}{30}.$$

6.5.5. Checking of element stiffness in extreme conditions

The derived element stiffness is checked against several extreme conditions by assigning a very large and a very small stiffness to the section springs and comparing against the theoretical stiffness expressions. In total, three conditions will be checked; no plastic hinge formed at element ends, one plastic hinge formed at one element end and two plastic hinges formed at two element ends.

6.5.5.1. No plastic hinge

If no plastic hinge is formed at nodes 1 and 2 of an element, the second and third rows in Eq. (6.49) can be simplified by assigning a very large number for both S_{s1} and S_{s2} and hence we have the entries from Eq. (6.49) as

$$k'_{22} = \lim_{\substack{S_{s1}\to\infty \\ S_{s2}\to\infty}} \left\{ \frac{S_{s1}-S_{s1}^2(K_{22}+S_{s2})}{\beta_s} \right\}$$

$$= \lim_{\substack{S_{s1}\to\infty \\ S_{s2}\to\infty}} \frac{\left(\dfrac{K_{22}}{S_{s2}}+1\right)K_{11} - \dfrac{K_{12}K_{21}}{S_{s2}}}{\left(\dfrac{K_{11}}{S_{s1}}+1\right)\left(\dfrac{K_{22}}{S_{s2}}+1\right) - \dfrac{K_{12}K_{21}}{S_{s1}S_{s2}}} = K_{11}, \tag{6.51a}$$

$$k'_{33} = \lim_{\substack{S_{s1} \to \infty \\ S_{s2} \to \infty}} \left\{ \frac{S_{s2} - S_{s2}^2(K_{11} + S_{s1})}{\beta_s} \right\}$$

$$= \lim_{\substack{S_{s1} \to \infty \\ S_{s2} \to \infty}} \frac{\left(\dfrac{K_{11}}{S_{s1}} + 1\right) K_{22} - \dfrac{K_{12}K_{21}}{S_{s1}}}{\left(\dfrac{K_{11}}{S_{s1}} + 1\right)\left(\dfrac{K_{22}}{S_{s2}} + 1\right) - \dfrac{K_{12}K_{21}}{S_{s1}S_{s2}}} = K_{22}, \qquad (6.51b)$$

$$k'_{23} = \lim_{\substack{S_{s1} \to \infty \\ S_{s2} \to \infty}} \left\{ \frac{S_{s1}S_{s2}K_{12}}{\beta_s} \right\}$$

$$= \lim_{\substack{S_{s1} \to \infty \\ S_{s2} \to \infty}} \frac{K_{12}}{\left(\dfrac{K_{11}}{S_{s1}} + 1\right)\left(\dfrac{K_{22}}{S_{s2}} + 1\right) - \dfrac{K_{12}K_{21}}{S_{s1}S_{s2}}} = K_{12}, \qquad (6.51c)$$

$$k'_{32} = \lim_{\substack{S_{s1} \to \infty \\ S_{s2} \to \infty}} \left\{ \frac{S_{s1}S_{s2}K_{21}}{\beta_s} \right\}$$

$$= \lim_{\substack{S_{s1} \to \infty \\ S_{s2} \to \infty}} \frac{K_{21}}{\left(\dfrac{K_{11}}{S_{s1}} + 1\right)\left(\dfrac{K_{22}}{S_{s2}} + 1\right) - \dfrac{K_{12}K_{21}}{S_{s1}S_{s2}}} = K_{21}. \qquad (6.51d)$$

The final incremental force–displacement equation becomes

$$\begin{pmatrix} \Delta P \\ \Delta M_{s1} \\ \Delta M_{s2} \end{pmatrix} = \begin{bmatrix} EA/L & 0 & 0 \\ 0 & K_{11} & K_{12} \\ 0 & K_{21} & K_{22} \end{bmatrix} \begin{pmatrix} \Delta L \\ \Delta \theta_{s1} \\ \Delta \theta_{s2} \end{pmatrix}. \qquad (6.52)$$

The stiffness equation is finally transformed to a typical beam-column element.

6.5.5.2. Plastic hinge at node 1 only

For the case of a plastic hinge formed at node 1 of an element, the values of S_{s1} and S_{s2} are assigned to be a very small and a very large value, respectively. This simplifies the entries from Eq. (6.49) to

$$k'_{22} = \lim_{\substack{S_{s1} \to 0 \\ S_{s2} \to \infty}} \left\{ \frac{S_{s1} - S_{s1}^2(K_{22} + S_{s2})}{\beta_s} \right\}$$

$$= \lim_{\substack{S_{s1} \to 0 \\ S_{s2} \to \infty}} \frac{S_{s1}\left(K_{11} - \dfrac{K_{12}K_{21}}{S_{s2}}\right)}{(K_{11} + S_{s1})\left(\dfrac{K_{22}}{S_{s2}} + 1\right) - \dfrac{K_{12}K_{21}}{S_{s2}}} = 0, \qquad (6.53a)$$

$$k'_{33} = \lim_{\substack{S_{s1} \to 0 \\ S_{s2} \to \infty}} \left\{ \frac{S_{s2} - S_{s2}^2(K_{11} + S_{s1})}{\beta_s} \right\}$$

$$= \lim_{\substack{S_{s1} \to 0 \\ S_{s2} \to \infty}} \frac{(K_{11} + S_{s1})K_{22} - K_{12}K_{21}}{(K_{11} + S_{s1})\left(\frac{K_{22}}{S_{s2}} + 1\right) - \frac{K_{12}K_{21}}{S_{s2}}} = \frac{K_{11}K_{22} - K_{12}K_{21}}{K_{11}}, \quad (6.53b)$$

$$k'_{23} = \lim_{\substack{S_{s1} \to 0 \\ S_{s2} \to \infty}} \left\{ \frac{S_{s1}S_{s2}K_{12}}{\beta_s} \right\}$$

$$= \lim_{\substack{S_{s1} \to \infty \\ S_{s2} \to \infty}} \frac{S_{s1}K_{12}}{(K_{11} + S_{s1})\left(\frac{K_{22}}{S_{s2}} + 1\right) - \frac{K_{12}K_{21}}{S_{s2}}} = 0, \quad (6.53c)$$

$$k'_{32} = \lim_{\substack{S_{s1} \to 0 \\ S_{s2} \to \infty}} \left\{ \frac{S_{s1}S_{s2}K_{21}}{\beta_s} \right\}$$

$$= \lim_{\substack{S_{s1} \to 0 \\ S_{s2} \to \infty}} \frac{S_{s1}K_{21}}{(K_{11} + S_{s1})\left(\frac{K_{22}}{S_{s2}} + 1\right) - \frac{K_{12}K_{21}}{S_{s2}}} = 0. \quad (6.53d)$$

The final element stiffness equation becomes

$$\begin{pmatrix} \Delta P \\ \Delta M_{s1} \\ \Delta M_{s2} \end{pmatrix} = \begin{bmatrix} EA/L & 0 & 0 \\ 0 & (K_{11}K_{22} - K_{12}K_{21})/K_{11} & 0 \\ 0 & 0 & 0 \end{bmatrix} \begin{pmatrix} \Delta L \\ \Delta \theta_{s1} \\ \Delta \theta_{s2} \end{pmatrix}. \quad (6.54)$$

It is noted that the value of $(K_{11}K_{22} - K_{12}K_{21})/K_{11}$ will be equal to $3EI/L$ when axial force is absent (i.e. $P = 0$, $K_{11} = K_{22} = 4EI/L$, $K_{12} = K_{21} = 2EI/L$). This is identical to the beam stiffness with one end pinned and the other end fixed.

6.5.5.3. Plastic hinges at both nodes 1 and 2

When plastic hinges are formed at both ends of an element, both S_{s1} and S_{s2} are taken as very small numbers and, therefore, the entries from Eq. (6.49) become

$$k'_{22} = \lim_{\substack{S_{s1} \to 0 \\ S_{s2} \to 0}} \left\{ \frac{S_{s1} - S_{s1}^2(K_{22} + S_{s2})}{\beta_s} \right\}$$

$$= \lim_{\substack{S_{s1} \to 0 \\ S_{s2} \to 0}} \frac{(K_{22} + S_{s2})K_{11}S_{s1} - K_{12}K_{21}S_{s1}}{(K_{11} + S_{s1})(K_{22} + S_{s2}) - K_{12}K_{21}} = 0, \quad (6.55a)$$

$$k'_{23} = \lim_{\substack{S_{s1}\to 0 \\ S_{s2}\to 0}} \left\{ \frac{S_{s1} S_{s2} K_{12}}{\beta_s} \right\}$$

$$= \lim_{\substack{S_{s1}\to 0 \\ S_{s2}\to 0}} \frac{S_{s1} S_{s2} K_{12}}{(K_{11}+S_{s1})(K_{22}+S_{s2}) - K_{12}K_{21}} = 0, \tag{6.55b}$$

and the final equation is given by

$$\begin{pmatrix} \Delta P \\ \Delta M_{s1} \\ \Delta M_{s2} \end{pmatrix} = \begin{bmatrix} EA/L & 0 & 0 \\ 0 & 0 & 0 \\ 0 & 0 & 0 \end{bmatrix} \begin{pmatrix} \Delta L \\ \Delta \theta_{s1} \\ \Delta \theta_{s2} \end{pmatrix}. \tag{6.56}$$

In this case, no additional moments can be resisted by the member and further moments are transferred to other members. The expressions become the stiffness equation of a pin-jointed truss element.

6.5.6. Correction of force point movement on full yield surface

Once a plastic hinge is formed at one or two ends of an element, the equilibrium condition may be violated as the axial resistance and moment at the section are greater than the applied axial force and moment. This reflects the condition of the force-moment point lying outside the full yield surface indicated by point A in Fig. 6.13. There are an infinite number of paths to bring back this force-moment point onto the yield surface. In this book, the path normal to the yield surface is chosen as the recovery path. Referring to Fig. 6.13, the new equilibrium force-moment point will move from point A to B such that the resisting moment M is equal to the reduced plastic moment M_{pr} of section in the presence of axial force P (i.e. $M = M_{pr}$). Therefore, the gradient of the force versus moment coordinate is computed and used for bringing the point A onto the yield surface for the section.

6.6. Combined effect of joint flexibility and material yielding – the springs-in-series model

The modelling of joint flexibility and material yielding have been described separately in Sections 5.6 and 6.5, respectively. Each model is simulated by a pseudo-spring. In this Section, the element stiffness formulation accounting for these effects simultaneously will be presented. In fact, these two effects always coexist in collapse of real structures. Consequently, they should be incorporated into an analysis in order to simulate the actual structural response. A direct method to combine their effects is that their spring elements can be connected together to form a new resultant springs-in-series, as shown in Fig. 6.17. The inner pair of spring elements (i.e. section springs) represents the effect of section yielding, while the outer springs (i.e. connection springs) are inserted to account for the

Chapter 6. Non-linear static analysis allowing for plastic hinges and semi-rigid joints

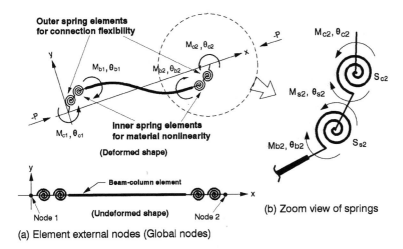

Fig. 6.17. Beam-column element attached with section and connection springs used in the present study.

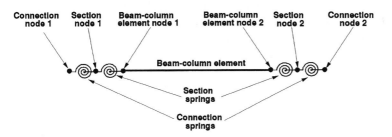

Fig. 6.18. Nodes of a hybrid element.

connection flexibility. Considering the force equilibrium and displacement compatibility conditions, the incremental moment–rotation relationship can be written as

$$\begin{pmatrix} \Delta M_{c1} \\ \Delta M_{s1} \\ \Delta M_{b1} \\ \Delta M_{b2} \\ \Delta M_{s2} \\ \Delta M_{c2} \end{pmatrix} = \begin{bmatrix} S_{c1} & -S_{c1} & 0 & 0 & 0 & 0 \\ -S_{c1} & S_{c1}+S_{s1} & -S_{s1} & 0 & 0 & 0 \\ 0 & -S_{s1} & S_{s1}+K_{11} & K_{12} & 0 & 0 \\ 0 & 0 & K_{21} & K_{22}+S_{s2} & -S_{s2} & 0 \\ 0 & 0 & 0 & -S_{s2} & S_{s2}+S_{c2} & -S_{c2} \\ 0 & 0 & 0 & 0 & -S_{c2} & S_{c2} \end{bmatrix} \begin{pmatrix} \Delta \theta_{c1} \\ \Delta \theta_{s1} \\ \Delta \theta_{b1} \\ \Delta \theta_{b2} \\ \Delta \theta_{s2} \\ \Delta \theta_{c2} \end{pmatrix}, \quad (6.57)$$

in which ΔM and $\Delta \theta$ are the incremental moments and the conjugate rotations; the subscripts 'c', 's' and 'b' refer to the moment and rotation at the connection node, the section node and the beam-column element node, as shown in Fig. 6.18. S_{c1} and S_{c2} are the tangent stiffnesses of the connection springs attached to element ends 1 and 2 respectively, and S_{s1} and S_{s2} are the section spring stiffness at ends 1 and 2.

Whilst all loads are assumed to be applied at global nodes, the internal incremental moments will be equal to zero (i.e. $\Delta M_{s1} = \Delta M_{s2} = \Delta M_{b1} = \Delta M_{b2} = 0$). Substituting the second and fifth rows into the remaining rows in Eq. (6.57) using the static condensation procedure, the equation becomes

$$\begin{pmatrix} \Delta M_{c1} \\ \Delta M_{b1} \\ \Delta M_{b2} \\ \Delta M_{c2} \end{pmatrix} = \begin{bmatrix} S_{cs1} & -S_{cs1} & 0 & 0 \\ -S_{cs1} & S_{cs1}+K_{11} & K_{12} & 0 \\ 0 & K_{21} & K_{22}+S_{cs2} & -S_{cs2} \\ 0 & 0 & -S_{cs2} & S_{cs2} \end{bmatrix} \begin{pmatrix} \Delta \theta_{c1} \\ \Delta \theta_{b1} \\ \Delta \theta_{b2} \\ \Delta \theta_{c2} \end{pmatrix}, \quad (6.58)$$

where

$$S_{cs1} = \frac{S_{c1} S_{s1}}{S_{c1}+S_{s1}}, \quad (6.59a)$$

$$S_{cs2} = \frac{S_{c2} S_{s2}}{S_{c2}+S_{s2}}. \quad (6.59b)$$

Equation (6.58) has the same form as Eq. (6.47) because the two springs (i.e. the connection and the section springs) connected in series in Eq. (6.57) have been combined to a single spring of stiffness S_{cs}. Similarly, the internal moments ΔM_{b1} and ΔM_{b2} can be condensed. The final force–displacement relationship of the element stiffness equation in the local coordinate axis can be formulated as

$$\begin{pmatrix} \Delta P \\ \Delta M_{c1} \\ \Delta M_{c2} \end{pmatrix} = \frac{1}{\beta_{cs}} \begin{bmatrix} EA/L & 0 & 0 \\ 0 & \beta_{cs}S_{cs1}-S_{cs1}^2(S_{cs2}+K_{22}) & S_{cs1}S_{cs2}K_{12} \\ 0 & S_{cs2}S_{cs1}K_{21} & \beta_{cs}S_{cs2}-S_{cs2}^2(S_{cs1}+K_{11}) \end{bmatrix} \begin{pmatrix} \Delta L \\ \Delta \theta_{c1} \\ \Delta \theta_{c2} \end{pmatrix}, \quad (6.60)$$

where

$$\beta_{cs} = \begin{vmatrix} S_{cs1}+K_{11} & K_{12} \\ K_{21} & S_{cs2}+K_{22} \end{vmatrix} = (S_{cs1}+K_{11})(S_{cs2}+K_{22}) - K_{12}K_{21}. \quad (6.61)$$

With the rotations at nodes computed as $\Delta \theta_{c1}$ and $\Delta \theta_{c2}$ in the solution for the structural stiffness equation, the rotations at the beam-column element nodes, $\Delta \theta_{b1}$ and $\Delta \theta_{b2}$, can be determined as

$$\begin{pmatrix} \Delta \theta_{b1} \\ \Delta \theta_{b2} \end{pmatrix} = \begin{bmatrix} t_{11} & t_{12} \\ t_{21} & t_{22} \end{bmatrix} \begin{pmatrix} \Delta \theta_{c1} \\ \Delta \theta_{c2} \end{pmatrix}, \quad (6.62)$$

and the rotations at the section nodes ($\Delta \theta_{s1}$ and $\Delta \theta_{s2}$) as

$$\begin{pmatrix} \Delta \theta_{s1} \\ \Delta \theta_{s2} \end{pmatrix} = \begin{bmatrix} \dfrac{S_{c1}+S_{s1}t_{11}}{S_{c1}+S_{s1}} & \dfrac{t_{12}S_{s1}}{S_{c1}+S_{s1}} \\ \dfrac{t_{21}S_{s2}}{S_{s2}+S_{c2}} & \dfrac{S_{c2}+S_{s2}t_{22}}{S_{s2}+S_{c2}} \end{bmatrix} \begin{pmatrix} \Delta \theta_{c1} \\ \Delta \theta_{c2} \end{pmatrix}, \quad (6.63)$$

where t_{ij} is defined as

$$\begin{bmatrix} t_{11} & t_{12} \\ t_{21} & t_{22} \end{bmatrix} = \frac{1}{\beta_{cs}} \begin{bmatrix} S_{cs1}(S_{cs2}+K_{22}) & -S_{cs2}K_{12} \\ -S_{cs1}K_{21} & S_{cs2}(S_{cs1}+K_{11}) \end{bmatrix}. \quad (6.64)$$

When the incremental deformations of connection and section springs are solved, the incremental resisting forces of element can be obtained.

Equation (6.60) is the final incremental force–deformation equation with the condensed section and connection springs. Therefore, it can be used directly in the conventional displacement-based finite element method without introducing additional degree-of-freedoms in the global structural stiffness matrix. The equation is simple but powerful for consideration of the combined effects of connection flexibility and material yielding since numerical instability due to a very large and a very small spring can be avoided.

For checking the validity of Eq. (6.60), the following cases can be considered by considering the extreme boundary conditions.

Case 1: No hinge

If both ends of an element are rigid joints and have no plastic hinges (i.e. $S_{c1} = S_{c2} = S_{s1} = S_{s2} = \infty$), Eq. (6.60) becomes

$$\begin{pmatrix} \Delta P \\ \Delta M_{c1} \\ \Delta M_{c2} \end{pmatrix} = \begin{bmatrix} EA/L & 0 & 0 \\ 0 & K_{11} & K_{22} \\ 0 & K_{21} & K_{22} \end{bmatrix} \begin{pmatrix} \Delta L \\ \Delta \theta_{c1} \\ \Delta \theta_{c2} \end{pmatrix}, \quad (6.65)$$

which is identical to the conventional elastic element stiffness equation.

Case 2: A hinge formed at end 1

If a fully plastic hinge or a pinned joint is assumed to be located at end 1 of the element (i.e. $[S_{c1} = 0$ or $S_{s1} = 0]$ and $S_{c2} = S_{s2} = \infty$), the equation is converted to the following:

$$\begin{pmatrix} \Delta P \\ \Delta M_{c1} \\ \Delta M_{c2} \end{pmatrix} = \begin{bmatrix} EA/L & 0 & 0 \\ 0 & (K_{11}K_{22} - K_{12}K_{21})/K_{22} & 0 \\ 0 & 0 & 0 \end{bmatrix} \begin{pmatrix} \Delta L \\ \Delta \theta_{c1} \\ \Delta \theta_{c2} \end{pmatrix}. \quad (6.66)$$

Ignoring the effect of axial force, we have $K_{11} = K_{22} = 4EI/L$ and $K_{12} = K_{21} = 2EI/L$ and the matrix in Eq. (6.66) can be converted to the followings:

$$\begin{pmatrix} \Delta P \\ \Delta M_{c1} \\ \Delta M_{c2} \end{pmatrix} = \begin{bmatrix} EA/L & 0 & 0 \\ 0 & 3EI/L & 0 \\ 0 & 0 & 0 \end{bmatrix} \begin{pmatrix} \Delta L \\ \Delta \theta_{c1} \\ \Delta \theta_{c2} \end{pmatrix}. \quad (6.67)$$

Case 3: Two hinges at both ends

If fully plastic hinges or pinned joints are formed at both ends 1 and 2 of the element (i.e. $[S_{c1} = 0$ or $S_{s1} = 0]$ and $[S_{c2} = 0$ or $S_{s2} = 0]$), Eq. (6.60) becomes

$$\begin{pmatrix} \Delta P \\ \Delta M_{c1} \\ \Delta M_{c2} \end{pmatrix} = \begin{bmatrix} EA/L & 0 & 0 \\ 0 & 0 & 0 \\ 0 & 0 & 0 \end{bmatrix} \begin{pmatrix} \Delta L \\ \Delta \theta_{c1} \\ \Delta \theta_{c2} \end{pmatrix}. \quad (6.68)$$

Stiffness Eq. (6.68) represents the pin-jointed truss element with zero bending stiffness.

From Eqs. (6.57) to (6.64), it can be seen that the developed spring-in-series model accounting for the joint flexibility and material yielding effects can automatically modify the element stiffness to the standard element stiffness matrix under the extreme conditions in Eqs. (6.65) to (6.68). In addition, for the occurrence of semi-rigid joint stiffness deterioration or/and partial material yielding at the ends of the element, the present element stiffness in Eq. (6.60) will automatically be adjusted in all these extreme cases. Thus, the spring-in-series model can be successfully employed in the conventional displacement-based finite element method of analysis because the widely used element stiffness formulation procedure can be retained in the analysis. The present element is only a natural extension of the conventional method to allow for connection stiffness and material yielding. This resultant spring-in-series model is simple and efficient in computational calculations.

From Eq. (6.60), the modified inelastic stiffness matrix of a flexibly connected beam-column element is given as

$$[_*\bar{k}]^e_{3\times 3} = \frac{1}{\beta_{cs}} \begin{bmatrix} EA/L & 0 & 0 \\ 0 & \beta_{cs}S_{cs1} - S_{cs1}^2(S_{cs2}+K_{22}) & S_{cs1}S_{cs2}K_{12} \\ 0 & S_{cs2}S_{cs1}K_{21} & \beta_{cs}S_{cs2} - S_{cs2}^2(S_{cs1}+K_{11}) \end{bmatrix}, \tag{6.69}$$

or in compact form as

$$[_*\bar{k}]^e_{3\times 3} = \begin{bmatrix} _*\bar{k}^a & 0 & 0 \\ 0 & _*\bar{k}_{11} & _*\bar{k}_{12} \\ 0 & _*\bar{k}_{21} & _*\bar{k}_{22} \end{bmatrix}, \tag{6.70}$$

where $_*\bar{k}_{ij}$ is the rotational stiffness of the element at node i due to rotation at node j.

The above derivation of the modified stiffness matrix is for plane beam-column elements. For three-dimensional elements, Eq. (6.60) can be extended and rewritten as

$$\{\Delta f\}^e_{6\times 1} = [_*\bar{k}]^e_{6\times 6}\{\Delta \bar{u}\}^e_{6\times 1}, \tag{6.71}$$

where

$$[_*\bar{k}]^e_{6\times 6} = \begin{bmatrix} _*\bar{k}^a & 0 & 0 & 0 & 0 & 0 \\ 0 & _*\bar{k}^y_{11} & 0 & 0 & _*\bar{k}^y_{12} & 0 \\ 0 & 0 & _*\bar{k}^z_{11} & 0 & 0 & _*\bar{k}^z_{12} \\ 0 & 0 & 0 & _*\bar{k}^x & 0 & 0 \\ 0 & _*\bar{k}^y_{21} & 0 & 0 & _*\bar{k}^y_{22} & 0 \\ 0 & 0 & _*\bar{k}^z_{21} & 0 & 0 & _*\bar{k}^z_{22} \end{bmatrix}, \tag{6.72}$$

$$\{\Delta f\}^e_{6\times 1} = \{\Delta P \quad \Delta M^y_{c1} \quad \Delta M^z_{c1} \quad \Delta M^x \quad \Delta M^y_{c2} \quad \Delta M^z_{c2}\}^{eT}, \tag{6.73}$$

$$\{\Delta \bar{u}\}^e_{6\times 1} = \{\Delta L \quad \Delta \theta^y_{c1} \quad \Delta \theta^z_{c1} \quad \Delta \theta^x \quad \Delta \theta^y_{c2} \quad \Delta \theta^z_{c2}\}^{eT}, \tag{6.74}$$

in which $\{\Delta f\}^e_{6\times 1}$ and $\{\Delta \bar{u}\}^e_{6\times 1}$ are the 6 × 1 incremental force vector and the corresponding 6 × 1 incremental displacement vector referred to the axis joining the two ends of a member; $_*\bar{k}^a$ and $_*\bar{k}^x$ are the axial and torsional stiffness; and the superscripts 'x', 'y' and 'z' refer to the local element axes.

In Eq. (6.71), the modified stiffness matrices derived are referred to the local element axis passing through its two ended nodes. In order to establish the tangent stiffness relationship for the equilibrium of a whole structure, the 6 member basic forces and displacements are first required to be transformed into 12 local nodal forces and displacements and then finally to a common and fixed reference axis system or the global coordinate system. The element tangent stiffness equation of equilibrium can be obtained by considering a variation of the equilibrium equation between the nodal and the member force vectors as

$$\delta\left(\{F\}^e_{12\times 1} - [T]_{12\times 6}\{f\}^e_{6\times 1}\right) = 0. \tag{6.75}$$

Expanding Eq. (6.75), we have

$$\{\Delta F\}^e_{12\times 1} - [T]_{12\times 6}\{\Delta f\}^e_{6\times 1} - [\Delta T]_{12\times 6}\{f\}^e_{6\times 1} = 0, \tag{6.76}$$

in which $\{F\}^e$ is the force vector of 12 degrees of freedom and referred to as the element local coordinate; $\{f\}^e$ is the force vector of dimension 6 and referred to as the member local axis and $[T]$ is the standard 12 × 6 transformation matrix given in Eq. (2.36) in Chapter 2.

From Eq. (6.76), it can be seen that the resistance of the structure corresponding to the incremental force vector $\{\Delta F\}^e$ is composed of two parts: the second term representing the variation of the equation with respect to the force, and the last term representing the work done created by the initial force and the nodal displacement. Assuming constant shear during the displacement and combining the change in forces of a displaced member as shown in Fig. 6.19, the relationship between the 12 nodal forces in ^tC and $^{t+\Delta t}C$ configurations can be expressed by a standard transformation process as

$$^{t+\Delta t}\begin{Bmatrix} F_{x1} \\ F_{y1} \\ F_{z1} \\ M_{x1} \\ M_{y1} \\ M_{z1} \\ F_{x2} \\ F_{y2} \\ F_{z2} \\ M_{x2} \\ M_{y2} \\ M_{z2} \end{Bmatrix}^e = \begin{bmatrix} 1 & -\psi_y & \psi_z & 0 & 0 & 0 & 0 & 0 & 0 & 0 & 0 & 0 \\ \psi_y & 1 & 0 & 0 & 0 & 0 & 0 & 0 & 0 & 0 & 0 & 0 \\ -\psi_z & 0 & 1 & 0 & 0 & 0 & 0 & 0 & 0 & 0 & 0 & 0 \\ 0 & 0 & 0 & 1 & 0 & 0 & 0 & 0 & 0 & 0 & 0 & 0 \\ 0 & 0 & 0 & 0 & 1 & 0 & 0 & 0 & 0 & 0 & 0 & 0 \\ 0 & 0 & 0 & 0 & 0 & 1 & 0 & 0 & 0 & 0 & 0 & 0 \\ 0 & 0 & 0 & 0 & 0 & 0 & 1 & -\psi_y & \psi_z & 0 & 0 & 0 \\ 0 & 0 & 0 & 0 & 0 & \psi_y & 1 & 0 & 0 & 0 & 0 \\ 0 & 0 & 0 & 0 & 0 & -\psi_z & 0 & 1 & 0 & 0 & 0 \\ 0 & 0 & 0 & 0 & 0 & 0 & 0 & 0 & 1 & 0 & 0 \\ 0 & 0 & 0 & 0 & 0 & 0 & 0 & 0 & 0 & 1 & 0 \\ 0 & 0 & 0 & 0 & 0 & 0 & 0 & 0 & 0 & 0 & 1 \end{bmatrix} {}^t\begin{Bmatrix} F_{x1} \\ F_{y1} \\ F_{z1} \\ M_{x1} \\ M_{y1} \\ M_{z1} \\ F_{x2} \\ F_{y2} \\ F_{z2} \\ M_{x2} \\ M_{y2} \\ M_{z2} \end{Bmatrix}^e \tag{6.77}$$

or in compact form as

$$^{t+\Delta t}\{F\}^e_{12\times 1} = [\tilde{B}]^e_{12\times 12}{}^t\{F\}^e_{12\times 1}, \tag{6.78}$$

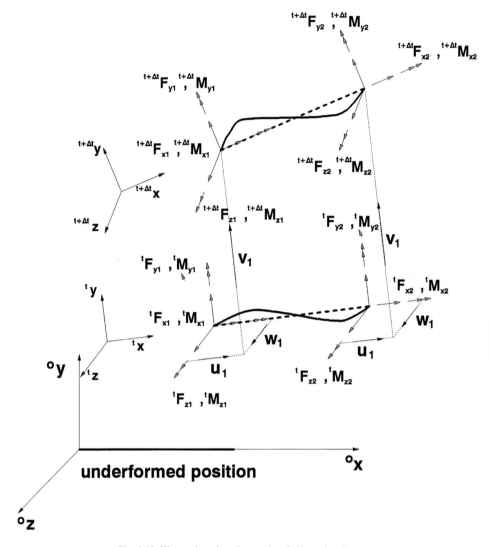

Fig. 6.19. Kinematics of an element in a 3-dimensional space.

in which ψ_y and ψ_z are the rotations due to nodal displacements between the two configurations given by

$$\begin{aligned}\psi_y &= \sin^{-1}\left[-\frac{(v_2 - v_1)}{L}\right] \approx -\frac{(v_2 - v_1)}{L}, \\ \psi_z &= \sin^{-1}\left[-\frac{(w_2 - w_1)}{L}\right] \approx \frac{(w_2 - w_1)}{L}.\end{aligned} \qquad (6.79)$$

The change of force due to the nodal displacements, therefore, can be obtained by subtracting $^t\{F\}^e$ from $^{t+\Delta t}\{F\}^e$ in Eq. (6.77). By expressing this incremental force in the last known ^tC configuration in terms of the 6×1 internal forces and after rearranging, the last term in Eq. (6.76) can be obtained as

$$[\Delta T]_{12 \times 6} \{f\}^e_{6 \times 1} = [N]_{12 \times 12} \{\Delta u\}^e_{12 \times 1}, \tag{6.80}$$

in which $\{\Delta u\}^e$ is the nodal displacement vector of the element with 12 degrees of freedom and referred to the local axis and $[N]$ is a 12×12 translational matrix allowing for change of stiffness due to rigid body motion and initial stress. Completed expression of $[N]$ is given in Eq. (3.18) in Chapter 3.

By the contragradient principle, the nodal displacement vector $\{\Delta u\}^e$ can be related to the member displacement vector $\{\Delta \bar{u}\}^e$ as

$$\{\Delta \bar{u}\}^e_{6 \times 1} = [T]^T_{6 \times 12} \{\Delta u\}^e_{12 \times 1}. \tag{6.81}$$

Substituting Eqs. (6.71), (6.80) and (6.81) into (6.76), the incremental equilibrium equation in local coordinates can be resulted as

$$\begin{aligned}
\{\Delta F\}^e_{12 \times 1} &= [T]_{12 \times 6} [*\bar{k}]^e_{6 \times 6} \{\Delta \bar{u}\}^e_{6 \times 1} + [N]_{12 \times 12} \{\Delta u\}^e_{12 \times 1} \\
&= [T]_{12 \times 6} [*\bar{k}]^e_{6 \times 6} [T]^T_{6 \times 12} \{\Delta u\}^e_{12 \times 1} + [N]_{12 \times 12} \{\Delta u\}^e_{12 \times 1} \\
&= \left([T]_{12 \times 6} [*\bar{k}]^e_{6 \times 6} [T]^T_{6 \times 12} + [N]_{12 \times 12} \right) \{\Delta u\}^e_{12 \times 1} \\
&= [k_T]^e_{12 \times 12} \{\Delta u\}^e_{12 \times 1},
\end{aligned} \tag{6.82}$$

in which $[k_T]^e$ is the 12×12 local tangent stiffness matrix of a beam-column element with connection springs and section springs attached to both element ends.

6.7. Numerical examples

The numerical procedures in Sections 6.5 and 6.6 have been incorporated into the computer program GMNAF (Non-linear Analysis of frames allowing for Geometric and Material Non-linearities). In this section, the computer program is verified via some benchmark problems such as European calibration frames. The numerical results predicted by the present theory are compared with results by others.

6.7.1. Linear analysis of space frame

The first example involves the linear analysis of a space frame. A rigid-jointed space frame with pinned diagonal bracing members is analysed and compared with the commercial program, microSTRAN-3D (Engineering Systems Limited, 1987). This is a 3-dimensional frame problem with commonly used rigid and pinned connection types. The geometry of the space frame is shown in Fig. 6.20. Each member is modelled by one element which is acceptable in a linear elastic analysis. At support 1, the translational

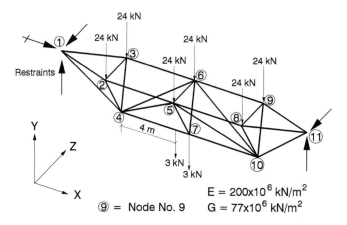

Fig. 6.20. Space frame.

movements in the global axis directions X, Y and Z, and the torsional rotation about the direction X are restrained. At support 11, only the translational movements in the directions Y and Z are fixed. Data relevant to the space frame is tabulated in Tables 6.1 to 6.3. A set of static vertical loads acting in the negative direction of the global Y-axis are applied. These include static loads of 24 kN placed at each of nodes 2, 3, 5, 6, 8, and 9, and a vertical load of 3 kN at node 7. In addition, a vertical point load of 3 kN is applied at the position of 4 meters from node 4 of member 10.

In Table 6.4, the print-out results of member forces computed by the present theory are compared with the values extracted from the microSTRAN user's manual. It was found that the two sets of results are very close. This example illustrates the accuracy of the present developed computer program for the simple linear elastic analysis of space frames.

6.7.2. Effect of semi-rigid connections in truss analysis

Conventionally, skeletal structures are analyzed based on the assumption that all connections are either frictionless pinned or fully rigid. For most trusses composed of relatively slender members, the joints are normally assumed to be pinned and only the axial stiffness is considered. However, when determining the ultimate axial capacity of individual members, an effective length ratio not necessarily equal to unity is commonly adopted. This implies a certain restraint exists at member ends. This use of different assumptions for connection stiffness in the design and analysis process is inconsistent. Furthermore, truss analysis based on the pinned joint assumption is sometimes found to be over-simplified and cannot reflect the actual structural behaviour. Recent limit state specifications such as AISC-LRFD (1986), Eurocode 3 (1988) and BSI-BS5950 (1990) permit the concept of semi-rigidity to be employed in the determination of member forces.

The measurements of the joint stiffness of various types of connections for beam-column types of frames have been conducted extensively by a number of researchers (Refer to Chapter 5). However, stiffnesses for the truss type of member have not been cal-

Chapter 6. *Non-linear static analysis allowing for plastic hinges and semi-rigid joints*

Table 6.1
Member connectivity of space frame

Member No.	Section group No.	Node 1	Node 2
1	1	1	2
2	1	2	5
3	1	5	8
4	1	8	11
5	1	1	3
6	1	3	6
7	1	6	9
8	1	9	11
9	2	1	4
10	2	4	7
11	2	7	10
12	2	10	11
13	3	2	3
14	3	3	4
15	3	2	4
16	3	5	6
17	3	6	7
18	3	5	7
19	3	8	9
20	3	9	10
21	3	10	8
22	4	4	5
23	4	5	10
24	4	4	6
25	4	6	10

Table 6.2
Nodal coordinates of space frame

Node No.	X-coordinate (m)	Y-coordinate (m)	Z-coordinate (m)
1	0	2	1
2	4	2	0
3	4	2	2
4	4	0	1
5	9	2	0
6	9	2	2
7	9	0	1
8	14	2	0
9	14	2	2
10	14	0	1
11	18	2	1

ibrated extensively. Taniguchi et al. (1993) recently measured the joint stiffness of truss members of circular cross-section but the information on this type of joint stiffness is still very scarce. At present, a simple method of determining the stiffness of a connection in a column from the effective length ratios is being discussed. The value of effective

Table 6.3
Section group properties of space frame

Section group No.	Area (m^2)	I_y (m^4)	I_z (m^4)	J (m^4)
1	2270.0 × 10^{-6}	5.13 × 10^{-6}	5.13 × 10^{-6}	10.30 × 10^{-6}
2	2510.0 × 10^{-6}	8.37 × 10^{-6}	8.37 × 10^{-6}	16.70 × 10^{-6}
3	1080.0 × 10^{-6}	1.00 × 10^{-6}	1.00 × 10^{-6}	1.95 × 10^{-6}
4	1080.0 × 10^{-6}	1.00 × 10^{-6}	1.00 × 10^{-6}	1.95 × 10^{-6}

length ratio can be obtained from the recommendation of various design codes of practice (BSI-BS5950, 1990) for steel frames.

Using the modified beam-column element described in Section 6.6, the elastic buckling load of a flexibly connected column can be evaluated by an ordinary procedure for bifurcation analysis. Shown in Fig. 6.21 shows the variation of the elastic buckling load against the joint stiffness. Once the effective length factor is assumed, the corresponding stiffness for the end connections, which gives the equal value of Euler's buckling load based on an assumed effective length, can be determined. The analysis based on the joint stiffness will then be compatible with the design code requirement and the corresponding prediction on the behaviour of the truss can be made.

The arch truss shown in Fig. 6.22 is selected for analysis and demonstration of the effects of assuming different joint stiffness for the web members. Before studies of the effect of the joint stiffness are made, the result of analysis for the truss given by Kondoh and Atluri (1985) is compared to check the validity of the computer program used. In their analysis, all the joints are assumed to be pinned and only the in-plane behaviour is considered. For ease of reference, the coordinates of the joints as well as the member size making up the truss are tabulated in Tables 6.5 and 6.6. The Young's modulus of elasticity for the material is taken as 7.03×10^5 kg/cm^2 and the cross section of all members is solid circular.

The load versus deflection curves by the developed computer program GMNAF, which possesses the described element, are shown in Fig. 6.23. The curve with higher buckling load corresponds to the case where the asymmetrical buckling mode is suppressed. This can be achieved by restraining the lateral movement (x-axis) of the centre node where the load is applied. This curve is coincident with the results by Kondoh and Atluri (1985) who incidentally, omitted the lower buckling mode in their analysis.

When the lateral displacement is allowed to move freely by releasing the restraint, the truss buckles asymmetrically before the symmetrical buckling load is attained. The discrepancy in behaviour represents the shallow and the deep arch types of buckling. The buckled mode shapes are also plotted behind the curves.

With the exception of space trusses connected by ball joints, top and bottom chord members are generally continuous and made up of several rigidly joined members. This arrangement will increase the global buckling capacity of the truss substantially, as demonstrated in the following example.

The truss studied previously is reanalyzed by assuming various fixity to the joints of the web members connected to continuous top and bottom chord members. Figure 6.24

Chapter 6. *Non-linear static analysis allowing for plastic hinges and semi-rigid joints* 163

Table 6.4
Computer results of linear analysis of space frame

Member No.	End node 1	End node 2	Axial force P (kN)	Moment M_{c1}^y (kN m)	Moment M_{c1}^z (kN m)	Torsion M^x (kN m)	Moment M_{c2}^y (kN m)	Moment M_{c2}^z (kN m)
1	1	2	-0.775×10^2 (−77.469)	-0.880×10^{-1} (−0.088)	0.213×10^0 (0.213)	-0.430×10^{-1} (−0.043)	-0.592×10^{-1} (−0.059)	0.242×10^0 (0.242)
2	2	5	-0.752×10^2 (−75.165)	0.622×10^{-1} (0.062)	-0.247×10^0 (−0.247)	0.169×10^{-1} (0.017)	0.317×10^{-1} (0.032)	0.815×10^0 (0.815)
3	5	8	-0.746×10^2 (−74.645)	-0.522×10^{-1} (−0.052)	-0.805×10^0 (−0.805)	-0.102×10^{-1} (−0.010)	-0.628×10^{-1} (−0.063)	0.216×10^0 (0.216)
4	8	11	-0.769×10^2 (−76.932)	0.740×10^{-1} (0.074)	-0.218×10^0 (−0.218)	0.439×10^{-1} (0.044)	0.951×10^{-1} (0.095)	-0.944×10^{-1} (−0.094)
5	1	3	-0.775×10^2 (−77.469)	0.880×10^{-1} (0.088)	0.213×10^0 (0.213)	0.430×10^{-1} (0.043)	0.592×10^{-1} (0.059)	0.242×10^0 (0.242)
6	3	6	-0.752×10^2 (−75.165)	-0.622×10^{-1} (−0.062)	-0.247×10^0 (−0.247)	-0.169×10^{-1} (−0.017)	-0.317×10^{-1} (−0.032)	0.815×10^0 (0.815)
7	6	9	-0.746×10^2 (−74.645)	0.522×10^{-1} (0.052)	-0.805×10^0 (−0.805)	0.102×10^{-1} (0.010)	0.628×10^{-1} (0.063)	0.216×10^0 (0.216)
8	9	11	-0.769×10^2 (−76.932)	-0.740×10^{-1} (−0.074)	-0.218×10^0 (−0.218)	-0.439×10^{-1} (−0.044)	-0.951×10^{-1} (−0.095)	-0.944×10^{-1} (−0.094)
9	1	4	0.168×10^3 (167.976)	0.726×10^{-6} (0.000)	-0.393×10^0 (−0.393)	0.323×10^{-6} (0.000)	0.886×10^{-6} (0.000)	-0.482×10^0 (−0.482)
10	4	7	0.216×10^3 (215.661)	-0.657×10^{-6} (0.000)	0.491×10^0 (0.491)	0.619×10^{-6} (0.000)	-0.436×10^{-6} (0.000)	0.542×10^0 (0.542)
11	7	10	0.216×10^3 (215.661)	0.144×10^{-6} (0.000)	-0.563×10^0 (−0.563)	0.619×10^{-6} (0.000)	-0.396×10^{-6} (0.000)	0.316×10^0 (0.316)
12	10	11	0.167×10^3 (166.928)	0.618×10^{-6} (0.000)	-0.309×10^0 (−0.309)	0.379×10^{-6} (0.000)	0.878×10^{-6} (0.000)	0.162×10^0 (0.162)
13	2	3	0.307×10^2 (30.737)	0.579×10^{-10} (0.000)	0.203×10^{-9} (0.000)	0.229×10^{-7} (0.000)	-0.579×10^{-10} (0.000)	-0.203×10^{-9} (0.000)

Table 6.4
(Continued)

Member No.	End node 1	End node 2	Axial force P (kN)	Moment M_{c1}^y (kN m)	Moment M_{c1}^z (kN m)	Torsion M^x (kN m)	Moment M_{c2}^y (kN m)	Moment M_{c2}^z (kN m)
14	3	4	-0.268×10^2 (−26.837)	0.837×10^{-9} (0.000)	0.625×10^{-10} (0.000)	0.340×10^{-2} (0.003)	0.104×10^{-8} (0.000)	-0.119×10^{-9} (0.000)
15	2	4	-0.268×10^2 (−26.837)	-0.837×10^{-9} (0.000)	0.625×10^{-10} (0.000)	-0.340×10^{-2} (−0.003)	-0.104×10^{-8} (0.000)	-0.119×10^{-9} (0.000)
16	5	6	0.119×10^2 (11.926)	-0.164×10^{-10} (0.000)	0.967×10^{-10} (0.000)	-0.158×10^{-6} (0.000)	0.164×10^{-10} (0.000)	-0.967×10^{-10} (0.000)
17	6	7	0.288×10^1 (2.876)	-0.130×10^{-10} (0.000)	0.678×10^{-10} (0.000)	-0.243×10^{-1} (−0.024)	-0.694×10^{-9} (0.000)	-0.187×10^{-10} (0.000)
18	5	7	0.288×10^1 (2.876)	0.130×10^{-10} (0.000)	0.678×10^{-10} (0.000)	0.243×10^{-1} (0.000)	0.694×10^{-9} (0.000)	-0.187×10^{-10} (0.000)
19	8	9	0.306×10^2 (30.617)	-0.423×10^{-10} (0.000)	0.161×10^{-9} (0.000)	-0.260×10^{-7} (0.000)	0.423×10^{-10} (0.000)	-0.161×10^{-9} (0.000)
20	9	10	-0.269×10^2 (−26.880)	-0.806×10^{-9} (0.000)	0.252×10^{-10} (0.000)	0.125×10^{-1} (0.013)	-0.556×10^{-9} (0.000)	-0.119×10^{-9} (0.000)
21	8	10	-0.269×10^2 (−26.880)	0.806×10^{-9} (0.000)	0.252×10^{-10} (0.000)	-0.125×10^{-1} (−0.013)	0.556×10^{-9} (0.000)	-0.119×10^{-9} (0.000)
22	4	5	-0.358×10^2 (−35.784)	-0.485×10^{-10} (0.000)	-0.459×10^{-9} (0.000)	-0.167×10^{-1} (−0.017)	0.322×10^{-10} (0.000)	0.543×10^{-9} (0.000)
23	5	10	-0.364×10^2 (−36.353)	-0.421×10^{-10} (0.000)	-0.528×10^{-9} (0.000)	0.130×10^{-1} (0.013)	0.325×10^{-10} (0.000)	0.237×10^{-9} (0.000)
24	4	6	-0.358×10^2 (−35.784)	0.485×10^{-10} (0.000)	-0.459×10^{-9} (0.000)	0.167×10^{-1} (0.017)	-0.322×10^{-10} (0.000)	0.543×10^{-9} (0.000)
25	6	10	-0.364×10^2 (−36.353)	0.421×10^{-10} (0.000)	-0.528×10^{-9} (0.000)	-0.130×10^{-1} (−0.013)	-0.325×10^{-10} (0.000)	0.237×10^{-9} (0.000)

Note: Values shown in brackets ( ) are extracted from the microSTRAN user's manual remainder have been computed by this study.

Chapter 6. *Non-linear static analysis allowing for plastic hinges and semi-rigid joints*

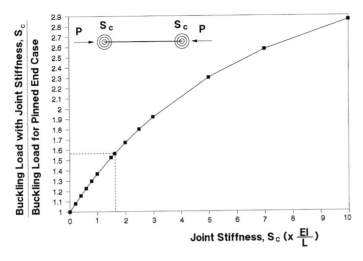

Fig. 6.21. Variation of column buckling load with end connection stiffness.

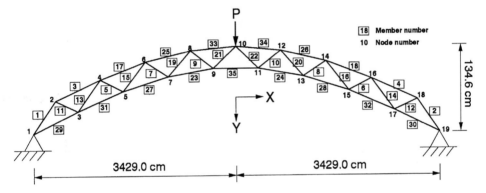

Fig. 6.22. Arch-truss structure.

Table 6.5
Coordinates of the arch-truss

Node number	X coordinate (cm)	Y coordinate (cm)
1, 19	∓3429	0.00
2, 18	∓3048	50.65
3, 17	∓2667	34.75
4, 16	∓2286	83.82
5, 15	∓1905	65.30
6, 14	∓1524	110.85
7, 13	∓1143	87.99
8, 12	∓762	128.50
9, 11	∓381	100.05
10	0	134.60

Table 6.6
Cross sectional area of the members

Member number	Cross sectional area (cm^2)
1–10, 35	51.61
11, 12	64.52
13–16	83.87
17, 18	96.77
19–22	103.23
23, 24	161.29
25, 26	193.55
27, 28	258.06
29–32	290.32
33, 34	309.68

shows the curves obtained by assuming the joint stiffness of web members equal to 0, $1.56 EI/L$ and ∞. The joint stiffness represents the pinned, semi-rigid (effective length factor, $k = 0.8$) and the rigid connection types respectively. It can be seen in Fig. 6.24 that increasing the joint stiffness for the web members has minimal effect on the snap-through buckling resistance of the truss. This observation is somewhat different to the case for the design of a web member as a strut using the Perry–Robertson formula in which the capacity of the member is sensitive to the effective length ratio and thus the connection stiffness. However, the increase in elastic capacity of the truss is significant if the continuity of the top and bottom chord members is assumed. It should further be noted that when comparing the case for all pinned connections with the present case of pinned web members, the buckling load differs considerably (see Figs. 6.23 and 6.24).

6.7.3. Buckling loads of semi-rigid simple portal frame

The elastic buckling load of a simple semi-rigid portal frame with geometry in Fig. 6.25 has been investigated by Lui and Chen (1988). The same frame is re-analysed by GMNAF for validation. The beam is assumed to be jointed to the columns through two connections of constant joint stiffness, S_c. To produce an instability sway, a small horizontal distributing force of $0.001 P$, where P is the vertical force placed at the top of columns, is applied to the frame. In the analysis, the values of E, I and L are assumed to be constant for all members. For the particular case of $S_c = 10 EI/L$, the theoretical bifurcation buckling load of the elastic and semi-rigid frame is determined as $P_{cr} = 1.56 EI/L^2$. The numerical elastic load-deflection curves traced by Lui and Chen (1988) and this study are plotted in Fig. 6.25. From the Figure, it can be seen that the present results are very similar to those obtained by Lui and Chen (1988).

In order to predict a more realistic buckling load of the frame, the plasticity effect due to gradual material yielding of members is considered in this study. The discussed refined-plastic hinge (R-P-H) and the elastic–plastic hinge (E-P-H) method are used. Since the loads are applied at the ends of the beam, plastic hinges will form at these locations where maximum moment occurs. According to the recommendations by ECCS (1983),

Chapter 6. *Non-linear static analysis allowing for plastic hinges and semi-rigid joints*

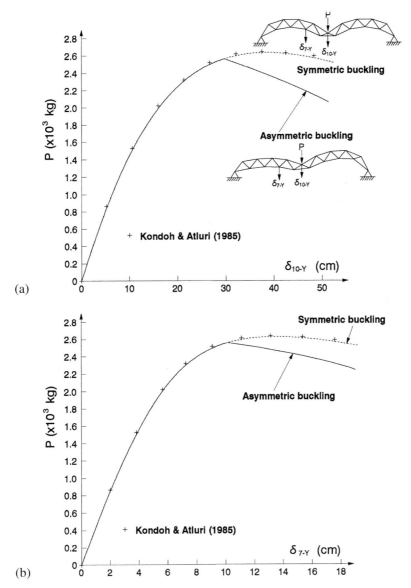

Fig. 6.23. Symmetric and asymmetric buckling of arch truss. (a) At node No. 10 in the Y-direction; (b) At node No. 7 in the Y-direction.

a maximum residual stress of 50% of yield stress is assumed to be present in hot-rolled members during the manufacturing process. The predicted load–deformation curves and the corresponding limit loads are shown in Fig. 6.26. It can be seen in the Figure that the buckling load is significantly reduced by the combined effects due to connection flexibility and member plastification. The buckling load varies from 1.82 for elastic and rigid frame

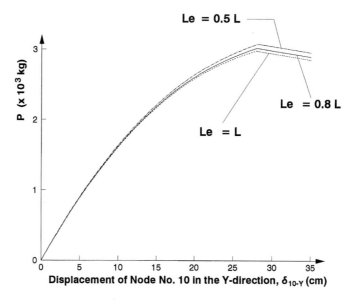

Fig. 6.24. Influence of effective length factor of web members of arch frame-truss.

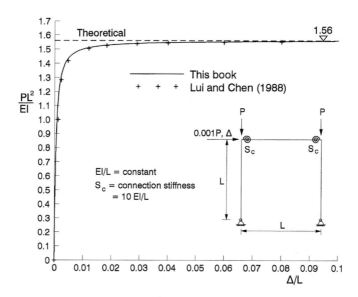

Fig. 6.25. Elastic buckling analysis of semi-rigid simple portal frame.

to 1.28 for inelastic and semi-rigid frame. It is noted that the limit load of the elastic frame with rigid joints is limited to 1.28 because geometric instability due to the P-Δ effect occurs. Since the early material yielding occurs due to the presence of residual stress in the members in the R-P-H method, its buckling load is lower in the case for the frame

Chapter 6. *Non-linear static analysis allowing for plastic hinges and semi-rigid joints* 169

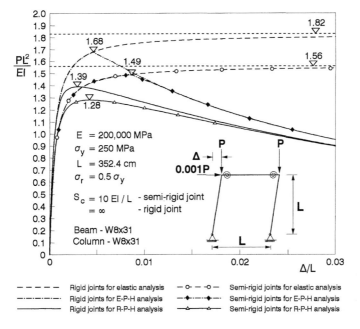

Fig. 6.26. Load–deflection curves for simple portal frame.

with members made of elastic–perfectly plastic material. In addition, the load-deflection traced by the R-P-H method is smoother than that by the E-P-H method because gradual yielding is allowed for in the R-P-H method but not in the E-P-H method.

In this example, it has been illustrated that geometric instability, connection flexibility and member plastification are significant to the overall buckling capacity of the structure studied.

6.7.4. *Fixed-end beam*

The inelastic behaviour of a fixed beam member with residual stress of 60% of the yield stress has been studied by Liew et al. (1992). This problem is reanalysed to validate the proposed plastic hinge method. A concentrated load is applied at position B and the paths of normalized load–deformation curves by elastic–plastic hinge (i.e. with the elastic–perfectly plastic material) and refined-plastic hinge analysis methods are plotted in Fig. 6.27. It can be seen from Fig. 6.27 that the yielding sequence starts at section A, then section B and finally section C. When the section at C is completely yielded, a collapse mechanism is formed and the beam fails by forming a mechanism with three plastic hinges. It can also be seen that the present results are in agreement with those obtained by Liew et al. (1992). It is noted that the paths of curves traced by the refined-plastic hinge methods are smooth while the paths by the elastic–plastic hinge methods have abrupt changes in slope. This is because gradual yielding of section stiffness from elastic to plastic status occurs in the refined-plastic analysis whilst sharp change of sec-

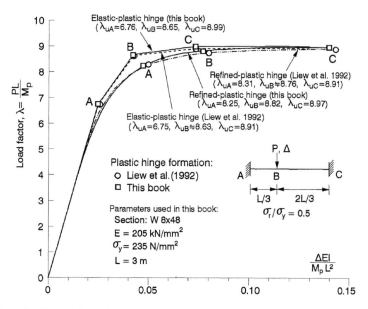

Fig. 6.27. Normalized load–deformation curves of a fixed-end beam for elastic–perfectly plastic hinge vs. plastic hinge methods.

tion stiffness occurs at the plastic hinges in the elastic–plastic hinge method. In addition, in the refined-plastic hinge analysis, the first formation of plastic hinge at section A is delayed because of the redistribution of internal inelastic forces before the attainment of fully plastic moment occurs at A. The ultimate load predicted by Liew et al. (1992) is $P = 8.91\,M_p/L$ which is slightly lower than the present limit load of $P = 8.97\,M_p/L$. This is due to the difference in yield function and numerical methods used. The advantage of the present full-yield surface is that it has a smoother shape than the method from AISC-LRFD (1986) based on Eq. (6.40). The present method is automatically determined from the basic sectional parameters (depth, breadth, flange and web thicknesses) of I- or H-sections without any curve-fitting procedure. The present method is, therefore, more general and rational in determining full-yield surfaces. The traces of force-moment points at these sections are plotted in Fig. 6.28. It is observed that the traces of force point at the right-hand corner in Fig. 6.28 gradually move up, because a significant tensile force is developed in the beam when the applied load increases.

6.7.5. Single storey braced frame

Two identical braced frames but with different supports have been studied by Liew et al. (1993). The geometry and loading pattern of the frames are shown in Fig. 6.29. The lateral sway is prevented in both the frames by effective bracing or tying to more rigid adjacent structures such as concrete core walls. Geometrical imperfection in the columns is assumed to be in a half sinusoidal shape of maximum in-plane deflection of $L/1000$ at mid-height. The beam is assumed to be ideally straight. In this study, the frames under a

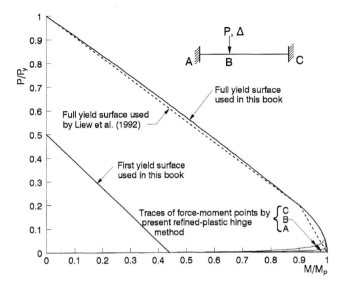

Fig. 6.28. Traces of force-moment points.

load factor of $\beta = 0.34$ are considered. In the plastic zone analysis used by Chen et al. (1990), all the cross-sections in the beam and columns were assumed to have an initial residual stress of maximum magnitude of 30% of the yield stress. In the plastic hinge analysis, based on the ECCS recommendation (1983) shown in Fig. 6.14, the maximum residual stress in the beam and columns are equal to 30% and 50% of the yield stress, respectively. In the frame model, the column is divided into four elements whilst the beam is discretized into eight elements and the loading is assumed to be applied at the element nodes. Material strain-hardening is not considered.

The traces of force points predicted by the present refined-plastic hinge method are compared with those results from Liew et al. (1993) who employed the LRFD bilinear interaction equations (AISC-LRFD, 1986) as the cross-sectional strength. The numerical results are plotted on the normalized axial-force and moment graph in Fig. 6.29. The shape of the LRFD bilinear interaction equations are assumed to be constant for all sections regardless of the geometrical configuration of cross-sections. These equations are generally conservative. In the present study, the full-yield surfaces for the beam and column sections are computed individually by the proposed method of section assemblage. The additional input data of the depth, breadth, and web and flange thickness of a section required can be found in any section tables saved in the data bank in a computer analysis, which leads to much convenience. The present method gives various yield surfaces for different sections instead of being governed approximately by the LRFD bilinear equations. The paths of force points traced by the present plastic hinge method are basically in good agreement to the results obtained by other researchers (see Fig. 6.29). As the axial force in columns increases, the force points of the column tops will move upwards and approach the full yield surface until the interaction of axial force and bending moment reaches the fully yielded status with the force point lying on the full yield surface. It is

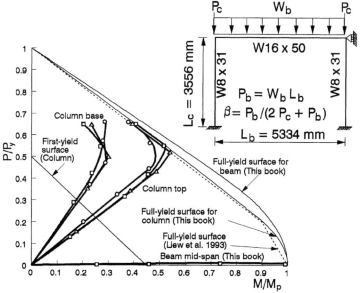

(a) Frame A: Braced frame with fixed supports

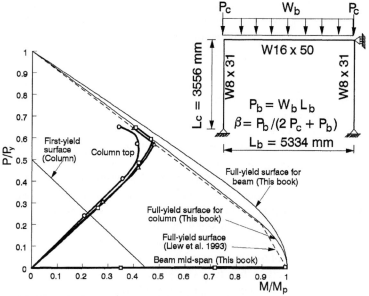

(b) Frame B: Braced frame with pinned supports

—□— Refined-plastic hinge (This book)
—△— Plastic hinge (Liew et al. 1993)
—○— Plastic zone (Chen et al. 1990)

Fig. 6.29. Trace of force points for braced frames.

Chapter 6. *Non-linear static analysis allowing for plastic hinges and semi-rigid joints*   173

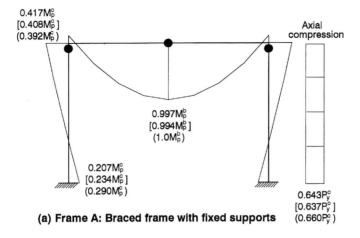

(a) **Frame A: Braced frame with fixed supports**

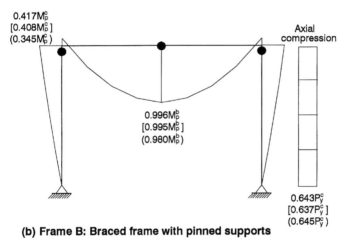

(b) **Frame B: Braced frame with pinned supports**

Value: Refined-plastic hinge, this book
[Plastic hinge, Liew et al. 1993]
(Plastic zone, Chen et al. 1990)

Symbol: ● Common plastic hinge location

Fig. 6.30. Axial force and bending moment diagrams of braced frames.

shown that force redistribution occurs in the structures after the formation of the plastic hinges at the column tops. When the plastic moment capacity at the mid-span of the beam is reached, a collapse mechanism in the beam with three pinned hinges will be formed. Bending moment and axial force diagrams of the braced frames by Chen et al. (1990), Liew et al. (1993) and the present method are compared and depicted in Fig. 6.30. Locations of the three plastic hinges (two at the column tops and one at the beam mid-span)

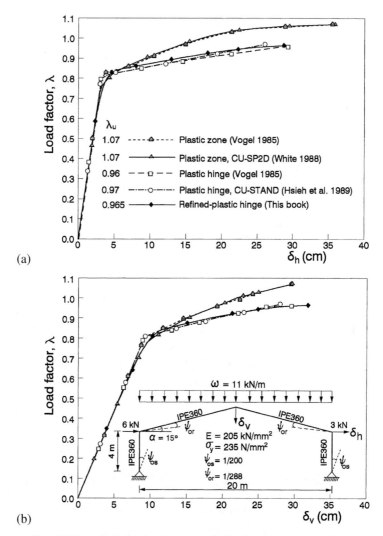

Fig. 6.31. Inelastic load-deflection curves of gable frame by various methods.

predicted by these studies are the same. The superscripts 'b' and 'c' refer to the beam and column member, respectively. The analysis results computed by the present study and by other researchers are generally in good agreement.

6.7.6. Gable frame

The gable frame is one of the European calibration frames which is used as a benchmark for verifying the reliability and accuracy of inelastic analysis programs (Vogel, 1985). Description of the gable frame with initial imperfections of the columns and roofs is given in Fig. 6.31. The same frame has also been studied by the Cornell inelastic programs

Chapter 6. *Non-linear static analysis allowing for plastic hinges and semi-rigid joints* 175

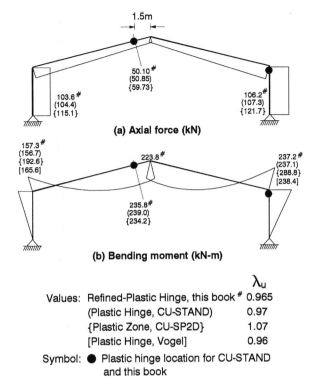

Fig. 6.32. Axial force and bending moment diagrams of gable frame.

at Cornell University, USA; the CU-STAND (Hsieh et al., 1989) for the plastic-hinge method and the CU-SP2D (White, 1988) for the plastic-zone method. Residual stress is considered in all members and its pattern is based on the ECCS recommendation (1983).

The load versus vertical and lateral displacements at the roof top and the right knee joints are plotted in Fig. 6.31. The load–deformation curve traced by the present refined-plastic hinge method is in close agreement with the plastic hinge analysis results by Vogel (1985) and the CU-STAND (Hsieh et al., 1989). In this problem, it is observed that the difference in the shape of the curves obtained by the plastic zone and plastic hinge approaches is due to the effect of spread of plasticity along the member in the plastic zone analysis. However, kinks occur at the fully yielded sections where the section stiffness diminishes rapidly in the plastic hinge analysis, resulting in an abrupt changes in the overall behaviour. Consequently, the ultimate loads predicted by the plastic zone analysis are larger than the plastic hinge analysis. The internal forces at the failure predicted by Vogel (1985), CU-STAND (Hsieh et al., 1989), CU-SP2D (White, 1988) and the proposed method are compared in Fig. 6.32. In the CU-STAND (Hsieh et al., 1989) and the present studies, two plastic hinges are predicted and developed at the same locations shown in the Fig. 6.32. At this instance, a collapse mechanism is formed and the structure reaches its ultimate load carrying capacity.

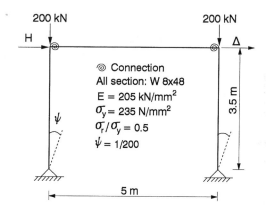

Fig. 6.33. Loading pattern and geometric configuration of portal frame.

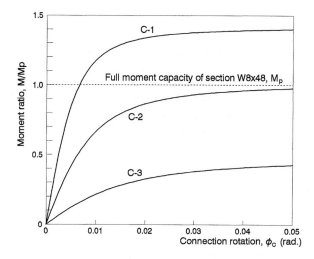

Fig. 6.34. Moment–rotation curves of connection types used for portal frame.

6.7.7. Inelastic response of simple semi-rigid jointed portal frame

To investigate the combined effect of connection flexibility and member plasticity on the overall structural behaviour, a simple portal frame with semi-rigid joints is studied. Figure 6.33 shows the loading pattern and the geometry of the portal frame with initial geometric imperfections. To account for the instability effect, a set of constant vertical loads each of 200 kN is applied at the top of columns. A horizontal load, H, then increases gradually until the ultimate load-carrying capacity of the structure is reached. In the analysis, three connection types with various moment capacities shown in Fig. 6.34 are considered Connection type C-1 has the largest moment capacity, type C-3 has the smallest moment capacity, and type C-2 has an intermediate capacity which is comparable to

Chapter 6. *Non-linear static analysis allowing for plastic hinges and semi-rigid joints* 177

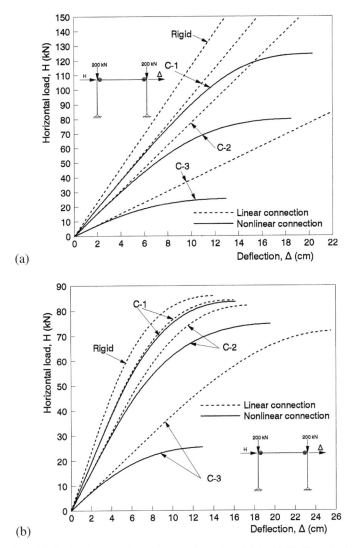

Fig. 6.35. Elastic and inelastic analyses of portal frame with various joints. (a) Elastic analysis; (b) inelastic analysis.

the plastic moment capacity of the member section. The moment–rotation relationships of these connections are modelled by the Richard–Abbott function described in Chapter 5.

The static load–deformation curves of the portal frame for various analyses are plotted in Fig. 6.35. In the elastic analysis shown in Fig. 6.35(a), connection non-linearity affects the ultimate load-carrying capacity significantly because the connection moment capacity limits the moment developed in connections. The ultimate loads of the frame are determined as 124.3, 79.9 and 26.6 kN for connection types C-1, C-2 and C-3, respectively. To further investigate the effect of gradual material yielding of members on the ultimate

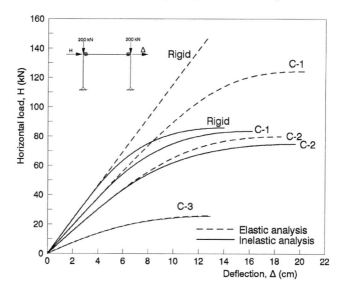

Fig. 6.36. Results comparison of elastic and inelastic analysis of portal frame with non-linear connection types.

load, the developed refined-plastic hinge method based on the section assemblage concept and with yield stress equal to 250 MPa is adopted. The inelastic load–deformation curves for various connections are plotted in Fig. 6.35(b). It can be observed from this Figure that the difference between the load–deformation paths of the linear and the non-linear connections is small for relatively stiff joint type C-1 but becomes large for weak joint type C-3. This is because joint C-1 is relatively linear before reaching the plastic stage in members, whilst joint C-3 is non-linear even at small loads. The behaviour of connection type C-2 is between these two cases.

For comparison, the results of inelastic analysis for the portal frame with non-linear joints are plotted in Fig. 6.36. It can be seen that the effect of member plasticity can significantly affect the ultimate load-carrying capacity of the frame. This effect is more pronounced in rigid jointed frames but less obvious in more flexible jointed frames (i.e. type C-3 case). This is expected as joint non-linearity controls the non-linear response at a load before material yields.

6.7.8. Inelastic limit load of a two-storey braced and unbraced semi-rigid frame with various support conditions

A single-bay two-storey frame with different support conditions as shown in Fig. 6.37 is analysed. The elastic load–deformation curve and the elastic load-carrying capacity of the frame have been computed by Lui and Chen (1988). The beams and columns are wide flange shaped sections of W 14×48 and W 12×96, respectively, and both the braced and unbraced frame cases have been considered. For the braced case, diagonal braces of angles L $3 \times 3 \times 1/2$ with pinned ends are used for both storeys. Five types of beam-column connections are considered. They are labelled as connections A, B, C, D and rigid as in

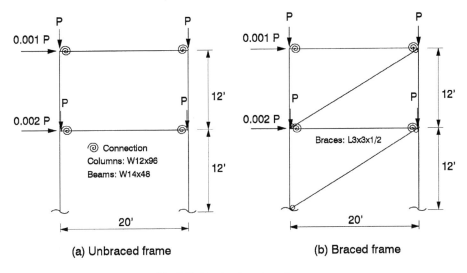

Fig. 6.37. One-bay two-storey frame.

Table 6.7
Connection parameters of the Chen–Lui exponential model (Lui and Chen, 1988)

	Connection types			
	A Single web angle	B Top and seated angle	C Flush end plate	D Extended end plate
Tested by	Richard et al. (1982)	Azizinamini et al. (1985)	Ostrander (1970)	Johnson and Walpole (1981)
M_0 (kip-in)	0	0	0	0
R_{kf} (kip/rad)	0.47104×10^2	0.43169×10^3	0.96415×10^3	0.41193×10^3
α	0.51167×10^{-3}	0.31425×10^{-3}	0.31783×10^{-3}	0.67083×10^{-3}
C_1	-0.43300×10^2	-0.34515×10^3	-0.25038×10^3	-0.67824×10^3
C_2	0.12139×10^4	0.52345×10^4	0.50736×10^4	0.27084×10^4
C_3	-0.58583×10^4	-0.26762×10^5	-0.30396×10^5	-0.21389×10^5
C_4	0.12971×10^5	0.61920×10^5	0.75338×10^5	0.78563×10^5
C_5	-0.13374×10^5	-0.65114×10^5	-0.82873×10^5	-0.99740×10^5
C_6	0.52224×10^4	0.25506×10^5	0.33927×10^5	0.43042×10^5
S_c^o (kip/rad)	0.48000×10^5	0.95219×10^5	0.11000×10^6	0.30800×10^6

Fig. 5.6. Connections A to D are realistic connection types obtained from experimental results. Connection A is a single web angle connection tested by Richard et al. (1982). Connection B is a top and seated angle connection with double web cleats examined by Azizinamini et al. (1985). C is a flush end plate connection tested by Ostrander (1970) and D is an extended end plate connection tested by Johnson and Walpole (1981). Their moment–rotation curves are represented by the Chen–Lui exponential model (Lui and Chen, 1988) and the required model parameters are listed in Table 6.7. The moment–

rotation relationship and the joint stiffness–rotation behaviour for these connections are plotted in Fig. 5.6. As illustrated in the Figure, the strength of connection A is the weakest while D is the stiffest. To investigate the effect of support conditions to the structural behaviour, the pinned and the fixed supports are considered. The beams are modelled by two elements, the columns by one element and the braces by one element in the structural model.

In the report by Lui and Chen (1988), the material is assumed to be elastic throughout the analysis. To simulate a more realistic structural behaviour, member plastification is considered in the present study. The proposed refined-plastic hinge method is employed. The elastic and inelastic load–deformation curves of the frame with various support conditions and braced/unbraced cases are traced in Figs. 6.38 to 6.41. The elastic and the inelastic stability limit loads of the load–deformation curves are summarized and tabulated in Table 6.8.

From Table 6.8, it can be observed that the present elastic analysis results are in good agreement with those obtained by Lui and Chen (1988). From the results, several observations can be made as follows.

(1) As expected, the limit load increases as the stiffness of connections increases. The ratio of the limit load value for the flexible connection case compared to that for the rigid joint case, (P_{cr})flexible connection$/(P_{cr})$rigid connection, is calculated and listed in Table 6.9. It is obvious that the value increases rapidly from connection A to rigid connections in unbraced cases. However, the sensitivity of the value to difference in connection flexibility is greatly reduced if bracing members are provided as in the braced cases. In the inelastic analysis results, the ratio is almost equal to one, except for the pinned support and unbraced case. This is because the limit load is totally dominated by the plastic moment capacity of the column members at the supports. This plastic strength of column members in the inelastic analysis are much less than the elastic moment of column in the elastic analysis. Consequently, the limit load of inelastic frames is basically restricted to a value and is non-sensitive to the joint stiffness.

(2) The bracing system is very effective in increasing the instability limit loads of the semi-rigid frames, P_{cr}. It is specially significant for more flexible connection types. This means that the strength of a structure with more flexible joints can be effectively increased by adopting a bracing system. In addition, it has also been demonstrated that the provision of bracings reduces the lateral drift greatly. Table 6.10 illustrates the ratio of the limit loads for a bracing system compared to that for an unbraced system (i.e. $(P_{cr})$with bracings$/(P_{cr})$without bracings) for all different frame cases. From the Table 6.10, the value of $(P_{cr})$with bracings$/(P_{cr})$without bracings increases significantly as more flexible connections are used. The range of the values for elastic frames with pinned supports varies from 62.86 for connection A to 7.62 for rigid connections. The bracing system is, therefore, suggested to be used because of its economical cost, rather than choosing larger sized members for the unbraced system. It is noted that all values are equal to one for inelastic semi-rigid frames with fixed supports. This is because the limit loads of the frames are completely controlled by the effect of member plasticity, and not by the connection flexibility. As illustrated in Figs. 6.40 and 6.41, the load–deformation curves for various connection types in the bracing system case are very close to each other. This phenomenon is due to the transformation of the frame action to the truss action. Therefore,

Chapter 6. *Non-linear static analysis allowing for plastic hinges and semi-rigid joints*   181

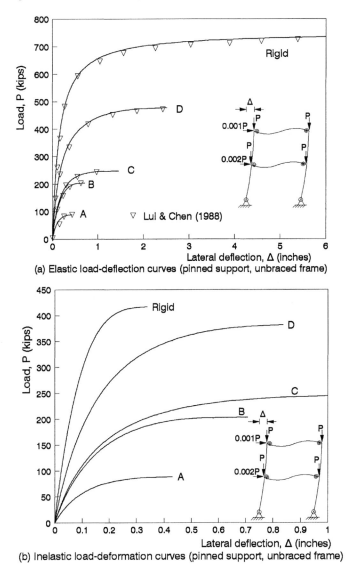

Fig. 6.38. One-bay two-storey unbraced frame with pinned supports.

if the bracing members remain effective, the frame behaviour is dominated by the truss action.

(3) From Figs. 6.38 to 6.41, the effect of member plasticity reduces the instability limit loads of the frames significantly, especially for the bracing system cases. Member yielding limits the applied load P to 480 kips, regardless of the connection type. The performance of material yield plays a more important role on the response of a stiffer structure such as heavily braced frames. For the unbraced frames, the structural behaviour is dominated by

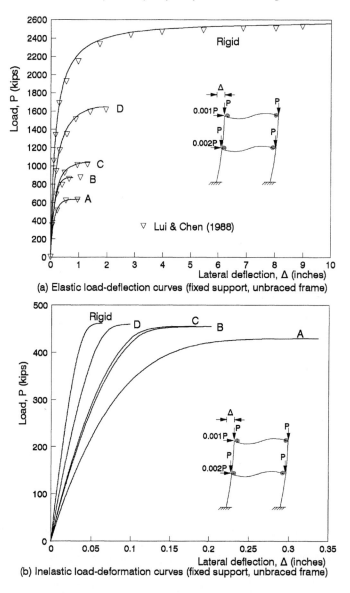

Fig. 6.39. One-bay two-storey unbraced frame with fixed supports.

the combined effect of member plasticity and connection flexibility. In determination of the ultimate load-carrying capacities of steel frames, therefore, both the semi-rigid joint behaviour and the member material yielding are important factors in an analysis.

(4) Two types of supports are considered in this example. They are the pinned and the fixed support conditions. The effect of support conditions on the limit load of flexibly connected frames is less important on braced frames than for unbraced frames. In cases

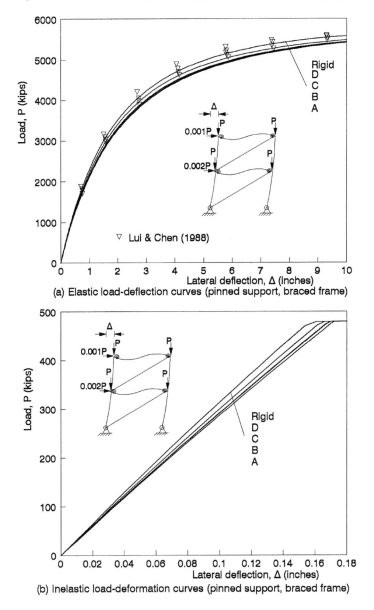

Fig. 6.40. One-bay two-storey braced frame with pinned supports.

of unbraced frames, fixed supports can improve the limit load of the frames significantly, especially for more flexible joint cases. This is because the overall structural stiffness will be increased as fixed supports are used.

From this example, the most effective and economical way to increase the overall structural strength is to use bracing members. In addition, in cases of braced frames, more

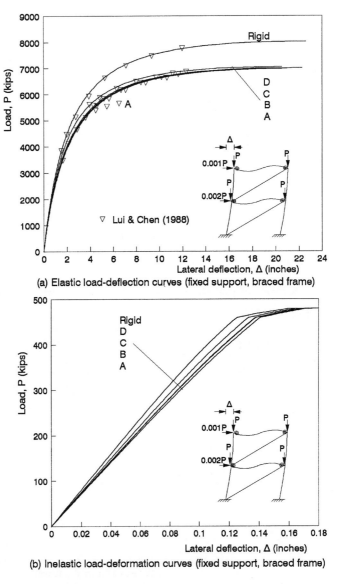

Fig. 6.41. One-bay two-storey braced frame with fixed supports.

flexible beam-to-column connections can be adopted because they affect very slightly the stability of the structure. The lateral deflections of braced frames can also be reduced significantly by bracings. Furthermore, the support conditions for braced frames can be either pinned or fixed because their contribution to the overall stiffness is not very significant. However, fixed supports and stiffer connections should be used in unbraced frames to increase the limit load and to reduce lateral sway of the frames.

Table 6.8
Stability limit loads of the two-storey frame, P_{cr}, in kips by present refined-plastic hinge method and the value in bracket [ ] by Lui and Chen 1988)

	Elastic analysis				Inelastic analysis			
	Pinned support		Fixed support		Pinned support		Fixed support	
Connection	Without bracings	With bracings	Without bracings	With bracings	Without bracings	With bracings	Without bracings	With bracings
A	89 [90]	5595 [5560]	638 [630]	6995 [5680]	89	479	480	480
B	204 [205]	5600 [5620]	875 [860]	7005 [6680]	204	479	480	480
C	249 [242]	5605 [5620]	1038 [1030]	7020 [6850]	247	479	480	480
D	477 [475]	5645 [5630]	1648 [1625]	7060 [6940]	382	479	480	480
Rigid	746 [725]	5690 [5640]	2560 [2530]	8035 [8000]	417	479	480	480

Table 6.9
Values of $(P_{cr})_{\text{flexible connection}}/(P_{cr})_{\text{rigid connection}}$ for two-storey frame by present refined-plastic hinge method

	Elastic analysis				Inelastic analysis			
	Pinned support		Fixed support		Pinned support		Fixed support	
Connection	Without bracings	With bracings	Without bracings	With bracings	Without bracings	With bracings	Without bracings	With bracings
A	0.119	0.983	0.249	0.870	0.213	1	1	1
B	0.273	0.984	0.341	0.872	0.489	1	1	1
C	0.333	0.985	0.405	0.874	0.592	1	1	1
D	0.639	0.992	0.643	0.879	0.916	1	1	1
Rigid	1	1	1	1	1	1	1	1

Table 6.10
Values of $(P_{cr})_{\text{with bracings}}/(P_{cr})_{\text{without bracings}}$ for the two-storey frame by the present refined-plastic hinge method

	Elastic analysis		Inelastic analysis	
Connection	Pinned support	Fixed support	Pinned support	Fixed support
A	62.86	10.96	5.38	1
B	27.45	8.00	2.35	1
C	22.51	6.76	1.94	1
D	11.83	4.28	1.25	1
Rigid	7.62	3.13	1.14	1

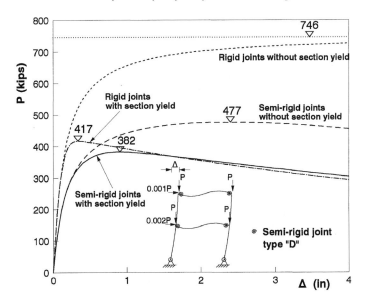

Fig. 6.42. Load–deformation curves for unbraced two-storey frame with pinned supports.

Referring to the predicted limit loads given in Table 6.8, Fig. 6.42 illustrates a graphical presentation on the combined effect of non-linear flexible joint and material yielding for the unbraced and pin-supported frame with joint type D. The limit load ranges from 746 kN for the elastic rigid joint case to 382 kN for the semi-rigid inelastic analysis.

6.7.9. Vogel six-storey frame

The two-bay six-storey European calibration frame subjected to proportionally applied distributed gravity loads and concentrated lateral loads has been reported by Vogel (1985). The frame is assumed to have an initial out-of-plumb straightness and all the members are assumed to possess the ECCS residual stress distribution (ECCS, 1983). The paths of load–deformation curves shown in Fig. 6.43 are primarily the same by the plastic-zone and the plastic-hinge analyses. The maximum capacity is reached at a load factor of 1.11 for Vogel's plastic-zone method (Vogel, 1985), 1.12 for Vogel's plastic-hinge method (Vogel, 1985), and 1.125 for the proposed refined-plastic hinge method. The maximum difference between these limit loads is less than 1.4%. This example shows the adequacy of the plastic hinge method for large deflection and inelastic analysis of steel frames.

The same frame has also been studied by the Cornell University inelastic programs: the CU-SP2D (White, 1988) and the CU-STAND (Hsieh et al., 1989). The force diagrams of the frame with key values at specified locations and at the maximum load of the frame are plotted in Fig. 6.44. The ultimate load factors are 1.18 for the CU-SP2D (White, 1988), 1.13 for the CU-STAND (Hsieh et al., 1989) and 1.125 for the present study. The Cornell's CU-SP2D plastic-zone result of 1.18 is slightly higher than the limit load of 1.11 by Vogel's plastic-zone method. The force distribution and the plastic hinge location

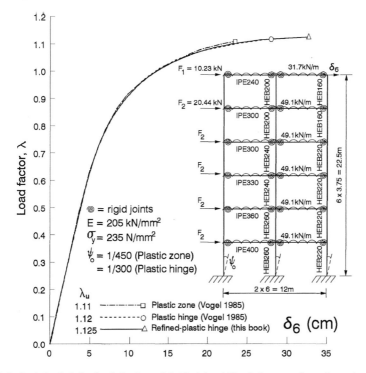

Fig. 6.43. Inelastic load–deflection behaviour of rigidly jointed Vogel six-storey frame for various studies.

obtained by the analyses are essentially similar. The CU-STAND hinge analysis (Hsieh et al., 1989) detects a total of 19 plastic hinges while the present study detects 16 plastic hinges. The difference may be explained by the fact that the present limit load, which is less than that obtained by Hsieh et al. (1989), is not high enough to produce further fully plastic hinges at these three locations. Referring to the Figure, the present bending moments at the three locations are very close to the fully plastic moment capacity of section just before structural collapse.

In order to reflect a more realistic behaviour of the Vogel frame, four semi-rigid connection types and traditional rigid joint type are assumed. They are the beam-to-column joints labelled as connections A, B, C and D used in the previous example. Details of the characteristics of these semi-rigid joints are referred to Fig. 5.6. For comparison, the linear and non-linear behaviour for each semi-rigid joint is also investigated. In the present analysis, the Chen–Lui exponential function (1991) is used to represent the non-linear moment–rotation curve of semi-rigid joints and the proposed refined-plastic hinge method is adopted to simulate the plastic deformation of members.

The load–deformation curves of the Vogel frame traced by the elastic and inelastic analyses are depicted in Fig. 6.45. From the graphs, it can be seen that the ultimate strength of the frame is greater for the linear joint cases than for the non-linear cases. This is because no deterioration occurs in linear joint stiffness. As expected, the ultimate limit load of the frame is larger for the elastic case than the inelastic case. This is the result of softening in

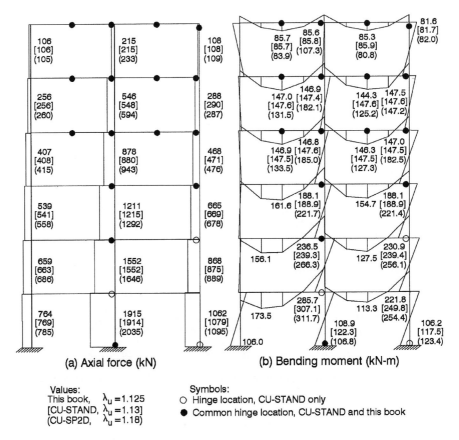

Fig. 6.44. Comparison of member forces of rigidly jointed Vogel frame by Cornell studies and this book.

structural stiffness due to partial yielding when the plastic hinge is initiated. For a frame with very flexible joints (e.g. connection types A and B), the overall performance of the frame is primarily dominated by the connection flexibility, because the connection ultimate moment is far smaller than the initial plastic moment of sections and thus moments at joints are limited by the moment capacity at connections. On the contrary, the ultimate limit load is affected by the member yield for a frame with very stiff joints (e.g. D and rigid connections). For the frame with semi-rigid connection C or D type, the behaviour of the frame is controlled by the combined effect of the connection flexibility and member plasticity. In fact, as illustrated from the elastic and rigid frame case in Fig. 6.45(a), the geometric non-linearity due to the P-Δ effect in column members reduces the buckling load of the system in all cases.

Figure 6.46 shows the elastic and inelastic load–deformation curves for the Vogel frame with linear and non-linear connection type B. It is obvious that the curves using linear and non-linear models for joint stiffness type B are identical. The reason for this result is that the non-linear type B has a very weak connection stiffness and therefore it totally governs

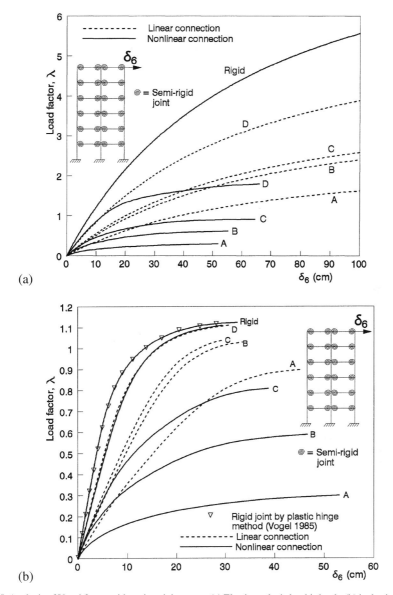

Fig. 6.45. Analysis of Vogel frame with various joint types. (a) Elastic analysis by this book; (b) inelastic analysis by present refined-plastic hinge method.

the frame behaviour. From the elastic analysis results, since there is no degradation of connection stiffness in the rigid joint and type B linear model, the ultimate limit load of the frame increases. However, in the inelastic analysis, the cross-sectional plastification of members reduces significantly the limit load of a frame with type B linear joints or rigid joints.

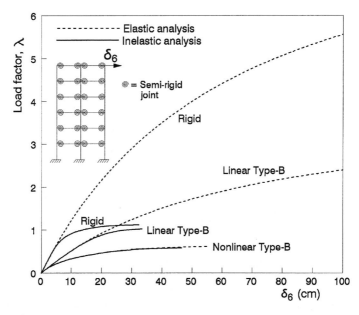

Fig. 6.46. Comparison of elastic and inelastic analysis results of Vogel frame with rigid and Type-B joints.

6.8. Conclusions

In this Chapter, a refined-plastic hinge method using the plate assemblage concept for analysis of semi-rigid steel frames is described. The method is based on the lumped plasticity approach. It has the merit of computational efficiency and accuracy comparable to the plastic zone model. A section spring of degradable stiffness is assumed to model the formation of plastic hinge. Partial yielding of the section is modelled by gradually reducing the stiffness of the section spring. In the inelastic analysis, the full-yield surface or the strength of a cross-section is determined by the proposed section assemblage concept based on the simple idea of the central core of web taking axial force and the remaining web and flange taking moments. The concept is derived from the basis of plate assemblage in section and is a more consistent and applicable assessment of section strength than other existing methods because only the fundamental engineering sectional parameters of depth, breadth, web and flange thicknesses are used for deriving the moment capacity of a section. Other methods usually employ some bilinear or artificially prescribed shapes of yield surfaces determined for all section types by curve-fitting or similar procedures less physical appreciation. The proposed full-yield strength surface formulation is more acceptable to practicing engineers for section capacity checks because the relationship between moment capacity and axial force is rationally established. To define the condition of initial yielding of section, a linear initial-yield surface is proposed. The equation of this surface includes for the effects of residual stress and axial force. When the force-point lies between the initial-yield and full-yield surfaces, the section is partially yielded and the stiffness of the section spring is finite to simulate this phenomenon.

As described in Chapter 4, the modelling of a connection component is represented by a connection spring element. This connection spring is then combined with a section together to form a springs-in-series model which is capable of automatically simulating the combined effect of material yielding and connection flexibility. The resultant spring can be easily incorporated into the conventional element stiffness formulation using the static condensation procedure without introducing extra degree-of-freedoms to the structural model. Additional time required for the procedure to deal with the spring-in-series model is minimal when compared with the total computational time. The concept of the springs-in-series model was found to be simple, reliable and efficient for inelastic analysis of semi-rigid frames. As illustrated from results of the numerical examples in Section 6.7, the computer program developed from the described method has been tested and verified to be valid, accurate, robust and efficient through the comparison with other analysis results and the benchmark solutions of some well-known problems, i.e. the European calibration frames. From these examples, the collapse load of structures predicted by the proposed refined-plastic hinge method is found to be slightly lower than that by the elastic–plastic hinge method. This is because, in the elastic–plastic hinge method, a section still behaves elastically prior to the attainment of fully plastic moment. The proposed refined-plastic hinge method can give a smooth transition from initial yield to full yield states by a gradual cross-sectional plastification process, avoiding an abrupt change at the formation of a plastic hinge.

From the numerical results, it was noted that the load-carrying capacity of an inelastic steel frame with semi-rigid joints can significantly affect connection flexibility and material yielding. The lesser of the connection moment capacity and the member cross-section strength dominates the overall response and the load capacity of the structure. If the moment capacity of a connection and the plastic moment strength of a cross-section are comparable, the structural response will be governed simultaneously by these two factors. Generally speaking, advanced analysis for steel structures due to static loads should consider three major non-linear factors, the geometric second-order effects, semi-rigid connections and plastic hinges.

References

AISC (1986): *Load and Resistance Factor Design Specification for Structural Steel Buildings*, 1st edn., American Institute of Steel Construction, AISC, Chicago, IL.

Al-Mashary, F. and Chen, W.F. (1991): Simplified second-order inelastic analysis for steel frames. *Struct. Engineer* **69**(23), 395–399.

Alvarez, R.J. and Birnstiel, C. (1969): Inelastic analysis of multistory multibay frames. *J. Struct. Div. ASCE* **95**(ST11), Proc. Paper 6922, 2477–2503.

Australian Standard (1990): AS4100-1990, Steel Structures, Sydney, Australia.

Azizinamini, A., Bradburn, J.H. and Radziminski, J.B. (1985): Static and Cyclic Behavior of Semi-Rigid Steel Beam-Column Connections, Technical Report, Dept. of Civil Engineering, Univ. of South Carolina, Columbia, SC.

Basu, P.K. (1990): Draft guidelines for evaluating forces caused by gravity loads acting on initially out-of-plumb members in structures, SSRC (Structural Stability Research Council) 1990 Annual Meeting TG28, Chairman's Report.

British Standard Institution (1990): BS5950, Part 1, Structural Use of Steelwork in Buildings, BSI, London, England.

Bridge, R.Q., Clarke, M.J., Hancock, G.J. and Trahair, N.S. (1991): Verification of approximate methods of structural analysis, ASCE Structural Congress '91 Compact Papers, Indianapolis, 29 April–1 May, pp. 753–756.
Chan, S.L. (1990): Strength of cold-formed box columns with coupled local and global buckling. *Struct. Engineer* **68**(7), 125–132.
Chan, S.L. and Chui, P.P.T. (1997): A generalised design-based elasto-plastic analysis of steel frames by section assemblage concept. *J. Eng. Struct.* **19**(8), 628–636.
Chen, P.F.S. and Powell, G.H. (1982): Generalized Plastic Hinge Concepts for 3-D Beam-Column Elements, Report No. UCB/EERC-82/20, Earthquake Engineering Research Center, Univ. of California, Berkeley, CA.
Chen, W.F. and Chan, S.L. (1995): Second-order inelastic analysis of steel frames using element with midspan and end springs. *J. Struct. Eng. ASCE* **121**(3), 530–541.
Chen, W.F. and Lui, E.M. (1991): *Stability Design of Steel Frames*, CRC Press, Boca Raton, FL.
Chen, W.F., Duan, L. and Zhou, S.P. (1990): Second-order inelastic analysis of braced portal frames – Evaluation of design formulae in LRFD and GBJ specification, *J. Singapore Struct. Steel Soc., Steel Struct.* **1**(1), 5–15.
Chu, K.H. and Pabarcius, A. (1964): Elastic and inelastic buckling of portal frames. *J. Mech. Div. ASCE* **90**(EM5), Proc. Paper 4094, 221–249.
Cohn, M.Z. and Abdel-Rohman, M. (1976): Analysis up to collapse of elasto-plastic arches. *Comput. Struct.* **6**, 511–517.
Duan, L. and Chen, W.F. (1989a): Design interaction equation for steel beam-columns. *J. Struct. Eng. ASCE* **115**(5), 1225–1243.
Duan, L. and Chen, W.F. (1989b): A Yield Surface Equation for Doubly Symmetrical Sections, Structural Engineering Report CE-STR-89-19, Purdue Univ., W. Lafayette, IN.
El-Zanaty, M.H., Murray, D.M. and Bjorhovde, R. (1980): Inelastic Behaviour of Multistory Steel Frames, Report No. 83, Dept. of Structural Engineering, Univ. of Alberta, Edmonton, AB, Canada.
Engineering Systems Limited (1987): microSTRAN-3D User's Manual, Vol. 3.30, Watford Herts, England, pp. Examples 35–40.
Eurocode 3 (1988): Common Unified Rules for Steel Structures, Commission of the European Communities, Brussels, Redraft.
European Convention for Constructional Steelwork (1983): Ultimate Limit State Calculation of Sway Frames with Rigid Joints, ECCS, Technical Working Group 8.2, Systems, Publication No. 33.
Freitas, J.A.T. and Ribeiro, A.C.B.S. (1992): Large displacement elastoplastic analysis of space trusses. *Comput. Struct.* **44**(5), 1007–1016.
Galambos, T.V. (1988): *Guide to Stability Design Criteria for Metal Structures*, 4th edn., Structural Stability Research Council, John Wiley and Sons, New York, NY.
Galambos, T.V. and Ketter, R.L. (1961): Columns under combined bending and thrust. *Trans. ASCE* **126**(1), 1–25.
Goto, Y. and Chen, W.F. (1987): Second-order elastic analysis for frame design. *J. Struct. Eng. ASCE* **113**(7), 1501–1519.
Guralnick, S.A. and He, J. (1992): A finite element method for the incremental collapse analysis of elastic–perfectly plastic framed structures. *Comput. Struct.* **45**(3), 571–581.
Harstead, G.A., Birnstiel, C. and Leu, K.C. (1968): Inelastic H-columns under biaxial bending. *J. Struct. Div. ASCE* **94**(ST10), Proc. Paper 6173, 2371–2398.
Hsiao, K.M., Hou, F.Y. and Spiliopoulos, K.V. (1988): Large displacement analysis of elasto-plastic frames. *Comput. Struct.* **28**(5), 627–633.
Hsieh, S.H., Deierlein, G.G., McGuire, W. and Abel, J.F. (1989): Technical Manual for CU-STAND, Structural Engineering Report No. 89-12, School of Civil and Environmental Engineering, Cornell University, Ithaca, NY.
Izzuddin, B.A. and Smith, D.L. (1996a): Large-displacement analysis of elastoplastic thin-walled frames. I: Formulation and implementation. *J. Struct. Eng. ASCE* **122**(8), 905–914.
Izzuddin, B.A. and Smith, D.L. (1996b): Large-displacement analysis of elastoplastic thin-walled frames. II: Verification and application. *J. Struct. Eng. ASCE* **122**(8), 915–925.
Johnson, N.D. and Walpole, W.R. (1981): Bolted End-Plate Beam-to-Column Connections Under Earthquake Type Loading, Research Report 81-7, Dept. of Civil Engineering, Univ. of Canterbury, Christchurch, New Zealand.

Johnston, B.G. and Cheney, L. (1942): Steel Columns of Rolled Wide Flange Section, Progress Report No. 2 of the AISC.

Karamanlidis, D. (1988): Alternative FEM formulations for the inelastic large deflection analysis of plane frames. *Comput. Struct.* **28**(6), 691–698.

Kassimali, A. (1983): Large deformation analysis of elastic–plastic frames. *J. Struct. Eng. ASCE* **109**(8), 1869–1886.

Kato, S., Mutoh, I. and Shomura, M. (1998): Collapse of semi-rigidly jointed reticulated domes with initial geometric imperfections. *J. Construct. Steel Res.* **48**(2/3), 145–168.

Ketter, R.L. (1955): Plastic deformation of wide-flange beam columns. *Trans. ASCE* **120**, 1028.

Ketter, R.L., Beedle, L.S. and Johnston, B.G. (1952): Column strength under combined bending and thrust. *Welding J. Res. Suppl.* **31**(12), 607.

Ketter, R.L., Kaminsky, E.L. and Beedle, L.S. (1955): Plastic deformation of wide-flange beam columns. *Trans. ASCE* **120**, 1028.

King, W.S., White, D.W. and Chen, W.F. (1992): On second-order inelastic analysis methods for steel frame design. *J. Struct. Eng. ASCE* **118**(2), 408–428.

Kitipornchai, S., Al-Bermani, F.G.A. and Chan, S.L. (1990): Elasto-plastic finite element models for angle steel frames. *J. Struct. Eng. ASCE* **116**(10), 2567–2581.

Kondoh, K. and Atluri, S.N. (1985): Influence of local buckling on global instability: Simplified, large deformation, post-buckling analysis of plane trusses. *Comput. Struct.* **21**(4), 613–627.

Li, Y. and Lui, E.M. (1995): A simplified plastic zone method for frame analysis. *Microcomput. Civil Eng.* **10**, 51–62.

Liapunov, S. (1974): Ultimate strength of multistory steel rigid frames. *J. Struct. Div. ASCE* **100**(ST8), Proc. Paper 10750, 1643–1655.

Liew, J.Y.R. (1992): Advanced Analysis for Frame Design, Ph.D. Thesis, Purdue University, West Lafayette, IN.

Liew, J.Y.R., White, D.W. and Chen, W.F. (1992): Second-Order Refined Plastic Hinge Analysis of Frames, Structural Engineering Report, CE-STR-92-12, Purdue University, West Lafayette, IN.

Liew, J.Y.R., White, D.W. and Chen, W.F. (1993): Second-order refined plastic-hinge analysis for frame design: Part 1 and 2. *J. Struct. Div. ASCE* **119**(11), 3196–3237.

Livesley, R.K. and Chandler, D.B. (1956): *Stability Functions for Structural Frameworks*, Manchester Univ. Press.

Lu, L.W. (1965): Inelastic buckling of steel frames. *J. Struct. Div. ASCE* **91**(ST6), Proc. Paper 4577.

Lui, E.M. and Chen, W.F. (1988): Behavior of braced and unbraced semi-rigid frames. *Int. J. Solids Struct.* **24**(9), 893–913.

Mason, R.E., Fisher, G.P. and Winter, G. (1958): Eccentrically loaded, hinged steel columns. *Proc. ASCE* **84**(EM4).

Massonnet, C. and Campus, F. (1956): Stanchion problem in frame structures designed according to ultimate carrying capacity. *Proc. Inst. Civil Engineers* **5**(August), 558.

Meek, J.L. and Loganathan, S. (1990): Geometric and material nonlinear behaviour of beam-columns. *Comput. Struct.* **34**(1), 87–100.

Nedergaard, H. and Pedersen, P.T. (1985): Analysis procedure for space frames with material and geometrical nonlinearities, Finite Element Methods for Nonlinear Problems, Europe–US Symposium, Trondheim, Norway, pp. 211–230.

Orbison, J.G., McGuire, W. and Abel, J.F. (1982): Yield surface applications in non-linear steel frame analysis. *Comput. Methods Appl. Mech. Eng.* **33**, 557–573.

Ostrander, J.R. (1970): An Experimental Investigation of End-Plate Connections, Master's Thesis, Univ. of Saskatchewan, Saskatoon, SK, Canada.

Powell, G.H. and Chen, P.F.S. (1986): 3-D beam-column element with generalized plastic hinges. *J. Eng. Mech. ASCE* **112**(7), 627–641.

Richard, R.M., Kreigh, J.D. and Hormby, D.E. (1982): Design of single plate framing connections with A307 bolts. *Eng. J. AISC* **19**(4), 209–213.

SCI (1987): *Steelwork Design Guide to BS5950*, Part 1, Vol. 1, Section Properties and Member Capacities, 2nd edn., The Steel Construction Institute, 1985.

Taniguchi, Y., Saka, T. and Shuku, Y. (1993): Buckling behaviour of space trusses constructed by a bolted jointing system, Space Structures – Proceeding of the Fourth International Conference on Space Structures, Vol. 1, pp. 89–98.

Toma, S. and Chen, W.F. (1992): European calibration frames for second-order inelastic analysis. *Eng. Struct.* **14**(1), 7–14.

Toma, S. and Chen, W.F. (1994): Calibration frames for second-order inelastic analysis in Japan. *J. Construct. Steel Res.* **28**, 51–77.

Toma, S., Chen, W.F. and White, D.W. (1995): A selection of calibration frames in North America for second-order inelastic analysis. *Eng. Struct.* **17**(2), 104–112.

Vogel, U. (1985): Calibrating frames. *Stahlbau* **54**(October), 295–311.

White, D.W. (1988): Analysis of Monotonic and Cyclic Stability of Steel Frame Subassemblages, Ph.D. Thesis, Cornell University, Ithaca, NY.

White, D.W. (1993): Plastic-hinge methods for advanced analysis of steel frames. *J. Construct. Steel Res.* **24**, 121–152.

Wong, M.B. and Tin-Loi, F. (1990): Analysis of frames involving geometrical and material nonlinearities. *Comput. Struct.* **34**(4), 641–646.

Yang, T.Y. and Saigal, S. (1984): A simple element for static and dynamic response of beams with material and geometric nonlinearities. *Int. J. Num. Methods Eng.* **20**, 851–867.

Yau, C.Y. and Chan, S.L. (1994): Inelastic and stability analysis of flexibly connected steel frames by springs-in-series model. *J. Struct. Eng. ASCE* **120**(10), 2803–2819.

Ziemian, R.D. (1993): Examples of frame studies used to verify advanced methods of inelastic analysis. In: *Plastic Hinge Based Methods for Advanced Analysis and Design of Steel Frames*, Structural Stability Research Council, SSRC, Lehigh Univ., Bethlehem, PA.

Chapter 7

CYCLIC BEHAVIOUR OF FLEXIBLY CONNECTED ELASTIC STEEL FRAMES

7.1. Introduction

Experimental results show that practical connections under cyclic loads behave non-linearly due to gradual yielding of bolts, connection plates, cleats, and other components. Flexible connections in low to medium-rise moment-resisting frames significantly affect the overall structural response which includes deflections, force distributions and dynamic response. In general, a flexibly connected steel frame experiences larger lateral drift but attracts smaller storey shear forces when compared with a rigidly jointed frame. Furthermore, this non-linear behaviour of connections in a ductile steel frame will exhibit non-recoverable hysteretic damping which is a primary source of passive damping during the vibration of a structure. It is interesting to note that this damping behaviour cannot be modelled in an analysis for frames made of connections with linear moment versus rotation stiffness, including rigid and pinned connections. This is because the reloading path overlaps the unloading path without generating enclosed areas in moment–rotation loops. On the other hand, for non-linear connection types, the amount of energy dissipated at a joint in each unloading and reloading cycle is equal to the area enclosed in the moment–rotation hysteretic loop.

In the dynamic analysis, there are three primary sources of damping. They are the viscous damping for other structural and non-structural components, the hysteretic damping at non-linear connections and the hysteretic damping at plastic hinges. In this Chapter, in order to focus attention on the influence of hysteretic damping at connections to the overall structural response, the material is assumed to be elastic, with viscous damping ignored. The combined effect of material yielding and connection flexibility in static problems has been discussed in Chapter 6 and the dynamic analysis allowing for these effects will be described in next chapter.

Under severe earthquakes, structures are designed against catastrophic collapse. From the global structural point, ignoring local instability, a structure will not collapse when the energy absorption capacity is larger than the input energy. The contribution of hysteric loop at connections in absorbing energy is considerable, which helps to improve structural safety. In the 7.8 magnitude Northridge earthquake in California San Fernando Valley, USA on 17th January 1994 (AISC, 1994), no steel buildings collapsed with only some local damage occurring at joints, indicating an inherent safety in steel structures. To further utilize the advantage of steel buildings in seismic zones a thorough study and analysis of the buildings under dynamic loads is needed. This reduces the cost of using steel in seismic design and thus increases the competitiveness of steel as a building material. In seismic analysis semi-rigid connections reduce the structural stiffness and the structure exhibits a larger drift than the rigid joint case. However, when under cyclic seismic loads, the hystereic loop damping at connections increases the energy-absorption capacity of a structure. Also, the storey shear forces may be reduced because of smaller inertial forces induced in a softer structure. This characteristic will add an advantage to the resistance of a structure against seismic loads. The dynamic response of flexibly connected steel structures is considerably affected by the connections when compared to static analysis and therefore they deserve a careful investigation.

Unlike brittle materials such as concrete, steel has a high ductility and thus a stronger resistance to cyclic and repeated loads. In addition, if a connection is properly detailed, its properties including stiffness can be retained with the hysteretic moment–rotation M–ϕ_c loops under cyclic loads. Furthermore, experiments (Popov and Pinkney, 1969; Vedero and Popov, 1972; Tsai and Popov, 1990; Azizinamini and Radziminski, 1989) demonstrated that the hysteretic M–ϕ_c loop at connections is stable and reproducible under cyclic loads. To make use of many previous results on static analysis, the moment–rotation curves are adopted using a simple reproducible assumption or model.

Computer simulation of a steel frame under seismic loads is referred in the design. Based on tested moment–rotation curves, an efficient and robust three-dimensional non-linear geometric large deflection dynamic analysis of steel structures accounting for non-linear and hysteretic beam-to-column connections is described in this chapter. Material yielding, however, is considered in Chapter 8, in order to focus on the effect of joint stiffness on the elastic behaviour of steel structures. The non-linear behaviour of connections was found to affect significantly the overall structural performance and is therefore suggested for inclusion in the dynamic analysis of steel structures for accurate prediction of their dynamic response.

In this Chapter, a literature review of research work on semi-rigid jointed structures is given in Section 7.2 and some commonly employed methods for the modelling of connection behaviour under cyclic loads are described in Section 7.3. A simple but powerful algorithm for simulating hysteretic moment–rotation looping path is proposed. The algorithm can be easily incorporated into computer program by a simple subroutine and can be directly applied to all kinds of connection models without the need of expensive curve-fitting procedure for each type of connections. In Sections 7.6 and 7.7, details of the numerical procedures for the present non-linear transient and vibration analysis of semi-rigid steel frames are described. The equations of motions are discretized in the time-domain and solved using the incremental-iterative Newmark integration method. The analysis procedure is coded to the developed computer program GMNAF/D (Geometric and Material Non-linear Analysis of Semirigid Frames – Dynamics). In Section 7.8, some experimental and numerical dynamic examples originally carried out by other researchers are reanalysed by the proposed method to verify and examine the validity of the program. From these numerical results, the non-linear dynamic and cyclic behaviour of flexibly connected steel frames is investigated and discussed.

7.2. Research work on semi-rigid connections

Compared to static tests of connection behaviour, cyclic tests on connections are rare. In the late 1960s and early 1970s, the cyclic behaviour of scaled connections was studied by Popov and Pinkney (1969). A comprehensive study of connection behaviour by one-third scale interior column sub-assemblages were reported by Krawinkler et al. (1971) and Carpenter and Lu (1973). Later, Clough and Tang (1975) carried out shaking table test for scaled moment resisting frames with practical joint details. In order to obtain more realistic results of connection behaviour, full-scale experiments on welded beam-to-column moment connections and bolted web and welded flanges connections were further tested by Popov and his co-workers (Popov and Stephen, 1970; Popov and Bertero, 1973; Popov, 1983).

More recently, Stelmack et al. (1986) conducted ten full-scale tests on steel frames with top and seat angle connections under cyclic loads. A further 14 specimens of 1/2 inch thick angle connections were tested under monotonic loads. The experimental response of load–deflection curves and moment–rotation curves were obtained. A trilinear function for moment–rotation $M-\phi_c$ curves was also proposed to model the angle connection, and comparison between the analytical and the experimental results was reported.

The hysteretic behaviour of top and seat flange angles with double web angle connection for various geometric parameters such as connection thickness, bolt size and spacing were investigated under constant and variable amplitude cyclic loads by Azizinamini and Radziminski (1989).

Portal frames with top and seat bottom flange and double bolted angles at web and with pinned (double bolted angle web) and rigid (welded flange and angle bolted web) connection types were analysed by Nader and Astaneh (1991). The frames were subjected to various earthquake forces in the shaking table. A total of 44 shaking table tests were carried out, and it was observed that the semi-rigid frames reduced the base shear, but the lateral drift was increased under ground excitation when compared to the rigid frames.

Behaviour of the commonly used, extended end-plate connections under cyclic loading was reported by Tsai et al. (1990) and Korol et al. (1990). Tsai et al. (1990) performed three cyclic tests on a full-size beam with extended end-plate connections and it was observed that the specimen with larger connection bolts and a slightly thicker end plate exhibited the superior ductility preferred in the design of structures against seismic loads. Korol et al. (1990) carried out tests on seven full-scale extended end-plate connections under cyclic loading and concluded that sufficient energy dissipation capability without substantial loss of strength can be achieved by proper detailing of the connection.

In static linear and non-linear analysis, the methods used for analyzing steel frames with flexible connections include linear analysis, bifurcation analysis (Cosenza et al., 1989) and its associated method for ultimate load determination based on the Merchant–Rankise formula (Jaspart, 1988), semi-bifurcation analysis (Ho and Chan, 1991), inelastic analysis (Al-Bermani and Kitipornchai, 1992; Chen and Zhou, 1987; Savard et al., 1994), second order non-linear analysis using the total and the incremental secant stiffness matrix (Goto and Chen, 1987; Anderson and Benterkia, 1991; Ho and Chan, 1993) and the ultimate analysis (Lui and Chen, 1987; Yau and Chan, 1994).

Vibration of semi-rigid frames is of interest in a preliminary and serviceability study of a structure against dynamic loads. Kawashima and Fujimoto (1984) conducted experimental investigations on the vibration of steel frames with semi-rigid connections. The connection stiffness was assumed to remain constant and the instability effects were ignored in their analysis. The natural frequencies measured from their experiments were compared with their analytical solutions. The transient response, however, was not investigated.

The seismic behaviour of two full-scale realistic 10 and 20 storey bare steel moment-resistant office buildings located in Canada was studied by Sivakumaran (1988). The connection stiffness was modelled by a bilinear moment–rotation function which is only accurate for the small deflection range. No details for the method of analysis is given by Sivakumaran (1988).

Using a developed computer package program, DRAIN-2D, a single-bay two-storey unbraced frame with top and seat angle connections under cyclic load was investigated by Youssef-Agha et al. (1989). Based on the geometric properties and configuration of a connection, the moment–rotation curve was represented by a bilinear function. Only the basic modelling procedure of the top-and seat angle connection type was given.

Shi and Atluri (1989) proposed a numerical method based on the complementary energy principle to study the dynamic and large deflection response of semi-rigid steel frames. Gao and Haldar (1995) developed a numerical procedure for the tangent stiffness formulation using a stress-based finite element method. In their formulations (Shi and Atluri, 1989; Gao and Haldar, 1995), the generally accepted displacement-based finite element method is not used.

In this Chapter, a robust and efficient method based on the conventional beam–column element for large deflection elastic analysis allowing the flexibility of beam-to-column connections in steel frames under dynamic loads is described. It uses the displacement-based finite element numerical procedure which is more widely known to engineers and researchers and is compatible to the currently used stiffness matrix method of analysis. The method is just an extension of the linear and the second-order analysis for frames.

7.3. Modelling of connection behaviour under cyclic loading

In the structural dynamic analysis of semi-rigid frames, the modelling of cyclic joint behaviour must first be defined. From many cyclic tests on beam–column connections, it was observed that the moment–rotation loops are stable and reproducible (Popov and Pinkney, 1969; Vedero and Popov, 1972; Tsai and Popov, 1990; Azizinamini and Radziminski, 1989). On the other hand, Korol et al. (1990) reported from their observations of the test results that properly detailed connections have a very ductile behaviour on moment–rotation loops. With reference to these experimental observations, studies in this Chapter assume that the static monotonic moment–rotation (M–ϕ_c) curve can be used for cyclic and dynamic analysis via a simple procedure. Basically, there are three types of methods to simulate the cyclic behaviour of connections.

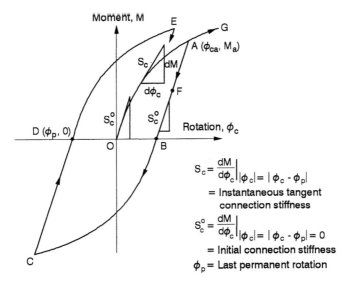

Fig. 7.1. Independent hardening model.

7.3.1. Independent hardening method

The independent hardening method is a simple method to describe the cyclic behaviour of connections under dynamic loading. In this method, the deterioration of connection property, such as the initial stiffness and the initial yield moment of joints, is not considered. In other words, the characteristic of connections is assumed to be unchanged, following the virgin moment versus rotation (M–ϕ) curve under the loading and reloading condition. Since the loops of the moment–rotation in each cycle are independent, no hardening effect is considered. The loop of connection moment–rotation curve is shown in Fig. 7.1. An algorithm for this method to describe the hysteretic behaviour of connections subjected to the dynamic loading is adopted in the present study and described as follows.

Under cyclic loads, the hysteretic loop and the instantaneous tangent connection stiffness on the moment–rotation (M–ϕ_c) curve will change in accordance with the current value of moment M, the relative slip rotation ϕ_c and the increment/decrement of moment ΔM at connection. An algorithm is described to monitor the non-linear moment–rotation relation of a connection looping path and the connection stiffness in the computer analysis (Fig. 7.1). The main objective of the algorithm is to produce a simple and generalized numerical procedure capable of modelling the non-linear behaviour of various type of connection models under dynamic loads.

Firstly, the virgin M–ϕ_c relation is assumed to be symbolically defined by

$$M = f(\phi_c), \tag{7.1}$$

in which $f(\phi_c)$ is a function of ϕ_c. Referring to Fig. 7.1, this algorithm is detailed as follows.

[1] If the path starts from or passes through a point $(\phi_p, 0)$ located on the x-axis or the line $M = 0$ and follows the virgin curve OA, BC, or DE, etc., it will be in the loading condition $(M \cdot \Delta M > 0)$ and follow the curve of the mathematical model by taking the point $(\phi_p, 0)$ as the new or updated origin, where ϕ_p is the last permanent rotation of the connection under cyclic loading. The M–ϕ_c relation is therefore written as

$$M = f(\phi_c - \phi_p), \tag{7.2a}$$

and the instantaneous tangent connection stiffness S_c is

$$S_c = \left. \frac{dM}{d\phi_c} \right|_{|\phi_c| = |\phi_c - \phi_p|}. \tag{7.2b}$$

[2] If the path is undergoing unloading $(M \cdot \Delta M < 0)$ for portion AB or CD it will move towards the x-axis along the straight line of slope equal to initial joint stiffness S_c^o and the reversal point (ϕ_{ca}, M_a) will be stored in the computer analysis for later use. The moment–rotation relation is simply expressed by

$$M = M_a - S_c^o (\phi_{ca} - \phi_c), \tag{7.3a}$$

and the tangent connection stiffness S_c is given as

$$S_c = \frac{dM}{d\phi_c} = S_c^o \tag{7.3b}$$

in which the reversal moment M_a is determined by

$$M_a = f(\phi_{ca} - \phi_p). \tag{7.3c}$$

[3] However, if the connection is unloaded from A to F and then reloaded from F to A, the path will follow a line parallel to the initial stiffness S_c^o until it reaches the last reversal moment M_a previously stored. When the connection is further reloaded $(M \cdot \Delta M > 0)$, it will follow the virgin curve AG which originates at the previous permanent rotation point $(\phi_p, 0)$. The moment at the connection is therefore given by

$$\begin{aligned} M &= M_a + S_c^o (\phi_c - \phi_{ca}) && \text{for } |M| < |M_a|, \\ M &= f(\phi_c - \phi_p) && \text{for } |M| \geqslant |M_a|, \end{aligned} \tag{7.4a}$$

and the corresponding tangent stiffness is computed by

$$\begin{aligned} S_c &= S_c^o && \text{for } |M| < |M_a|, \\ S_c &= \left. \frac{dM}{d\phi_c} \right|_{|\phi_c| = |\phi_c - \phi_p|} && \text{for } |M| \geqslant |M_a|. \end{aligned} \tag{7.4b}$$

Using the proposed algorithm from step [1] to step [3], stable and reproducible hysteretic loops of a non-linear connection can be traced. Provided that the force versus time function can be defined, this method performs well for structures subjected to dynamic loads. The present algorithm is simple, logical and can be applied to all types of connection models. This is more general and powerful than some existing methods which are restricted to one specified type of connection models for dynamic analysis. Furthermore, some of the the existing methods are too coarse for some highly non-linear connections, which include the linear and the bilinear moment–rotation functions.

A merit of the proposed algorithm is that numerous existing valuable equations for moment–rotation relations based on test data (Nethercot, 1985; Kishi and Chen, 1986a and 1986b; Lui and Chen, 1988) can be directly employed and incorporated into the present computer code, without the need of using another curve-fitting procedure to obtain the required parameters of a connection model. This reduces the effort and simplifies considerably the practical analysis and design. In addition, the new algorithm requires only a simple subroutine in the computer program for modelling and tracing of the hysteretic connection behaviour. As illustrated from the numerical examples in Section 7.8, the analysis results based on the above algorithm were found to be in good agreement with the experimental and numerical results obtained by others.

7.3.2. Kinematic hardening method

The kinematic hardening method is a modified independent hardening method allowing for the effect of material hardening. This is represented by the hardening line ($M = S_h \phi_c$) with slope S_h on the moment–rotation diagram shown in Fig. 7.2. In the case of reversal unloading, the path of the moment–rotation curve moves along the line with the slope of the initial connection stiffness S_c^o (i.e. for line AB or CD) until it reaches the hardening line. For further reversal unloading, the path follows the virgin non-linear M–ϕ_c curve of connections under the monotonic static loading (i.e. for curve BC or DE). The connections therefore follow the virgin curve earlier in this method than in the independent hardening method. Mathematically, if the hardening line has a zero slope (i.e. $S_h = 0$), the kinematic hardening method is exactly the same as the independent hardening method.

7.3.3. Bounding surface method

In the aforementioned independent and kinematic hardening methods, the moment–rotation relation of connections is first assumed to follow the straight line with the slope of the initial connection stiffness under the reversal unloading, and then obeys the virgin moment–rotation relation under the subsequent reloading. Alternatively, in order to simplify the procedure for handling the cyclic reversal connection behaviour, the moment–rotation path follows a non-linear curve directly when the reversal unloading occurs.

According to the Masing rule, if the virgin loading curve is defined by

$$f(M, \phi_c) = 0, \tag{7.5}$$

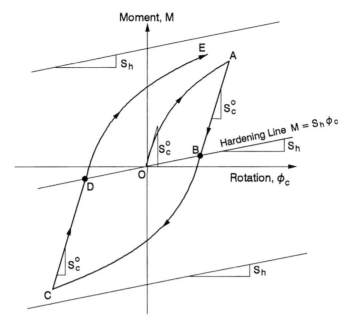

Fig. 7.2. Kinematic hardening model.

the unloading and reloading curve can be assumed by the following equation:

$$f\left(\frac{M - M^*}{2}, \frac{\phi_c - \phi_c^*}{2}\right) = 0, \tag{7.6}$$

where (ϕ_c^*, M^*) is the point at which the load reversal occurs. Therefore, the corresponding instantaneous connection stiffness of the virgin curve is given by

$$S_c = \frac{dM}{d\phi_c} = g(M, \phi_c), \tag{7.7}$$

and the instantaneous stiffness of the unloading and reloading curve is expressed by

$$S_c = \frac{dM}{d\phi_c} = g\left(\frac{M - M^*}{2}, \frac{\phi_c - \phi_c^*}{2}\right). \tag{7.8}$$

Gao and Haldar (1995) adopted the Masing rule and chose the Richard–Abbott model (Richard and Abbott, 1975) to represent the virgin connection moment–rotation relationship for curve OA given by

$$M = \frac{(k - k_p)\phi_c}{\left(1 + \left|\frac{(k - k_p)\phi_c}{M_o}\right|^n\right)^{1/n}} + k_p\phi_c, \tag{7.9}$$

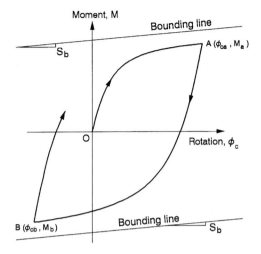

Fig. 7.3. Bounding surface model.

in which k is the initial connection stiffness, k_p is the strain-hardening stiffness, M_o is a reference moment and n is a parameter defining the sharpness of the curve. Hence, the expression of the M–ϕ_c curve for modelling the unloading and reloading part AB can be written as

$$M = M_a - \frac{(k - k_p)(\phi_{ca} - \phi_c)}{\left(1 + \left|\frac{(k - k_p)(\phi_{ca} - \phi_c)}{2M_o}\right|^n\right)^{1/n}} - k_p(\phi_{ca} - \phi_c), \qquad (7.10a)$$

and the tangent joint stiffness is

$$S_c = \frac{dM}{d\phi_c} = \frac{(k - k_p)}{\left(1 + \left|\frac{(k - k_p)(\phi_{ca} - \phi_c)}{2M_o}\right|^n\right)^{(n+1)/n}} + k_p, \qquad (7.10b)$$

in which (ϕ_{ca}, M_a) is the reversal point as shown in Fig. 7.3. If (ϕ_{cb}, M_b) is the next reversal point, the reloading M–ϕ_c relation is obtained by replacing (ϕ_{ca}, M_a) with (ϕ_{cb}, M_b) in Eqs. (7.10a) and (7.10b).

7.4. Connection models used in this study

The M–ϕ_c mathematical function between the moment and rotation is essential in the application of the modelling technique in Section 7.3. Since the experimental data for the behaviour of connections subjected to cyclic loading is limited, it is quite difficult to model the actual response of the hysteretic moment–rotation looping process in strict accordance with the test results. Popov and Pinkney (1969), Azizinamini and Radziminski

(1989), Tsai and Popov (1990), and Korol et al. (1990) showed from the tests that properly detailed and fabricated connections perform well and are stable and reproducible under cyclic loading. For simplicity, the cyclic behaviour of connections with the independent hardening rule method is used in this study (i.e. using Eqs. (7.1) to (7.4)). The virgin curves of connections obtained from monotonic loading tests are referred to in the present dynamic analysis.

As discussed in the last Chapter, numerous types of mathematical functions for modelling the non-linear behaviour of connections have been developed. They include the bilinear model (Sivakumaran, 1988; Youssef-Agha, 1989), the trilinear model (Gerstle, 1988), the polynomial model (Frye and Morris, 1975), the cubic B-spline model (Cox, 1972; Jones et al., 1980), the bounding-line model (Al-Bermani et al., 1994; Zhu et al., 1995), the power model (Batho and Lash, 1936; Colson and Louveau, 1983; Kishi and Chen, 1987a, 1987b; King and Chen, 1993), the Ramberg–Osgood model (Shi and Atluri, 1989), the Chen–Lui exponential model (Lui and Chen, 1988), the modified Chen–Lui exponential model (Kishi and Chen, 1986a, 1986b), and the Richard–Abbott model (Gao and Haldar, 1995).

For analysis of frames with semi-rigid connections, a $M-\phi_c$ function must be first selected. Since the Chen–Lui exponential, the Ramberg–Osgood and the Richard–Abbott models can produce a smooth $M-\phi_c$ curve and give a good fit to experimental data, they are used in the present transient analysis of flexibly jointed structures. Furthermore, these models can always guarantee a positive connection stiffness which is desirable for modelling of realistic joint behaviour. The description of various functions for modelling of connection stiffness under monotonic load has been given in Chapter 5. When the connection reverses in rotation, the initial slope parallel to the virgin moment versus rotation curve is followed. The condition is sensed by the product of total moment and incremental moment being less than zero.

7.5. Hybrid element with connection springs

To consider connection flexibility in the present analysis, a connection element is mathematically idealized as a pseudo-rotational-spring or the connection spring element, which is attached to the two ends of a beam–column element to form a hybrid element. Figure 7.4 shows the connection between the hybrid element and its adjacent column. One side of a connection spring is connected to the beam–column element and the other side is connected to a global node. In fact, for the column elements in Fig. 7.4, connection springs are jointed at their ends. However, since nearly all steel columns are designed and built as continuous columns with heavily welded joints at the junction between columns, their end connection springs are usually assumed to be rigid and therefore not illustrated in the Figure. The deflection configuration of a hybrid element with end-springs is shown in Fig. 7.5. The element is initially straight before being loaded, and then deforms to a curved shape by translational and rotational movements. The internal forces and deformations at a connection spring are graphically illustrated in the Figure. The rotation of a connection is defined as the difference between the angles on the two sides of the connection (i.e. $\phi_c = \theta_c - \theta_b$). Eliminating the additional degrees-of-freedom due to the rotational

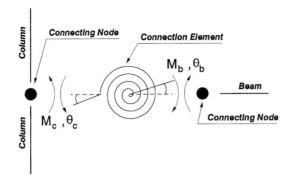

Fig. 7.4. Connection spring element.

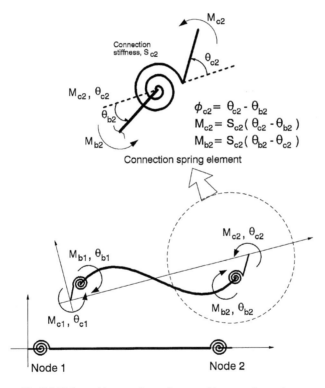

Fig. 7.5. Deformed beam–column element with connection springs.

deformations of connection spring elements, the transverse displacement function of the hybrid element can be derived by the conventional approach, which is described in Section 7.6.3. Hence, the linear stiffness, geometric stiffness, consistent mass, and viscous damping matrices for the hybrid element can be formulated in the procedures discussed in Sections 7.6.4 to 7.6.7.

7.6. Numerical procedure for present non-linear transient analysis

The following subsections introduce the numerical procedures in the developed computer program for transient analysis of semi-rigid frames.

7.6.1. Assumptions

The following assumptions are made for the analysis procedure in this Chapter.
(a) Bernoulli's assumption that a plane section normal to the centroidal axis before deformation remains plane after deformation and normal to the axis is made. Warping shear and cross section distortion are not considered.
(b) Small strains but arbitrarily large displacements and rotations are considered.
(c) Non-linearities due to joint flexibility and geometric change are considered. Material is assumed to remain elastic throughout the whole loading range.
(d) All connections are assumed sufficiently ductile and retain their connection stiffness and characteristic under cyclic loading.
(e) The hysteretic loop at a connection is based on its static monotonic moment–rotation curve which is fitted by a mathematical function.
(f) The connection element is of zero length.

7.6.2. Numerical solution method

In the present study, the widely used displacement-based finite element method is used. The equations of motions are based on the coordinate system of the updated Lagrangian description and are expressed in an incremental-iterative procedure. In the time-domain dynamic analysis, the time history is discreted into a number of equally spaced time steps. Based on the last known equilibrium configuration, the solution at the next time step is solved successively, starting from the initial condition to the desired time. The initial condition here is assumed to be stationary. By a step to step procedure, the unconditionally stable Newmark numerical integration method with the constant-average-acceleration assumption is employed (Newmark, 1959; Bathe, 1996). Iterations are performed for each time step to satisfy equilibrium before marching to the next step. Within a time step, the tangent connection stiffness is assumed constant in order to eliminate the possible connection stiffness oscillation under the iterative unloading/reloading process. Owing to the geometrical change, the tangent stiffness can be reformed and updated in each iteration in order to accelerate convergence within a time step. This procedure, however, requires a longer computational time for each iteration in forming the updated tangent stiffness. Alternatively, the constant stiffness iteration scheme using the initial tangent stiffness can be chosen. Although this scheme does not require such a tangent stiffness reforming procedure at every iteration, it needs more numbers of iterations for convergence. Therefore, a combined use of both methods by reforming the tangent stiffness covering every several iterations can be adopted.

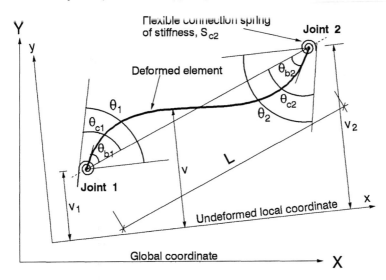

Fig. 7.6. The lateral deflections and rotations of a deformed hybrid element with flexible connection springs at the ends.

7.6.3. Derivation of shape function for beam–column element with flexible joints

In the displacement based finite element formulation, probably the most direct, but not necessarily the simplest, method to include the effect of connections modelled as springs is to adopt a shape function with this feature of having end-springs (see Fig. 7.6). For the beam–column type of one-dimensional element, the cubic Hermitian function is most commonly used because of its simplicity. For other types of shape functions, publications by So and Chan (1991) and Chan and Zhou (1994) should be referred to. It is noted that other shape functions can also be used in the stiffness matrix formulation of beam–column elements with end connection springs by the following procedure.

Referring to Fig. 7.6, the lateral deflection v on the y-direction at a displacement x along the centre line of a straight element can be written as

$$v = a_0 + a_1 x + a_2 x^2 + a_3 x^3, \qquad (7.11)$$

in which a_j is a coefficient. Denoting the end beam rotations by θ_{b1} and θ_{b2}, and the lateral deflections by v_1 and v_2 at the end nodes 1 and 2 of the element, we have the following boundary conditions as

$$\begin{aligned} \text{at } x = 0, \quad & v = v_1, \quad \frac{dv}{dx} = \theta_{b1}, \\ \text{at } x = L, \quad & v = v_2, \quad \frac{dv}{dx} = \theta_{b2}, \end{aligned} \qquad (7.12)$$

in which L is the length of the element.

The four unknowns (a_j, $j = 0, \ldots, 3$) in Eq. (7.11) can be solved by the four boundary conditions in Eq. (7.12), and we thus obtain

$$\begin{bmatrix} v_1 \\ \theta_{b1} \\ v_2 \\ \theta_{b2} \end{bmatrix} = \begin{bmatrix} 1 & 0 & 0 & 0 \\ 0 & 1 & 0 & 0 \\ 1 & L & L^2 & L^3 \\ 0 & 1 & 2L & 3L^2 \end{bmatrix} \begin{bmatrix} a_0 \\ a_1 \\ a_2 \\ a_3 \end{bmatrix}. \tag{7.13}$$

Rearranging the above equation, the unknowns are determined as

$$\begin{bmatrix} a_0 \\ a_1 \\ a_2 \\ a_3 \end{bmatrix} = \begin{bmatrix} 1 & 0 & 0 & 0 \\ 0 & 1 & 0 & 0 \\ -3/L^2 & -2/L & 3/L^2 & -1/L \\ 2/L^3 & 1/L^2 & -2/L^3 & 1/L^2 \end{bmatrix} \begin{bmatrix} v_1 \\ \theta_{b1} \\ v_2 \\ \theta_{b2} \end{bmatrix}. \tag{7.14}$$

Substituting the coefficients into Eq. (7.11), the lateral deflection function v becomes

$$v = \begin{bmatrix} \rho_1^2 \rho_2 L & -\rho_1 \rho_2^2 L \end{bmatrix} \begin{bmatrix} \theta_{b1} \\ \theta_{b2} \end{bmatrix} + \begin{bmatrix} (3 - 2\rho_1)\rho_1 & (3 - 2\rho_2)\rho_2 \end{bmatrix} \begin{bmatrix} v_1 \\ v_2 \end{bmatrix}, \tag{7.15}$$

in which

$$\rho_1 = 1 - \frac{x}{L}, \qquad \rho_2 = \frac{x}{L}. \tag{7.16}$$

The lateral deflection v in the above expression has not yet accounted for the effect of connection flexibility at the ends of the beam–column element.

Considering the in-plane bending of the inner prismatic beam–column within the connection springs at two ends of the hybrid element composed of the beam–column element with the two end connection springs, the element stiffness matrix of the Hermitian cubic element can be expressed in matrix form as

$$\begin{bmatrix} M_{b1} \\ M_{b2} \end{bmatrix} = \begin{bmatrix} K_{11} & K_{12} \\ K_{21} & K_{22} \end{bmatrix} \begin{bmatrix} \theta_{b1} \\ \theta_{b2} \end{bmatrix}, \tag{7.17}$$

in which the subscripts '1' and '2' are referred to as end nodes 1 and 2 of the beam–column element; and K_{ij} is the stiffness entries of the beam–column element.

Combining the two connection springs and the beam–column element together to form the complete system, the element stiffness matrix for the system can be written as

$$\begin{bmatrix} M_{c1} \\ M_{b1} \\ M_{b2} \\ M_{c2} \end{bmatrix} = \begin{bmatrix} S_{c1} & -S_{c1} & 0 & 0 \\ -S_{c1} & S_{c1} + K_{11} & K_{12} & 0 \\ 0 & K_{21} & S_{c2} + K_{22} & -S_{c2} \\ 0 & 0 & -S_{c2} & S_{c2} \end{bmatrix} \begin{bmatrix} \theta_{c1} \\ \theta_{b1} \\ \theta_{b2} \\ \theta_{c2} \end{bmatrix}. \tag{7.18}$$

For simplicity, the loads are assumed to be applied only at the global nodes and hence both M_{b1} and M_{b2} are equal to zero. Thus, we obtain

$$\begin{bmatrix} \theta_{b1} \\ \theta_{b2} \end{bmatrix} = \begin{bmatrix} S_{c1}+K_{11} & K_{12} \\ K_{21} & S_{c2}+K_{22} \end{bmatrix}^{-1} \begin{bmatrix} S_{c1} & 0 \\ 0 & S_{c2} \end{bmatrix} \begin{bmatrix} \theta_{c1} \\ \theta_{c2} \end{bmatrix}. \quad (7.19)$$

In the assemblage of the element stiffness matrix, it is essential to express the matrix in the global coordinate system for the consideration of geometrical compatibility. Substituting Eq. (7.19) into Eq. (7.15) and transforming the external nodal rotations, θ_{c1} and θ_{c2}, about the convected axis along the ends of the element to the global nodal rotations, θ_1 and θ_2, the displacement function v can be finally written as

$$v = \begin{bmatrix} \rho_1^2 \rho_2 L & -\rho_1 \rho_2^2 L \end{bmatrix} \begin{bmatrix} K_{11}+S_{c1} & K_{12} \\ K_{21} & K_{22}+S_{c2} \end{bmatrix}^{-1} \begin{bmatrix} S_{c1} & 0 \\ 0 & S_{c2} \end{bmatrix}$$
$$\times \begin{bmatrix} 1/L & 1 & -1/L & 0 \\ 1/L & 0 & -1/L & 1 \end{bmatrix} \begin{bmatrix} v_1 \\ \theta_1 \\ v_2 \\ \theta_2 \end{bmatrix} + \begin{bmatrix} (3-2\rho_1)\rho_1 & (3-2\rho_2)\rho_2 \end{bmatrix} \begin{bmatrix} v_1 \\ v_2 \end{bmatrix}, \quad (7.20)$$

in which θ_1 and θ_2 are the nodal rotations of the element about the global axis.

The above expression represents the relationship between the lateral deflection, v, and the connection stiffness, S_{c1} and S_{c2}, and the nodal variables, v_1, v_2, θ_1 and θ_2. Since the expression does not contain additional degrees-of-freedom for the connection spring elements, this can be directly used in the displacement-based finite element method which is familiar to engineers and researchers. In other words, the numerical procedure of the conventional finite element method for frame analysis can be retained. Therefore, Eq. (7.20) can be directly used to formulate the linear, the geometric stiffness, the mass and the viscous damping matrices.

7.6.4. Formulation of the linear stiffness matrices

Based on the variational principle and the energy theorem, the linear stiffness matrix for a beam–column element can be expressed as

$$[K_L] = \int_0^L \left(\frac{\partial^2 v}{\partial x^2}\right)^T EI \left(\frac{\partial^2 v}{\partial x^2}\right) dx \quad (7.21)$$

in which EI is the flexural rigidity and L is the length of the element. The coefficients of the elastic stiffness matrix $[K_L]$ are given in Appendix A.1.

7.6.5. Formulation of the geometric stiffness matrices

In order to consider the instability effect due to initial stress of the element, the geometric stiffness matrix $[K_G]$ is included in the incremental equilibrium equation by addition to the linear stiffness matrix $[K_L]$ to form the instantaneous tangent stiffness matrix. Considering the instability effect due to the axial load P in the element, we have

$$[K_G] = \frac{P}{2} \int_0^L \left(\frac{\partial v}{\partial x}\right)^T \left(\frac{\partial v}{\partial x}\right) dx. \tag{7.22}$$

Evaluating Eq. (7.22), the geometric stiffness matrix can be obtained and the complete expression is detailed in Appendix A.2.

Therefore, using the updated Lagrangian approach, the tangent stiffness matrix of the beam–column element accounting for the connection spring stiffness can be evaluated directly by

$$[K_T] = [K_L] + [K_G]. \tag{7.23}$$

7.6.6. Formulation of consistent mass matrix

In most previous studies, the lumped mass matrix has been used for simplification of calculation by ignoring the distributed masses. However, based on the shape function of deflection of the element in Eq. (7.20), the consistent mass matrix allowing for the semi-rigid connections can be determined as

$$[M] = \int_0^L v^T \overline{m} v \, dx, \tag{7.24}$$

in which $\overline{m}$ is the mass per unit length of the element. The mass matrix $[M]$ for prismatic members can therefore be evaluated directly and explicitly. The complex derivation can be carried out conveniently by mathematical software, Maple V (Waterloo Maple Software, 1992), and the expressions are given in Appendix A3.

7.6.7. Formulation of viscous damping matrix

Similar to the consistent mass matrix, the viscous damping matrix may be written in the following form

$$[C] = \int_0^L v^T c(x) v \, dx, \tag{7.25}$$

where $c(x)$ is the distributed damping coefficient per unit length. However, in practice, the evaluation of the damping property $c(x)$ is complex and may be unreliable. For this reason, the viscous damping matrix is usually assumed to be proportional to the mass and stiffness matrices (Clough and Penzien, 1975) as

$$[C] = a[M] + b([K_L] + [K_G]), \tag{7.26}$$

in which a and b are proportional constants that can be evaluated from the natural frequencies of the structure (Leger and Dussault, 1992). In order to illustrate the effect of hysteretic damping at connections, the viscous damping $[C]$ is not considered here although it can be easily included in the analysis through this proportional damping assumption.

7.6.8. Equations of motions

In transient analysis, the equilibrium equation at time $t + \Delta t$ is sought on the basis of the last known equilibrium state at time t. The incremental equilibrium equation can be written as

$$[M]\{\Delta \ddot{u}\} + [C]\{\Delta \dot{u}\} + ([K_L] + [K_G])\{\Delta u\} = \{\Delta F\}, \tag{7.27}$$

in which $[M]$ and $[C]$ are the mass and the viscous damping matrices; $[K_L]$ and $[K_G]$ are the linear and the geometric stiffness matrices; $\{\Delta F\}$ is the incremental applied force vector; and $\{\Delta u\}$, $\{\Delta \dot{u}\}$ and $\{\Delta \ddot{u}\}$ are the incremental displacement, velocity and acceleration vectors, respectively.

In the case of seismic loads, the incremental load vector $\{\Delta F\}$ can be expressed as

$$\{\Delta F\} = -[M][E]\{\Delta \ddot{x}_g\}, \tag{7.28}$$

in which $[E]$ is an index vector of the inertia forces and $\{\Delta \ddot{x}_g\}$ is an increment of absolute ground acceleration. In this case, it is noted that $\{\Delta u\}$, $\{\Delta \dot{u}\}$ and $\{\Delta \ddot{u}\}$ are the incremental displacement, velocity and acceleration vectors relative to the ground, respectively.

Expressing the velocity and the acceleration vectors at time $t + \Delta t$ in terms of the known displacement, velocity and acceleration vectors at time t and the acceleration vector at time $t + \Delta t$ by the Newmark method (Newmark, 1959), we have

$$\{^{t+\Delta t}\dot{u}\} = \{^t\dot{u}\} + (1-\beta)\Delta t\{^t\ddot{u}\} + \beta\Delta t\{^{t+\Delta t}\ddot{u}\}, \tag{7.29}$$

$$\{^{t+\Delta t}\ddot{u}\} = \left(1 - \frac{1}{2\alpha}\right)\{^t\ddot{u}\} - \frac{1}{\alpha(\Delta t)}\{^t\dot{u}\} - \frac{1}{\alpha(\Delta t^2)}\{^t u\} + \frac{1}{\alpha(\Delta t^2)}\{^{t+\Delta t}u\}, \tag{7.30}$$

and hence the final incremental equilibrium equation for solving the unknown displacement increment $\{\Delta u\}$ at time $t + \Delta t$ can be expressed as

$$[K]_{\text{eff}}\{\Delta u\} = \{\Delta F\}_{\text{eff}}, \tag{7.31}$$

where

$$[K]_{\text{eff}} = [K_L] + [K_G] + \frac{1}{\alpha(\Delta t)^2}[M] + \frac{\beta}{\alpha\Delta t}[C], \tag{7.32}$$

$$\{\Delta F\}_{\text{eff}} = \{\Delta F\} + [M]\left(\frac{\{^t\dot{u}\}}{\alpha\Delta t} + \frac{\{^t\ddot{u}\}}{2\alpha}\right)$$
$$+ [C]\left[\frac{\beta}{\alpha}\{^t\dot{u}\} + \frac{\Delta t}{2}\left(\frac{\beta}{\alpha} - 2\right)\{^t\ddot{u}\}\right], \tag{7.33}$$

in which $[K]_{\text{eff}}$ and $\{\Delta F\}_{\text{eff}}$ are the effective tangent stiffness and the effective load increment; α and β are the Newmark's parameters taken as 0.25 and 0.5 for the constant-average-acceleration assumption within each time step and $\{^t\dot{u}\}$ and $\{^t\ddot{u}\}$ are the total velocity and acceleration at time t.

When the displacement increment $\{\Delta u\}$ at time $t + \Delta t$ in Eq. (7.31) is solved, it is used to update the geometry of the structure, the displacement, the acceleration and the velocity by the following equations:

$$\{^{t+\Delta t}x\} = \{^t x\} + \{\Delta u\}, \tag{7.34}$$

$$\{^{t+\Delta t}u\} = \{^t u\} + \{\Delta u\}, \tag{7.35}$$

$$\{^{t+\Delta t}\ddot{u}\} = \frac{1}{\alpha(\Delta t)^2}\{\Delta u\} - \frac{1}{\alpha(\Delta t)}\{^t\dot{u}\} - \left(\frac{1}{2\alpha} - 1\right)\{^t\ddot{u}\}. \tag{7.36}$$

For checking of equilibrium, the following equations can be adopted

$$\{^{t+\Delta t}\dot{u}\} = \{^t\dot{u}\} + \Delta t(1-\beta)\{^t\ddot{u}\} + \beta\Delta t\{^{t+\Delta t}\ddot{u}\}, \tag{7.37}$$

$$\{\Delta F\}_i = \{^{t+\Delta t}F\} - \left([M]\{^{t+\Delta t}\ddot{u}\} + [C]\{^{t+\Delta t}\dot{u}\}_i + \{^{t+\Delta t}R\}_i\right) \tag{7.38}$$

in which $\{\Delta F^*\}_i$ is the unbalanced residual force increment vector; $\{^{t+\Delta t}F_a\}$ is the total applied nodal force; $\{^{t+\Delta t}R\}_i$ is the resisting force of the complete structure and i is the number of iterations in a particular time increment. The residual displacement increment $\{\Delta u\}_i$ due to the unbalanced residual force $\{\Delta F\}_i$ is, therefore, given by

$$\{\Delta u\}_i = [K]_{\text{eff}}^{-1}\{\Delta F\}_i. \tag{7.39}$$

If the unbalanced force norm, being equal to the ratio of the Euclidean norm of $\{\Delta F\}_i$ to $\{^{t+\Delta t}F\}$, and the residual displacement norm, equal to the ratio of $\{\Delta u\}_i$ to $\{^{t+\Delta t}u\}$, are also less than a certain tolerance *TOLER*, the equilibrium condition is assumed to have been satisfied and no further equilibrium iteration is activated. These conditions are stipulated in Eq. (7.40). Under such a condition, the procedure in Eq. (7.31) for the next time step is repeated.

$$\frac{\{\Delta F\}_i^T\{\Delta F\}_i}{\{^{t+\Delta t}F\}_i^T\{^{t+\Delta t}F\}_i} < TOLER, \tag{7.40a}$$

$$\frac{\{\Delta u\}_i^T\{\Delta u\}_i}{\{^{t+\Delta t}u\}_i^T\{^{t+\Delta t}u\}_i} < TOLER. \tag{7.40b}$$

If the condition in Eq. (7.40) is not satisfied, the following iterative scheme for correction of equilibrium error is activated.

The residual displacement in Eq. (7.40) is then used to update the geometry, the displacement, the velocity and the acceleration by

$$\{^{t+\Delta t}x\}_{i+1} = \{^{t+\Delta t}x\}_i + \{\Delta u\}_i, \quad (7.41)$$

$$\{^{t+\Delta t}u\}_{i+1} = \{^{t+\Delta t}u\}_i + \{\Delta u\}_i, \quad (7.42)$$

$$\{^{t+\Delta t}\dot{u}\}_{i+1} = \{^{t+\Delta t}\dot{u}\}_i + \frac{\beta}{\alpha(\Delta t)}\{\Delta u\}_i, \quad (7.43)$$

$$\{^{t+\Delta t}\ddot{u}\}_{i+1} = \{^{t+\Delta t}\ddot{u}\}_i + \frac{1}{\alpha(\Delta t)^2}\{\Delta u\}_i. \quad (7.44)$$

In order to accelerate the convergence rate, if necessary, the connection stiffness can be updated and the effective structural stiffness matrix is reformed.

The procedure from Eqs. (7.38) to (7.44) is iterated until convergence is achieved. After satisfaction of equilibrium at a particular time instance, the next time increment is marched out from Eq. (7.31) and the process is repeated until the desired duration in the time history is tracked.

7.6.9. Determination of resistant member forces

In the Newton–Raphson type of iterative analysis, the resisting force vector $\{R\}$ of a structure under deformation must be evaluated to check whether or not convergence meets the convergence requirement dictated by Eq. (7.40). It is obtained by adding the incremental resisting member force, which is obtained from the displacement increment $\{\Delta u\}$, to the total resisting force in the last iteration. The procedure for determining $\{R\}$ is described as follows.

After solving the incremental nodal displacement $\{\Delta u\}$ about the global axis from Eq. (7.31) or (7.39), the nodal displacement increments $[\Delta \psi]$ about the last local coordinate axis can be obtained directly as

$$\{\Delta \psi\} = [L]^{\mathrm{T}}\{\Delta u\}$$
$$= \{\Delta u_1 \quad \Delta v_1 \quad \Delta w_1 \quad \Delta \theta_1^x \quad \Delta \theta_1^y \quad \Delta \theta_1^z \quad \quad (7.45)$$
$$\Delta u_2 \quad \Delta v_2 \quad \Delta w_2 \quad \Delta \theta_2^x \quad \Delta \theta_2^y \quad \Delta \theta_2^z\}^{\mathrm{T}},$$

in which $[L]^{\mathrm{T}}$ is the transformation matrix relating the 12-displacements from the global axis to the local axis; Δu, Δv and Δw are the translational displacement increments; $\Delta \theta$ is the rotational displacement increment; the subscripts '1' and '2' refer to nodes 1 and 2; and the superscripts 'x', 'y' and 'z' are the principal axes in the local coordinate system.

Hence, the external rotational increments $\Delta \theta_c$ about the axis passing through the end nodes in the updated coordinate system can be determined by subtracting the rigid body rotation increments $\Delta \mu$ from the total rotation increments $\Delta \theta$ about the last known axis. This task can be carried out by the following equations:

$$\Delta \theta_{c1}^y = \Delta \theta_1^y + \Delta \mu^y, \quad (7.46a)$$

$$\Delta\theta_{c2}^y = \Delta\theta_2^y + \Delta\mu^y, \tag{7.46b}$$

$$\Delta\theta_{c1}^z = \Delta\theta_1^z - \Delta\mu^z, \tag{7.46c}$$

$$\Delta\theta_{c2}^z = \Delta\theta_2^z - \Delta\mu^z, \tag{7.46d}$$

in which the incremental rotations due to translational displacements are given by

$$\Delta\mu^y = \sin^{-1}\left(\frac{\Delta w_2 - \Delta w_1}{L}\right), \tag{7.47a}$$

$$\Delta\mu^z = \sin^{-1}\left(\frac{\Delta v_2 - \Delta v_2}{L}\right). \tag{7.47b}$$

When the external rotation increments $\Delta\theta_c$ about the last known coordinate axes are evaluated, the corresponding internal rotations can be determined by Eq. (7.19) for y- and z-axes as

$$\begin{Bmatrix} \Delta\theta_{b1}^y \\ \Delta\theta_{b2}^y \end{Bmatrix} = \begin{bmatrix} S_{c1}^y + K_{11}^y & K_{12}^y \\ K_{21}^y & S_{c2}^y + K_{22}^y \end{bmatrix}^{-1} \begin{bmatrix} S_{c1}^y & 0 \\ 0 & S_{c2}^y \end{bmatrix} \begin{Bmatrix} \Delta\theta_{c1}^y \\ \Delta\theta_{c2}^y \end{Bmatrix}, \tag{7.48a}$$

$$\begin{Bmatrix} \Delta\theta_{b1}^z \\ \Delta\theta_{b2}^z \end{Bmatrix} = \begin{bmatrix} S_{c1}^z + K_{11}^z & K_{12}^z \\ K_{21}^z & S_{c2}^z + K_{22}^z \end{bmatrix}^{-1} \begin{bmatrix} S_{c1}^z & 0 \\ 0 & S_{c2}^z \end{bmatrix} \begin{Bmatrix} \Delta\theta_{c1}^z \\ \Delta\theta_{c2}^z \end{Bmatrix}. \tag{7.48b}$$

With the external and the internal rotations at the end connections about the local y- and z-axes, $\Delta_e\theta$ and $\Delta_i\theta$, the corresponding relative slip rotations $\Delta\phi_c$ at the connections can be calculated directly as

$$\Delta\phi_c = \Delta\theta_c - \Delta\theta_b. \tag{7.49a}$$

Applying Eq. (7.49a) to both nodes and about the y- and z-axes, we obtain

$$\Delta\phi_{c1}^y = \Delta\theta_{c1}^y - \Delta\theta_{b1}^y, \tag{7.49b}$$

$$\Delta\phi_{c2}^y = \Delta\theta_{c2}^y - \Delta\theta_{b2}^y, \tag{7.49c}$$

$$\Delta\phi_{c1}^z = \Delta\theta_{c1}^z - \Delta\theta_{b1}^z, \tag{7.49d}$$

$$\Delta\phi_{c2}^z = \Delta\theta_{c2}^z - \Delta\theta_{b2}^z. \tag{7.49e}$$

Equations (7.49a) to (7.49e) can be used to update the current connection stiffness.

On the other hand, the incremental moments ΔM at the connections about the last known axis can be evaluated in terms of the internal rotations $\Delta\theta_{b1}$ and $\Delta\theta_{b2}$ by

$$\Delta M_1^y = K_{11}^y \Delta\theta_{b1}^y + K_{12}^y \Delta\theta_{b2}^y, \tag{7.50a}$$

$$\Delta M_2^y = K_{21}^y \Delta\theta_{b1}^y + K_{22}^y \Delta\theta_{b2}^y. \tag{7.50b}$$

The incremental moments in Eq. (7.50) are referred to the local y-axis. Similarly, the incremental moments about the local z-axis can be obtained as

$$\Delta M_1^z = K_{11}^z \Delta\theta_{b1}^z + K_{12}^z \Delta\theta_{b2}^z, \tag{7.50c}$$

$$\Delta M_2^z = K_{21}^z \Delta\theta_{b1}^z + K_{22}^z \Delta\theta_{b2}^z. \tag{7.50d}$$

The incremental torsional moment ΔM_x and the incremental axial load ΔP can be evaluated as

$$\Delta M_x = \frac{GJ + P\rho_o^2}{L}\left[\Delta\theta_2^x - \Delta\theta_1^x\right], \tag{7.51a}$$

$$\Delta P = \frac{EA}{L}\bigg[(\Delta u_2 - \Delta u_1) + \frac{1}{30}L\big(4\Delta\theta_{b1}^y\theta_{b1}^y - \Delta\theta_{b1}^y\theta_{b2}^y - \Delta\theta_{b2}^y\theta_{b1}^y + 4\Delta\theta_{b2}^y\theta_{b2}^y\big)$$
$$+ \frac{1}{30}L\big(4\Delta\theta_{b1}^z\theta_{b1}^z - \Delta\theta_{b1}^z\theta_{b2}^z - \Delta\theta_{b2}^z\theta_{b1}^z + 4\Delta\theta_{b2}^z\theta_{b2}^z\big)\bigg], \tag{7.51b}$$

in which $(\Delta u_2 - \Delta u_1)$ and $(\Delta\theta_2^x - \Delta\theta_1^x)$ are the relative incremental axial deformation and torsional twist, where Δu_1, Δu_2, $\Delta\theta_1^x$ and $\Delta\theta_2^x$ are obtained from Eq. (7.45); P is the axial force in the element; A is the cross-sectional area; G is the shear modulus; J is the torsional constant; and ρ_o is the polar radius of gyration. The second and third terms in Eq. (7.51b) are to account for the bowing effect due to axial force.

Internal forces for each element will then be updated by the force increments in Eqs. (7.50) to (7.51) as

$$\{P\}_k = \{P\}_{k-1} + \{\Delta P\}, \tag{7.52}$$

in which $\{P\}_k$ and $\{P\}_{k-1}$ refer to the six independent total internal force vectors for the current and previous equilibrium states respectively; and $\{P\}$ is given by

$$\{P\} = \begin{Bmatrix} P & M_1^y & M_1^z & M_x & M_2^y & M_2^z \end{Bmatrix}^T. \tag{7.53}$$

The resistance force vector of the whole structure $\{R\}$ can then be computed as

$$\{R\} = \sum_{N_{\text{ele}}}[L][T]\{P\}, \tag{7.54}$$

in which $[T]$ and $[L]$ are the transformation matrices relating the 6 internal forces to the 12 forces in the local coordinate axes and then to the global coordinates, and N_{ele} is the total number of elements. $\{R\}$ is the resistant force for the complete structure and it is used for determination of the unbalanced force vector $\{\Delta F\}$. Matrices for $[L]$ and $[T]$ have been given in Eqs. (2.11) and (2.36).

The tedious computational work can be carried out by computers. The numerical procedure for the present non-linear dynamic finite element method described in this Chapter has been coded to the developed computer FORTRAN program GMNAF/D (Non-linear Analysis of Frames – Geometric and Material analysis in Dynamics). The program is verified in the next section by comparison with the numerical results by others through some benchmark problems. It was found that the present results are in good agreement with those by others. The developed program is, therefore, believed to a powerful software tool in predicting the non-linear transient response of steel structures with hysteretic semi-rigid joints under various types of dynamic loading such as earthquakes, wind, industrial machinery and blasting.

7.7. Procedure for present vibration analysis

The vibration frequencies and modes of a structure connected by flexible joints can be idealised as a set of simultaneous equations for the stiffness matrix relations. The computed eigenvalue and eigen-vector represent the frequency and the vibration mode of the structure, respectively.

The dynamic equilibrium equation of a free and undamped vibrating system can be simply written as

$$[M]\{\ddot{u}\} + [K]\{u\} = 0, \tag{7.55}$$

in which $\{u\}$ is the displacement vector and $\{\ddot{u}\}$ is the acceleration vector. The first term represents the inertia force and the second term is the stiffness resistance.

Substituting the displacement function in terms of a sine time function as $u = u_0 \sin \omega t$ into the above equation, we obtain the characteristic equation as

$$\left| -\omega^2 [M] + [K_L] \right| = 0, \tag{7.56}$$

in which ω is the frequency of the system.

In the presence of axial force, the stiffness of a member will be changed and this effect can be included in an analysis by the addition of the geometric stiffness or the initial stress matrix to the equilibrium Eq. (7.55) as

$$\left| [K_L(\lambda)] - \omega^2 [M] + \lambda [K_G] \right| = 0, \tag{7.57}$$

in which λ is the load intensity parameter. In cases where the change of connection stiffness due to the moments at joints is considered, the linear stiffness matrix $[K_L(\lambda)]$ will be formed in accordance with the current joint stiffness at the assumed or the design load level. It is, therefore, also a function of the load factor, λ. The proportional damping effect can also be included in the analysis through the following well known expression (Paz, 1991)

$$\omega_{\text{Damping}} = \omega\sqrt{1-\xi^2},\qquad(7.58)$$

in which ξ is the damping ratio.

In some structures, the variation of connection stiffness may be considerable due to the moment at joints and the initial axial force in columns. On the basis of the above theory, the numerical procedure for computation of the natural frequency of a flexibly jointed steel frames allowing for the effect of initial stress from axial load in members and of joint stiffness due to moment, is summarised as follows.

A linear analysis to determine the moments at joints and the initial stress in the elements is carried out, and the linear and geometric stiffness matrices are formed accordingly. If the joint stiffness is non-linear, it will be related to the bending moment in the frame and iterations may be necessary to achieve equilibrium in the structure. With the initial stress calculated according to the load distribution in the frame, the geometric stiffness matrix for the element and the tangent stiffness of the complete structure can subsequently be formed. The determinant of this tangent stiffness matrix is calculated and iteration is activated until this determinant is less than a certain tolerance.

In cases where the effect of initial stress are not considered and the joint stiffness is assumed constant, only the mass and the linear stiffness matrices are required to be formed. The process, therefore, of finding the non-linear joint stiffness under the assumed loads and the formation of the geometric stiffness can be omitted. The remaining part of the numerical scheme is unchanged.

7.8. Numerical examples

The following examples include vibration and transient analysis of steel frames subjected to cyclic, impact, sinusoidal and seismic loads with various types of connections. Rigid connections and linear stiffness connections are also included for comparison against non-linear connections to illustrate the importance of connection flexibility and non-linearity. In modelling of the M–ϕ_c curves of non-linear connections, the Chen–Lui exponential, the Richard–Abbott and the Ramberg–Osgood models are chosen because they require less parameters, guarantee positive first-derivative for connection stiffness and represent the curves very well. Static gravitational loads are also studied because they can reduce column stiffness via the P-Δ and the P-δ effects.

7.8.1. Natural frequencies of beams with varying connection stiffness

In this example, the natural frequency for a beam–column member with two ends flexibly connected is studied. The results of the change in natural frequency due to the variation of joint stiffness is shown in Fig. 7.7. The frequency of the beam is represented by the square root of its flexural rigidity, mass per unit length and the quadruple of length.

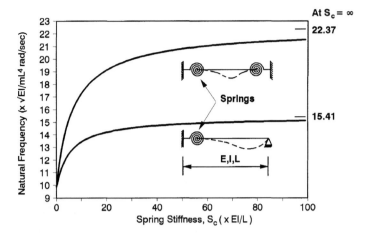

Fig. 7.7. Natural frequency of beam with various spring stiffness.

When the spring stiffness is zero or infinite, the solutions converge to the analytical results obtained in standard texts (Paz, 1991). When the joint stiffness is varied, the natural frequency changes and the result is depicted in Fig. 7.7. It can be seen that the change of the natural frequency is considerable when the joint stiffness varies and ignoring this joint flexibility may lead to a large error in vibration analysis of flexibly jointed structures.

7.8.2. Vibration analysis of a cantilever column subjected to a varying axial load

The presence of an axial load affects the stiffness of a beam–column. This effect can be included via the geometric stiffness matrix or by the stability function. It, in turn, affects the dynamical characteristics of a beam–column.

This example is aimed at the study of the effect of axial load on the vibration of a cantilever column shown in Fig. 7.8. In the Figure, the computed frequency is expressed in a non-dimensional form via the division by the natural frequency, ω_0 of the cantilever column free from any initial axial load. P_e is the Euler's buckling load of the cantilever. As seen in Fig. 7.8, an axial force alters the natural frequency significantly when the force is large. If the axial load is in tension, the frequency of the member will be increased and vice versa. When the load is close to the Euler's buckling load, the frequency tends to zero, indicating an unstable equilibrium state where frequency will be very low.

7.8.3. Vibration of two storey portals

The natural frequency of a two storey portal frame of height 7.32 m (288 inches) and width 6.10 m (240 inches) shown in Fig. 7.9 has been analyzed. To activate the non-linear characteristic of the joints, a set of point loads of P equal to 1601 kN (360 kips) is applied at the mid-span and the two ends of the beam. This arrangement will create moments at the two ends of the beam, which in turn will change the joint stiffness and the

Chapter 7. *Cyclic behaviour of flexibly connected elastic steel frames*

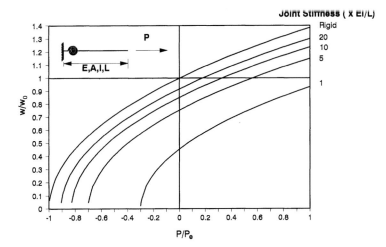

Fig. 7.8. Natural frequency of cantilever column with various axial load.

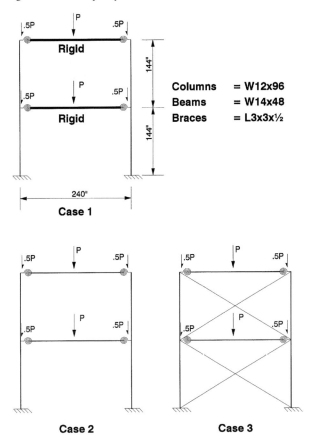

Fig. 7.9. Vibration analysis of portal frames.

Table 7.1
Natural frequency of one bay two storey frame with semi-rigid connections

Joint type	Natural frequency, f (Hz)		Natural frequency, f (Hz) (with axial effect)	
	Linear joint	Non-linear joint	Linear joint	Non-linear joint
Case 1: Without bracing (rigid floor assumption)				
A	6.57	6.57	4.32	4.33
B	7.49	7.50	5.26	5.28
C	7.73	7.73	5.51	5.50
D	9.94	9.94	7.60	7.60
Rigid		17.87		14.58
Case 2: Without bracing				
A	6.44	5.26	4.19	2.80
B	7.13	5.27	4.91	2.82
C	7.30	5.29	5.08	2.84
D	8.53	5.29	6.28	2.85
Rigid		10.35		7.99
Case 3: With bracing				
A	26.38	26.31	21.63	21.57
B	26.44	26.31	21.68	21.57
C	26.45	26.31	21.69	21.57
D	26.58	26.31	21.79	21.57
Rigid		26.85		22.05

structural frequency. In total, the three cases of rigid floor, unbraced and braced frames are studied in this example. In the rigid floor case, the slab is assumed to be one hundred times stiffer than the column. The flexural rigidity of the beams is therefore strengthened by the slab. The results of the analysis are tabulated in Table 7.1. The linear and non-linear joint stiffness and geometry have been assumed and their results are presented. Five types of joints have been considered in total; the single web angle connection (A), the top and seated angle connection with double web cleats (B), the flush end plate connection (C), the extended end plate connection (D) and the rigid joints. These typical joint rigidities are adopted from Lui and Chen (1988) and shown in Fig. 5.6 in Chapter 5.

Generally speaking, joint flexibility affects the vibration characteristics of a structure through its influence on the structural stiffness. It can also be seen in Table 7.1 that the non-linearity of joint stiffness may or may not have a notable influence over the vibration characteristics of a portal. The effects depend on the change of joint stiffness by the applied loads. Consequently, if the initial stiffness before loading differs significantly from the stiffness after loading, the frequency will be altered and vice versa.

The effect of axial load on the natural frequency of a structure depends on the magnitude of the change of the member stiffness by the load and cannot be over-generalized. For slender structures subjected to high axial load, this effect is significant and vice versa. In general, if columns of medium slenderness are designed without the consideration of axial load effect, the frequency of a structure will be over-estimated.

Table 7.2
Natural frequencies of six-storey frame

	Natural frequencies (Hz)		
	Mode 1	Mode 2	Mode 3
Present	2.832	7.003	12.073
COSMOS/M	2.832	7.003	12.074

7.8.4. Linear dynamic analysis of six-storey frame

For verification purposes, the linear dynamic analysis of a two-bay six-storey frame with rigid joints shown in Fig. 7.10 has been carried out by this study and the results compared with those obtained by a commercial computer analysis software package, COSMOS/M (1990). For simplicity, the lumped matrix is used. Using the present vibration analysis, the natural frequencies for the first three modes are determined and tabulated in Table 7.2. They match the results by COSMOS/M very well. In the next dynamic analysis, the frame is assumed to be subjected to a sinusoidal ground motion of magnitude 9.81 m/s^2 and of a period of 14 rad/s. Viscous damping is also assumed to be 0.25 of mass matrix. The displacement time-history at the top-right roof level is plotted in Fig. 7.10. It can be seen that the two sets of results are very close. From this example, the linear part of the present dynamic analysis program is validated.

7.8.5. Cantilever beam with uniformly distributed load

A cantilever with fixed end rigidity supported against deflection and rotation is re-analysed. The beam is subjected to an impact uniformly distributed load, as shown in Fig. 7.11. The beam is made of material with a Young's modulus of elasticity (E) of 1.2×10^4 psi, Poisson's ratio (v) of 0.2 and density (ρ) of 10^{-6} lb sec^2/in^4. Four elements are used and a time step of 15 micro-seconds is employed. In order to simulate the fixed support of the beam, the stiffness of the spring at the support is set to $10^{10} EI/L$. The response of the tip deflection by the developed program is shown in Fig. 7.11 and compares very well with Yang and Saigal (1984). It is also noted that the amplitude is very large and up to 60 percent of its length, indicating the validity of the method in the large deflection elastic range.

7.8.6. Clamped–clamped beam subjected to impact point load

A clamped–clamped beam under a concentrated load applied at the mid-span has been investigated by Mondkar and Powell (1977) and Yang and Saigal (1984) using the rigid joint assumption at supports. In the present study, the results for the rigid connection case are compared and the effect of the semi-rigid supports on the response is further investigated. The beam is divided into ten elements and its properties are shown in Fig. 7.12. For the condition of fully rigid joints at supports, the value of the connection stiffness, S_c, is assigned a very large value of $S_c = 10^{10} EI/L$. Two semi-rigid connections of

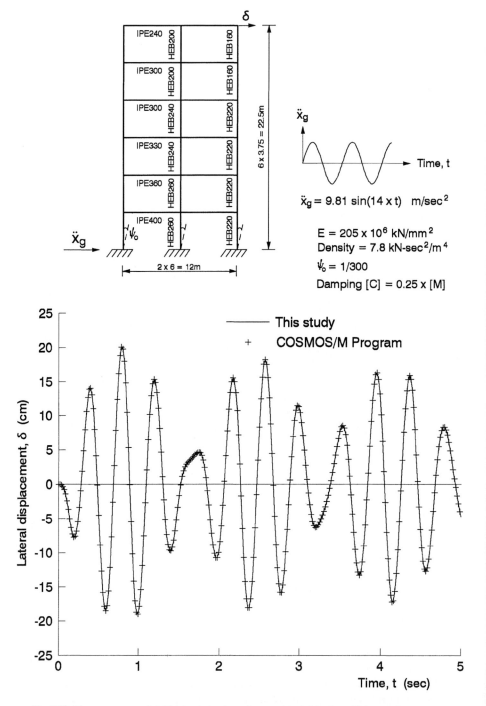

Fig. 7.10. Linear response of rigid-joined six-storey frame subjected to sinusoidal ground movement.

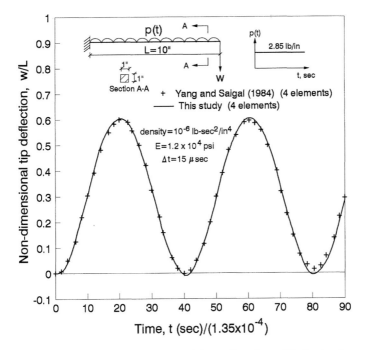

Fig. 7.11. Large displacement dynamic response of the cantilever beam fully fixed to the support.

stiffness $S_c = 10EI/L$ and $S_c = EI/L$ are further assumed in the analysis. The static load–deformation curves of the mid-span deflection are shown in Fig. 7.12. For rigid connections, the results were found to be very close to the results obtained by Mondkar and Powell (1977) and Yang and Saigal (1984). It was noted that the beam stiffness, equal to the slope of the curves, increases with the load due to the presence of a large axial tension and the subsequent catenary action in the beam in the large deflection range.

The dynamic behaviour of the beam with a finite stiffness for connections is further considered. An impact load of 640 lb is applied at the mid-span. The Newmark method with $\alpha = 0.25$ and $\beta = 0.5$ was employed and a time step of 10 microseconds was chosen. The results are shown in Fig. 7.13 and were found to be in close agreement with others. The initial responses for the semi-rigid joints are similar to the fully clamped case but subsequent paths deviate from the rigid joint case. As expected, the beam with softer connections has a longer period and a larger peak deflection.

7.8.7. Large deflection dynamic analysis of 45° curved cantilever beam

Static analysis of a cantilever 45° bend subjected to a concentrated end load has been studied by Bathe and Bolourchi (1979), Cardona and Geradin (1988) and Kouhia and Tuomala (1993). This example is a three-dimensional large deflection problem. To validate the proposed theory, this problem is reanalysed and semi-rigid connections at the support are further studied for comparison. The properties of the cantilever are shown in

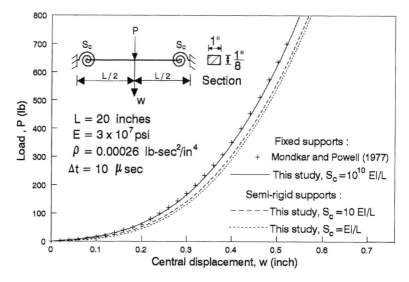

Fig. 7.12. Static analysis of clamped–clamped beam with semi-rigid supports.

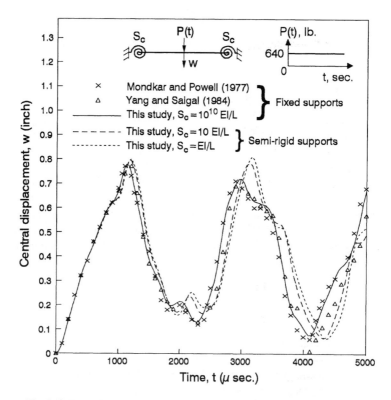

Fig. 7.13. Dynamic response of clamped–clamped beam with semi-rigid supports.

Chapter 7. *Cyclic behaviour of flexibly connected elastic steel frames*

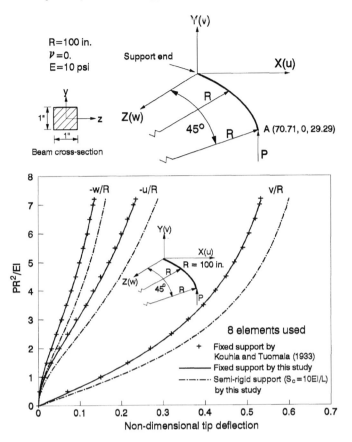

Fig. 7.14. Three-dimensional large deflection of 45° circular bend.

Fig. 7.14. Two rotational connection springs about the local axes, y–y and z–z, are assumed at the support ends. The semi-rigid connection of stiffness is taken as $S_c = 10EI/L$ and the bend is modelled by eight straight elements. The static analysis results are plotted in Fig. 7.14. They agree well with the results by Kouhia and Tuomala (1993). The tip positions at various load levels tabulated in Table 7.3 are also in close agreement with others.

Since its dynamic behaviour has not yet been studied, the problem is further extended to include dynamic analysis. Material density is assumed to be 2.54×10^{-4} lb sec^2/in^4. A concentrated load of 300 lb suddenly applied at the tip is studied and the time history of the tip is plotted in Fig. 7.15. From the results, the bend with semi-rigid connection has a larger deflection amplitude and a longer response period when compared to the response of the rigid connection case. It can be seen that semi-rigid connections can significantly affect the structural behaviour. The present developed computer program, therefore, is demonstrated to be capable of handling the transient response of structures undergoing large deflection in three dimensional space.

Table 7.3
Comparison of static results for cantilever 45° bend

Load	Tip position (inch)		
level (lb)	This study	Kouhia and Tuomala (1993)	Cardona and Geradin (1988)
0	(70.71, 0, 29.29)	(70.71, 0, 29.29)	(70.71, 0, 29.29)
300	(58.77, 40.17, 22.29)	—	(58.64, 40.35, 22.14)
450	(52.22, 48.48, 18.56)	—	(52.11, 48.59, 18.38)
600	(47.12, 53.45, 15.75)	(47.01, 53.50, 15.69)	(47.04, 53.50, 15.55)

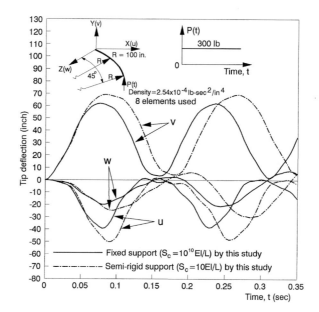

Fig. 7.15. Dynamic behaviour of the 45° circular bend.

7.8.8. Portal frame subjected to harmonic ground motion

An unbraced simple portal frame made of wide flange sections is shown in Fig. 7.16. A vertical load of 10 kips is applied at the mid-span of the beam. The static response of the semi-rigid frame has been studied by Lui and Chen (1986), Chen and Zhou (1987) and Al-Bermani and Kitiporchai (1992). The moment–rotation curve of semi-rigid connections is modelled by the Chen–Lui exponential model and the required parameters for the model are listed in Table 7.4. Al-Bermani et al. (1994) further investigated the dynamic behaviour of the same frame subjected to a sinusoidal ground motion of acceleration amplitude 150 inches/sec^2 using four different vibrating frequencies. These were 15.7, 12.56, 10.47 and 8.97 rad/sec. Lumped masses each of 5 kips due to the vertical static load are assumed to be attached at the two ends of the beam. The time histories of lateral deflection at the right top node for the rigid and the semi-rigid frames were compared.

Chapter 7. *Cyclic behaviour of flexibly connected elastic steel frames*

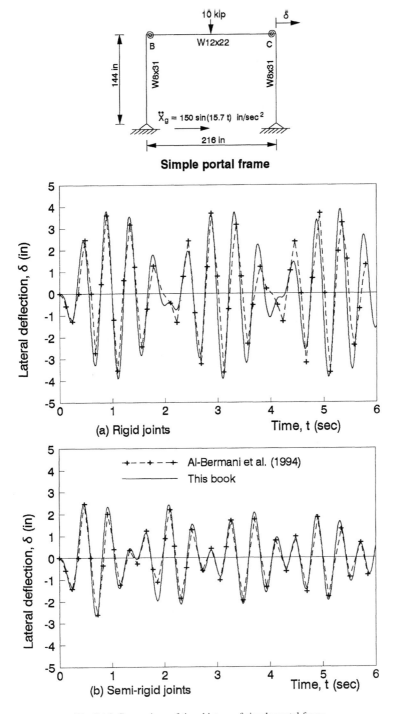

Fig. 7.16. Comparison of time history of simple portal frame.

Table 7.4
Required parameters for the Chen–Lui exponential model

Coefficients	Values
C_1	-0.37565×10^2
C_2	0.10758×10^4
C_3	-0.77282×10^4
C_4	0.22126×10^5
C_5	-0.28368×10^5
C_6	0.13544×10^5
α	0.10789×10^{-3}
M_o	0.0 kip in
R_{kf}	0.0 kip in/rad

It was found that non-linear joints dampen the deflection under the higher range of vibrating frequencies while, on the contrary, magnify under the lower range. For the case of the exciting frequency equal to 15.7 rad/sec, the results from the present analysis and from Al-Bermani et al. (1994) are plotted in Fig. 7.16. It can be seen that they are in good agreement.

7.8.9. Single-bay two-storey frame for experimental verification

This example is to verify the present proposed connection modelling algorithm (Chui and Chan, 1996) with experimental test results. A total of ten flexibly connected steel frames were tested by Stelmack et al. (1986). The geometrical and material properties of these frames is given in Fig. 7.17(a). Half inch thick angle connections were used in the test, and fourteen monotonic moment–rotation test results of such connections were reported by Stelmack et al. (1986). The upper and the lower bounds of the scattered data are shown in Fig. 7.17(c). The trilinear model was used by Stelmack et al. (1986) whilst the Richard–Abbott model is used in this study. Youssef-Agha et al. (1989) also studied the seismic response of a similar frame. In their paper, a bilinear model for the top and seat angle connection type was proposed and expressed in terms of the connection details such as the dimensions, geometries and arrangement of the connection components. In the test by Stelmack et al. (1986), the frame was subjected to cyclic lateral loads with load increment equal to ±1 kip and increased up to ±5 kips. For the second-storey, the load is always one-half of the load at the first storey. The load history is plotted on Fig. 7.17(b) for clarity.

The experimental and analytical results of the load–deflection curves and the connection moment–rotation loops are compared and plotted separately for each cycle in Figs. 7.18 and 7.19. It can be seen that the predicted results by the Richard–Abbott model are close to the experimental results and smoother than the results from the trilinear model. This shows that the proposed theory can generate accurate and smooth deflection curves for the frame.

The next study was to investigate the influence and numerical stability of various connection models. The frame with rigid, linear and non-linear connection types is assumed.

Chapter 7. *Cyclic behaviour of flexibly connected elastic steel frames*

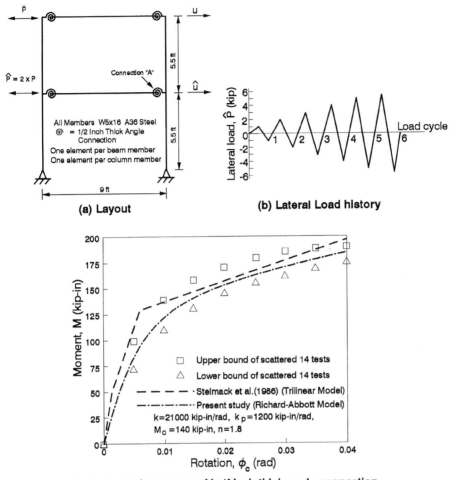

Fig. 7.17. Single-bay two-storey frame.

For the non-linear connections, three models, namely the Richard–Abbott model, the Ramberg–Osgood model and the Chen–Lui exponential model, are used. The required parameters and the resulting plots of these models on half inch angle connection are shown in Fig. 7.20. The frame is assumed to be subjected to a sudden load which pushes and pulls the frame in one second and then disappears. The time histories of the deflections at the second storey of the frames are shown in Fig. 7.21. It can be seen that the frame with semi-rigid connections has a larger amplitude and a longer period when compared with the rigid connection case. In addition, the frame with non-linear connections dampens and exhibits a non-recoverable deflection due to the presence of permanent rotations at connections. This feature, of course, was not observed in a linear joint model. In contrary to the non-linear connection case, the frame with either rigid or linear connections has

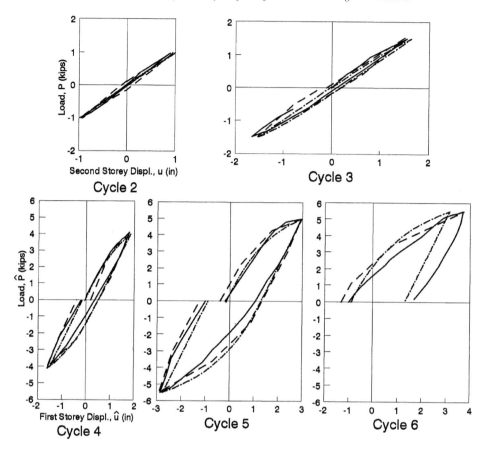

Fig. 7.18. Load–deflection curves.

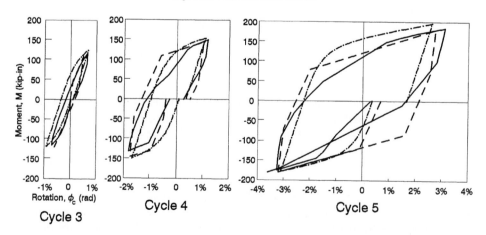

Fig. 7.19. Moment–rotation loops at connection A. ——— Tested results by Stelmack et al. (1986); – – – – Theory predicted by Stelmack et al. (1986); – · – · – · Predicted results by present theory.

Chapter 7. *Cyclic behaviour of flexibly connected elastic steel frames*

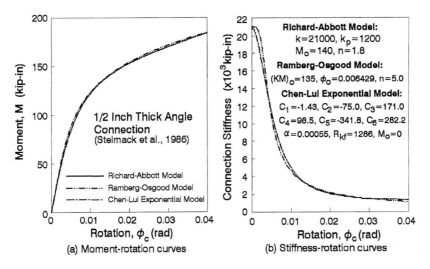

Fig. 7.20. Models representing half inch thick angle connection.

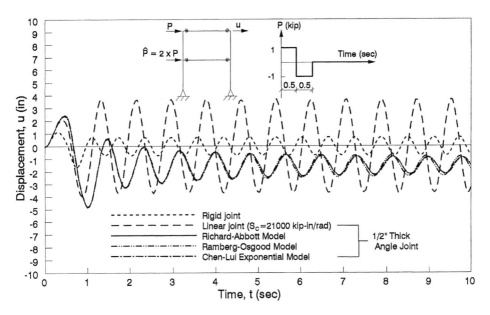

Fig. 7.21. Transient response of frames with different joints.

no damping. It was also noted that the response of the frame simulated by these three non-linear connection models are very close to each other. None of them experiences numerical problem such as divergence. The validity of using these models indicates that the choice of these three connection models is not important when the connection behaviour is represented appropriately by any one of them. Similar observation can be obtained in

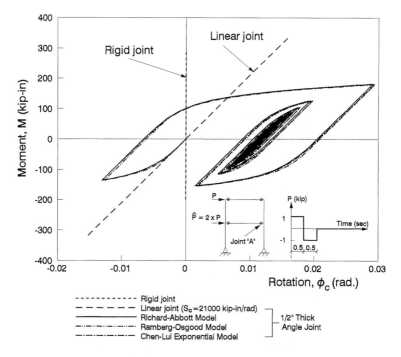

Fig. 7.22. Hysteretic $M-\phi_c$ loops traced at joint A using different models.

Fig. 7.22 which shows the hysteretic $M-\phi_c$ loops at connection A. It also demonstrates that the proposed algorithm for simulating hysteretic and cyclic moment–rotation loops at connections can be described for different connection models. As shown in the previous study, the coarse model based on the simple trilinear function produces a curve with kinks. This slope discontinuity is undesirable in a computer analysis and thus the linearized model is not recommended.

7.8.10. Dynamic response of L-shaped frame

The natural frequency of an L-shaped frame composed of two mild steel members was measured experimentally by Kawashima and Fujimoto (1984). The connection stiffness is adjusted to different values and the frequencies at the first four modes were recorded in the test. The length of the horizontal and the vertical members are equal to 150 cm and 100 cm, respectively, and their flexural rigidity (EI) is 42,300 kgf cm². Details of the frame are depicted in Fig. 7.23. The connection stiffness used in their test is given by the following relation:

$$S_c = \frac{3EI}{L} \frac{v}{1-v},$$

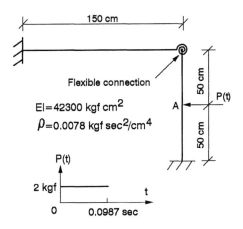

Fig. 7.23. L-shaped frame.

Table 7.5
Frequencies of the L-shaped frame

Vibration mode	Joint stiffness parameter, v	Frequency (Hz)	
		Experiment (Kawashima and Fujimoto, 1984)	Present theory
1	0.47	15.5	15.22
	0.307	14.1	14.76
	0.025	13.9	13.76
2	0.47	30.8	33.52
	0.307	30.2	32.59
	0.025	29.9	30.86
3	0.47	45.0	46.69
	0.307	44.1	45.64
	0.025	43.9	44.27
4	0.47	95.8	92.48
	0.307	89.1	92.07
	0.025	83.4	91.19

in which v is a parameter varies from zero for the pinned joint case to 1 for the rigid joint case. In tests carried out by Kawashima and Fujimoto (1984), the value of v is set to 0.025, 0.307 and 0.470, and frequencies correspondent to the joint stiffness were measured.

The experimental results together with the predicted frequencies by the present method are tabulated in Table 7.5. It can be seen that the discrepancy between the two sets of results is small, with an averaged difference of 4.3%. The source of error is mainly due to unavoidable experimental errors such as ignorance of the connection mass and the idealised theoretical finite element model for the structure.

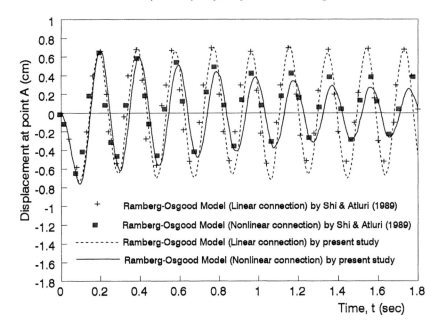

Fig. 7.24. Response of the L-shaped frame.

Shi and Atluri (1989) further studied the dynamic response of the same frame with an assumed header plate flexible connection type, as shown in Fig. 7.23. The frame was subjected to an impact load at the mid-height of the column. In the report of Shi and Atluri (1989), the tangent connection stiffness of the header plate connection type was represented by the Ramberg–Osgood model with the following parameters, $\phi_o = 0.00603$ rad, $M_o = 8.446$ kg cm and $n = 4.32$. Thus, the stiffness equation is expressed as

$$S_c = \frac{1400}{1 + 4.32 \, |0.1184 \, M|^{3.32}}.$$

Both the linear and the non-linear connections were considered by Shi and Atluri (1989). For the linear connection, the joint stiffness was taken as $S_c = S_c^o = 1400$ kg cm/rad. An impact load of 2 kgf for a short duration is applied at point A. The same problem is re-analysed by the present method and the displacement response at point A is depicted in Fig. 7.24. From the Figure, there is a slight difference between the curves by Shi and Atluri and in the current study. This may be due to different numerical methods used. Shi and Atluri (1989) adopted the stress-based tangent stiffness formulation while the current study used the more popular displacement-based finite element method. However, both the results indicate both an important character of hysteretic damping occurred at the non-linear connection. This hysteretic damping cannot be observed at a connection with linear stiffness.

The next study will investigate the influence of the use of connection models. This problem will be reanalysed for the same frame with three connection models, namely the

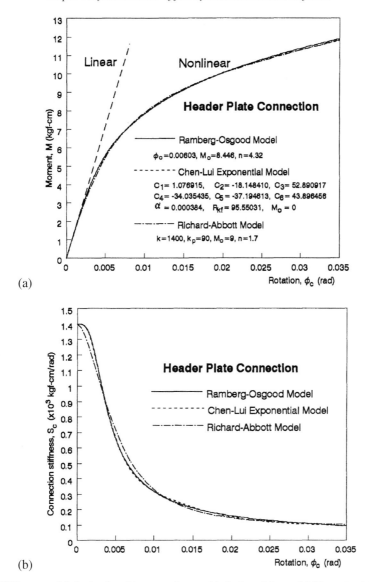

Fig. 7.25. Different models for header plate connection used in L-shaped frame. (a) Moment–rotation curves; (b) Stiffness–rotation curves.

Ramberg–Osgood, the Chen–Lui exponential and the Richard–Abbott models. Based on the previous M–ϕ_c curve using the Ramberg–Osgood model, the approximate parameters for describing the same curve by the Chen–Lui exponential and the Richard–Abbott models are obtained by a curve-fitting procedure. Their moment–rotation and stiffness–rotation curves are shown in Fig. 7.25(a) and (b). It can be seen that they simulate the connection quite well and are similar to each other. Under the same loading condition,

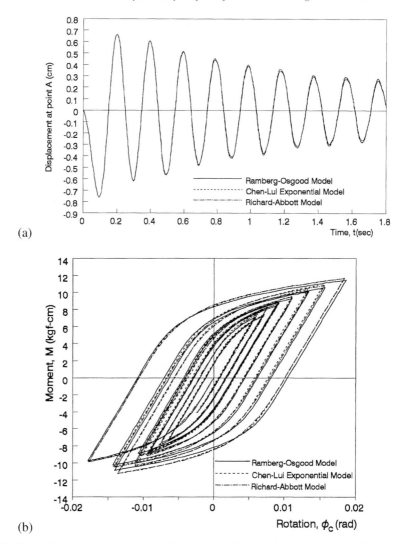

Fig. 7.26. Comparison of dynamic response of the L-shaped frame by different models. (a) Response of displacement vs. time; (b) Response of moment–rotation curves.

the displacement response at point A and the moment–rotation response at the connection are traced in Figs. 7.25(a) and (b), respectively. It is noted that the results are in close agreement. This indicates that the choice of these models is unimportant. In addition, the moment–rotation curve data from laboratory test results described by different models can be used and incorporated into the computer code directly without the need to re-fit the model. Furthermore, the algorithm described in Section 7.3.1 for modelling the hysteretic response of a semi-rigid joint is applicable to any one of these commonly used models. No numerical instability is observed and accurate solutions can be obtained.

Chapter 7. *Cyclic behaviour of flexibly connected elastic steel frames* 237

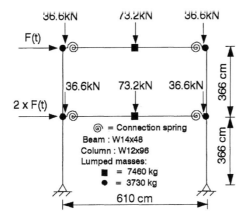

Fig. 7.27. Details of single-bay two-storey portal frame.

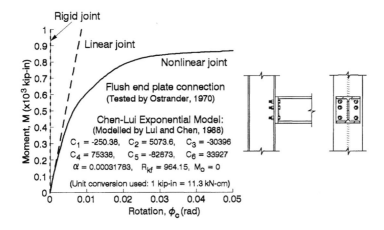

Fig. 7.28. Monotonic moment–rotation curve of flush end plate connection.

7.8.11. Gravitational load effects on single-bay two-storey frame

To predict a more realistic behaviour of a structure, gravitational loads (e.g. floor slabs and furniture) statically acting on the structure should be included in dynamic analysis. This is due to the fact that gravitational loads can induce axial forces on columns and, therefore, reduce the column stiffness. In addition, the additional moment due to the P-Δ effect is generated by the gravitational loads. In order to study these effects on structural response, gravitational loads are considered in this example.

An unbraced two-storey frame with static gravitational loads of, 36.6 kN and 73.2 kN, is studied and its geometry is shown in Fig. 7.27. These gravitational loads are considered as lumped masses and attached rigidly to the frame throughout the analysis. Linear and non-linear flush end plate and rigid connection types are assumed. The flush end plate connection modelled by the Chen–Lui exponential model plotted in Fig. 7.28 is assumed

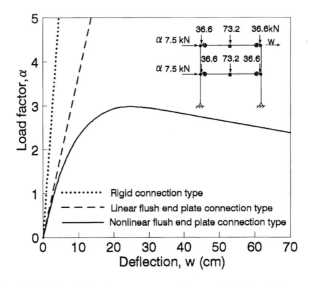

Fig. 7.29. Static analysis of two-storey portal frame with different joint types.

for the connection. Two elements per beam member and one element per column are used in the model and a time step of $\Delta t = 0.05$ second is chosen. The static load–deformation response of the frames is studied and plotted in Fig. 7.29. The load carrying capacity is considerably affected by the non-linear behaviour of connections. To test the reproducibility of hysteretic moment–rotation loops at connections, the frame is subjected to cyclic loads, and the response of displacement and connection M–ϕ_c loops are traced in Fig. 7.30. As shown in Fig. 7.30(a), the deflection initially oscillates between ∓ 7 cm at transient state but later oscillates steadily between ∓ 5 cm. It can be seen in Fig. 7.30(b) that the hysteretic loops are highly reproducible and stable. During this stable oscillation response, the amount of the input energy by the exciting loads and the energy dissipation at connections are found to be the same and approximately equal to 320 Joules per each cycle. In Fig. 7.31, the same frame is studied against the presence and the absence of gravitational loads and under impact loads. Various connection types are also considered. It was found that the deflection amplitude of the frames accounting for gravitational loads is larger than those without gravitational loads. This is due to the column softening by the P-Δ effect. Because of the formation of permanent rotations at the connections, there exists a shift of deflection response in the flexibly jointed frames. As indicated in the previous example, non-linear connections reduce the frequency whilst hysteretic damping induced at the connections decreases the successive amplitudes of the response due to dissipation of energy.

7.8.12. Buckling and vibration analysis of a portal

In the design of steel frames, vibration check may be necessary to ensure the structure will not violate the serviceability and to avoid resonance. Studies in this example are par-

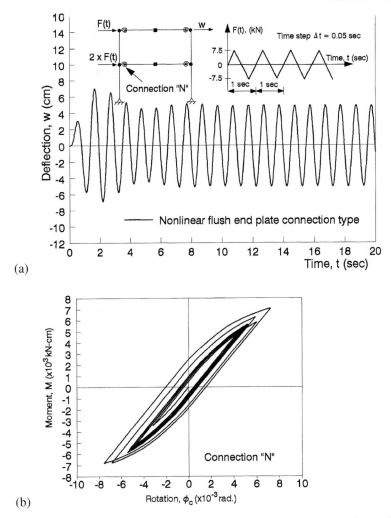

Fig. 7.30. Dynamic behaviour of two-storey frame with non-linear flush end plate connections under cyclic loads. (a) Displacement response; (b) Response of the hysteretic M–ϕ_c loops at connection N.

ticularly relevant to the design of portals and platforms supporting machines and human beings. As the vibration characteristic is sensitive to the joint stiffness and thus joint details, it is of interest to assess how the natural frequency of a frame varies against the joint stiffness.

The natural frequency of the simple portal shown in Fig. 7.32 is determined using various values for the joint stiffness. The mass along the beam is assumed to be equal to 10 times, 5 times and the same as the self mass of the beam in the study in order to investigate the influence of loading on the vibration behaviour of the portal with different joint stiffness.

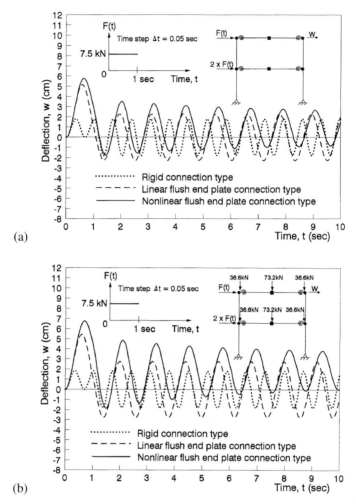

Fig. 7.31. Dynamic response of steel frame with various connection types. (a) Without gravity loads; (b) With gravity loads.

It can be seen in Fig. 7.32 that the connection stiffness has an important influence over the natural frequency of the portal. For practical extended end plate connections, the joint stiffness is approximately 1.3 times the flexural stiffness of the beam (EI/L) and the natural frequency in case 3 is 4.37 Hz which is 22% below the rigid joint case of 5.59 Hz. For the flexible joint detail of angle web cleats, the joint stiffness is about 0.1 of the beam stiffness and the natural frequency is 3.54 Hz, compared to the frequency for the idealised pinned joint case of 3.45 Hz. For other types of connections like the flush end plate and the top and bottom cleat connections, the dynamic characteristics of the portal is not close to either the rigid or the pinned joint cases. These observations show clearly the relationship between joint stiffness and the vibration frequency of steel frames.

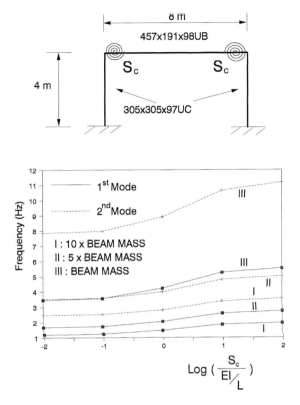

Fig. 7.32. Variation of frequency against joint stiffness.

7.8.13. Dynamic analysis of steel frames accounting for instability effect

The stiffness of an elastic beam will be increased by the presence of an applied tensile load and reduced when the load is compressive. The change in stiffness of a member in the presence of an axial load can also be allowed for in the present analysis through the adoption of Eqs. (7.57) instead of (7.56). The importance of considering this instability effect is studied by using the portal in the previous example but with an additional pair of applied loads acting on top of the columns (P in Fig. 7.33).

On the assumption that the applied loads equal 1000 kN for Case 1 and 2000 kN for Case 2 and the joint stiffness equals to the beam stiffness (EI/L), the frequencies of the portal are computed and compared with the results from Case 3 in the previous example. The calculated first mode frequencies are equal to 4.045 Hz and 3.841 Hz, respectively, for Cases 1 and 2, compared with the value of 4.239 Hz in the previous study where the instability effect due to axial loads is not considered. This effect will be more important in slender frames susceptible to instability and vice versa. Consequently, in the vibration check of slender structures, the unstable effect due to initial loads may be required in the analysis.

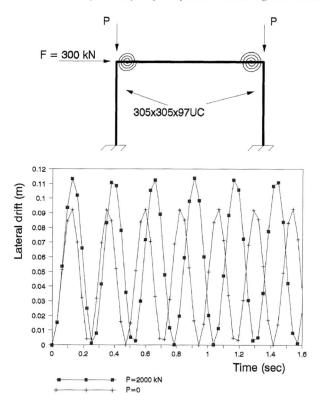

Fig. 7.33. The dynamic response of the semi-rigid jointed frame.

Using the obtained natural frequency and the mode superposition method, the dynamic response of the structure can be plotted. An impact force of 300 kN is applied horizontally to the top of the column which is further under the action of the vertical static loads, as shown in Fig. 7.33. The first three vibration frequencies are determined and used in the calculation of the deflection of the portal against time. Plotted in Fig. 7.33 is the response for the frame when each of the vertical loads is 2000 kN and zero, respectively. The difference in the response of the portal under the two loading cases is manifested clearly with respect to time. The maximum deflections under the dynamic loads are different, leading to a difference in calculated stress induced by the dynamic force. Note that the P-Δ effect has not been accounted for in this example because of the ignorance of the large displacement matrix in the incremental equilibrium equation. The effect may become significant when the lateral drift is large.

Non-linearity of joint stiffness due to bending moment at joints may be considerable in some cases. This effect can be included in the analysis by using the updated joint stiffness according to the magnitude of the moments at joints. The importance of this effect depends on the sensitivity of the connection stiffness against moments and also on the magnitude of the moments. Obviously the reduction of joint stiffness reduces the frame stiffness and thus the structural frequency.

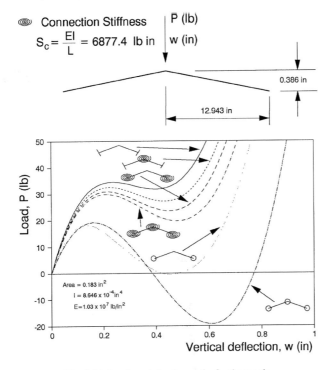

Fig. 7.34. Load vs. deflection paths for the toggle.

7.8.14. Static and dynamic analysis of the William's toggle frame with semi-rigid joints

The Williams' toggle frame (1964) with properties shown in Fig. 7.34 has been studied extensively. However, the behaviour of the frame with semi-rigid joint stiffness has only been investigated by Chan and Zhou (1994). This problem is selected for study in this example in order to understand the influence of semi-rigid connections in the pre- and post-buckling behaviour of this well-known frame. Roof structures like rafters may be susceptible to this snap-through buckling mode.

The equilibrium paths for the frames with different locations of the semi-rigid joints of stiffness equal to EI/L of the member are plotted in Fig. 7.34, together with the plots for the pinned and the rigid joint cases of which the result for the rigid joint case agrees with the experimental plot by Williams (1964). Each member was modelled by four elements. It can be seen from Fig. 7.34 that the locations as well as the magnitude of the joint stiffness are important in affecting the stability of the structure. An over-simplification of joint type in an analysis is dangerous in design.

The dynamic response of the toggle subjected to a suddenly applied point load of 15 lb at its top is further investigated. The joint stiffness is reduced to 0.1 of the beam stiffness (EI/L) in order to differentiate the curves for the semi-rigid and the rigid cases. The density of the material is taken as 2.54×10^{-4} lb sec^2/in^4. The vertical tip deflection is

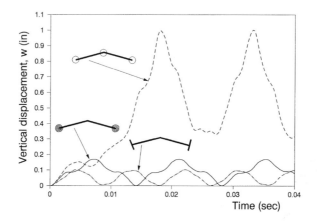

Fig. 7.35. The transient response of the toggle with various joint stiffness.

plotted against time and shown in Fig. 7.35. It is of interest to note that the load forces the pinned toggle to vibrate in the post-buckling range and the structure cannot return to its original state during vibration. The same load will only cause the toggle with semi-rigid or rigid joints to vibrate in its pre-buckling range. This discrepancy in the vibration characteristics for the two cases demonstrates clearly the interesting interaction of semi-rigid connections and the non-linear structural behaviour.

7.8.15. Analysis of the 3-dimensional hexagonal frame

The hexagonal frame shown in Fig. 7.36 has also been studied by a number of researchers (Papadrakakis, 1981; Meek and Tan, 1984) using the rigid joint assumption. In this case, it should be noted special care should be exercised in extending an analysis program for two dimensional frames to the three dimensional case in which the rotations are non-vectorial and non-accumulative (Chan, 1992). Adopting the incremental secant stiffness method (Chan, 1989), rotations can be arbitrarily large in both the static and dynamic analyses. In spite of previous research, static and dynamic analysis of the frame allowing for the effect of joint flexibility appears to have not been investigated before.

By assigning the joint stiffness a very large and a very small value, the equilibrium curves obtained are similar to the results by other researchers (Papadrakakis, 1981; Meek and Tan, 1984). When the joint stiffness used is set to 10 times the member stiffness, the equilibrium curve was found to be between the two extreme cases.

The transient response of the frame has been further investigated. The mass density is assumed to be 2.54×10^{-4} lb sec^2/in^4 and the time step used is 0.002 second. The point load applied at the top of the frame is assumed to be harmonic with an angular velocity of 50 radian/second. The joint stiffness is taken to be 366.5 lb in which is the same as the beam flexural stiffness (EI/L). Figure 7.37 shows the displacement response against time for the rigid and for the semi-rigid joint cases.

As expected, the semi-rigid joint response is more flexible than its rigid counterpart. The non-linear natural periods for the two frames undergoing large deflection are also

Chapter 7. Cyclic behaviour of flexibly connected elastic steel frames

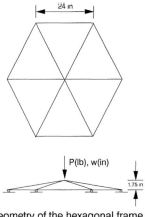

Geometry of the hexagonal frame

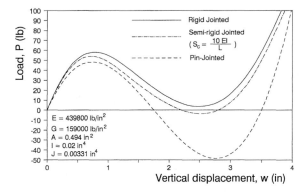

Fig. 7.36. Static response of the hexagonal frame with semi-rigid joints.

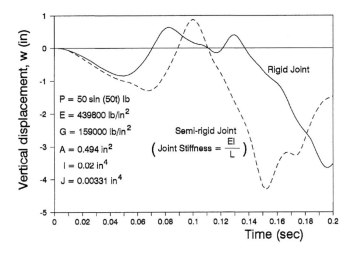

Fig. 7.37. Non-linear response of the hexagonal frame subjected to a harmonic point load at the top.

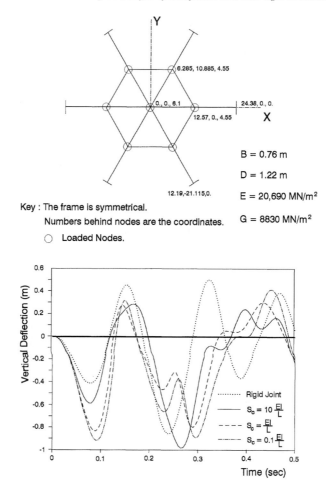

Fig. 7.38. Dynamic response of the dome frame with semi-rigid joints.

considerably different. From this example, the response of a space frame under the action of dynamic or earthquake loads will be better predicted if the joint stiffness is also considered in the analysis.

7.8.16. Large deflection analysis of 24 member dome frame with damping

The shallow dome with geometric properties shown in Fig. 7.38 has been analysed previously by Remseth (Remseth, 1979) dynamically. All members are 0.76 m wide, 1.22 m deep and of density 2400 kg/m^3. In his studies, the damping effect was included as 5% of the critical damping and the damping matrix is proportional to the mass matrix only. The point load of magnitude 34.4 MN applied at tip is assumed to be sinusoidal of period equal to 0.15 second.

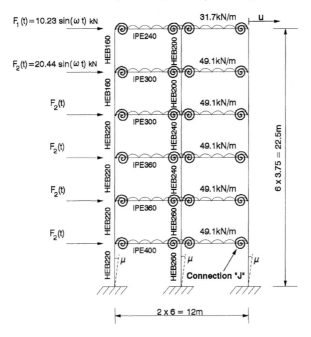

Fig. 7.39. Layout of the Vogel six-storey frame.

As shown in Fig. 7.38, the deflection response from Remseth (1979) agrees with the present results in the rigid joint case. For the semi-rigid joint case, no previous results are available for comparison.

The structure of the same material and geometrical properties but with different assumed joint stiffness is analysed and the resulting plots for the response of the tip deflection are depicted in Fig. 7.38. A connection stiffness of 10, 1 and 0.1 of the member stiffness (EI/L) of 188 MN m is assumed. Two elements were used to model each member and the lump mass was employed to formulate the mass matrix.

It can be seen from Fig. 7.38 that the response of the frame with different joint stiffness differs considerably. Generally speaking, the first local maximum displacement and the period increases with a decrease in joint stiffness.

7.8.17. Dynamic behaviour of the Vogel two-bay six-storey frame

The rigid jointed Vogel two-bay six-storey frame has been used as a calibration frame for static inelastic analysis by Vogel (1985), Mashary and Chen (1991), Toma and Chen (1992) and many others. The scope of calibration for the structure with semi-rigid joints was further extended by Yau and Chan (1994). The dynamic behaviour of the structure is studied in this paper. The arrangement of the Vogel frame is shown in Fig. 7.39. Assuming

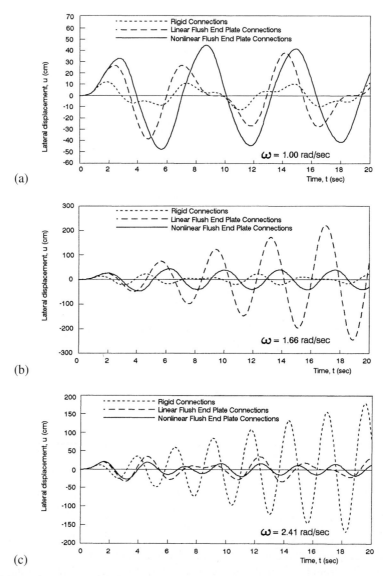

Fig. 7.40. Dynamic response of the Vogel frame with various connection types. (a) Displacement response for $\omega = 1.00$ rad/sec; (b) Displacement response for $\omega = 1.66$ rad/sec; (c) Displacement response for $\omega = 2.41$ rad/sec.

material to be elastic throughout the analysis and ignoring viscous damping (i.e. $[C] = 0$) for focusing attention on the effects of connection types, the frame with various types of connections under different horizontal loads, $F_1(t)$ and $F_2(t)$, is analysed. Four elements per beam member and one element per column member are used in the structural model. The beams are assumed to be connected to the columns by the semi-rigid joints. A flush

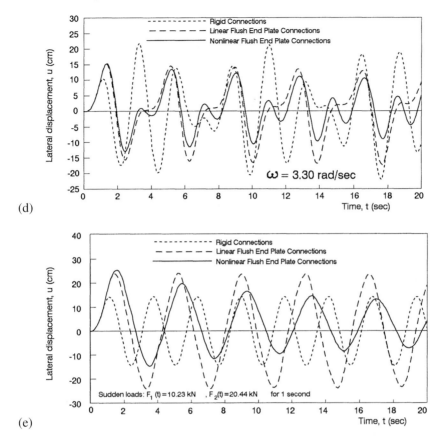

Fig. 7.40. (Continued). Dynamic response of the Vogel frame with various connection types. (d) Displacement response for $\omega = 3.30$ rad/sec; (e) Displacement response for impact loads ($F_1(t) = 10.23$ kN, $F_2(t) = 20.44$ kN for one second).

end plate connection tested by Ostrander (1970) and the curve fitted by Lui and Chen (1988) using the Chen–Lui exponential model is shown in Fig. 7.28. The linear and nonlinear type of the flush end plate connection as well as the rigid connection are assumed for the present study. The uniformly distributed static loads on the beams are considered as additional masses which, together with the self-weight density of 7.8 kN sec^2/m^4 for all members, are included in the lumped mass matrix. The fundamental natural frequencies for the cases of fully rigid and linear semi-rigid connections are found to be 2.41 rad/sec (period = 2.60 sec) and 1.66 rad/sec (period = 3.78 sec), respectively, by the present modal analysis.

Under the action of applied forces $F_1(t)$ and $F_2(t)$ with low frequencies, it can be seen in Fig. 7.40(a) that the frame with semi-rigid connections will magnify its lateral displacement at the top, u. On the contrary, when the structure is under loads with high frequency [see Fig. 7.40(d)], the response of the lateral displacement of the frame with the semi-rigid joints will be dampened.

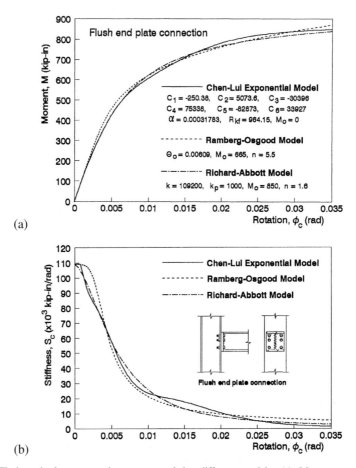

Fig. 7.41. Flush end plate connection represented by different models. (a) Moment–rotation curves; (b) Stiffness–rotation curves.

Although the dynamic response of the frame with the non-linear connection type can either be magnified or dampened under various harmonic vibration frequencies of the applied loads, it is interesting that, under two resonance cases for the linear and the rigid joint types in Fig. 7.40(a) and (b), it will not resonate due to the hysteretic damping induced by the energy dissipation of the non-linear hysteretic loops at the non-linear connections. This behaviour indicates the significance of the hysteretic damping effect due to non-linear connections in ductile steel frames.

In the case of the Vogel frame subjected to suddenly applied loads ($F_1(t) = 10.23$ kN and $F_2(t) = 20.44$ kN) for a duration of one second, it can be seen in Fig. 7.40(e) that the frame with the non-linear connection type will be dampened gradually with diminishing displacement whilst the other two cases (the linear and the rigid joint types) will not be dampened at all. This clearly shows that the conventional joint type of rigid or linear semi-rigid type cannot produce a damping effect which is an important characteristic for

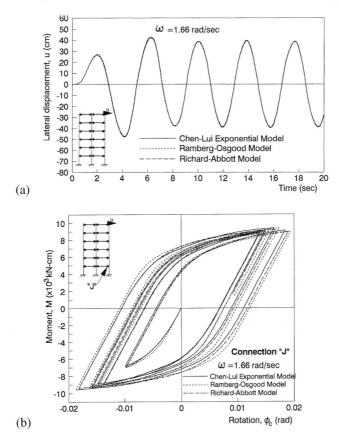

Fig. 7.42. Dynamic response of the Vogel frame with the non-linear flush-end-plate connection by different models (Load frequency, $\omega = 1.66$ rad/sec). (a) Displacement vs. time response; (b) Moment–rotation response.

moment resistant steel frames in an analysis. It was noted that the displacement response of the semi-rigid frame will be larger than that of the rigid frame because the stiffness of the semi-rigid frame is smaller. It was also noted that the displacement response in the positive direction is larger than in the negative direction because of the presence of the permanent rotational deformations at the non-linear connections.

The next study is to investigate the effect on the choice of connection models on the overall structural response. The flush end plate connection is proposed to be modelled by the Ramberg–Osgood, the Richard–Abbott and the Chen–Lui exponential models in this study in order to see the discrepancy amongst the models used. The parameters required by the models are determined and shown in Fig. 7.41. The frame with these three non-linear connection models under an excitation frequency of 1.66 rad/sec is analysed. The displacement response at the top, u, and the moment–rotation response at the connection J is traced in Fig. 7.42(a) and (b), respectively. It can be seen that there is a minimal difference between these models. It can, therefore, be concluded that the choice of models is unimportant in terms of accuracy and numerical efficiency.

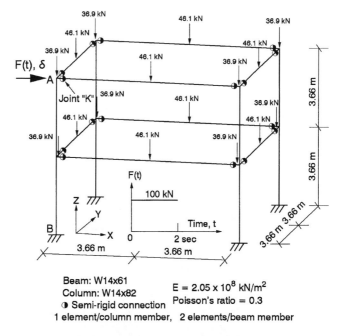

Fig. 7.43. Layout and properties of space frame.

7.8.18. Semi-rigid space frame under impulse force loading

The non-linear dynamic response of a three dimensional frame with various connection types (rigid, linear and non-linear joints for comparison) subjected to an impulse load of 100 kN is studied. The geometry of the frame is shown in Fig. 7.43. Static gravitational loads of 36.9 and 46.1 kN are applied before the application of dynamic load in order to study the P-Δ and the P-d effects. Their equivalent masses are lumped to nodes. A semi-rigid connection is assumed to be the flush end plate of which the moment–rotation curve was represented by the Chen–Lui exponential model shown in Fig. 7.28. For semi-rigid cases, it is assumed that the connection stiffness about the weak-axis of sections is one fifth of the stiffness about the strong-axis. The transient displacement response at point A and the hysteretic M–ϕ_c loops at joint K are plotted in Fig. 7.44. It can be seen that the semi-rigid frame has a larger displacement and a longer period because connection flexibility leads to a structure with a softer stiffness. For the non-linear joint case, the response has a positive displacement shift due to formation of permanent rotations at connections and there is hysteretic damping due to energy dissipation at connections. However, it is interesting that these effects will not be observed in both rigid and linear connection cases. The deformed shapes of the frame with the non-linear connections at the specified times are depicted in Fig. 7.45. The lateral drift envelopes of the column A–B of the frame with various connection types are also plotted in Fig. 7.46. As mentioned before, there exists a permanent drift in the non-linear joint case. It is also noted that the deflection in the rigid joint case is very large within the first two seconds and then oscillates periodically

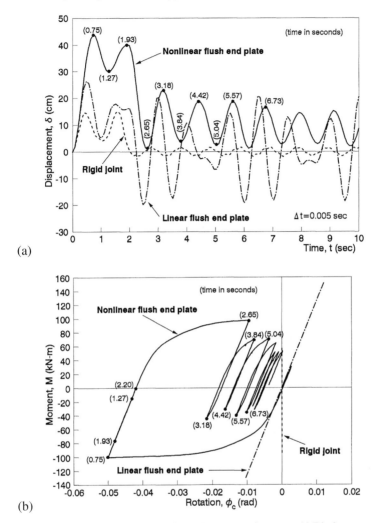

Fig. 7.44. Dynamic response of the space frame with various connection types. (a) Displacement response at point A; (b) Moment–rotation relationship at joint K about the strong-axis.

and stably with a small constant amplitude. In this example, the connection types can significantly change the dynamic response of structures and the effect of the connection flexibility is an important parameter affecting its structural response.

7.8.19. Seismic response of four-bay five-storey frame

In a seismic zone, it is necessary to ensure buildings possess adequate strength to prevent collapse under severe earthquakes or serious damage under moderate earthquakes. Since flexible connections can greatly alter the behaviour of structures, it is interesting to note their significance on seismic response.

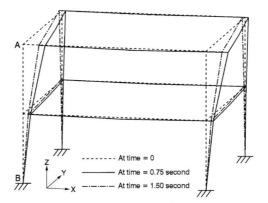

Fig. 7.45. Two times enlarged deformed shapes of space frame with the non-linear flush end plate connections at the specified time.

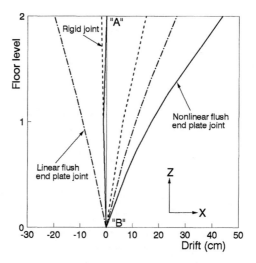

Fig. 7.46. Storey drift envelopes of the column A–B of space frame with various connection types on the X–Z plane in the first ten seconds.

The layout of a four-bay five-storey steel frame located in Shanghai, China, is examined. The geometric configuration of a similar frame is shown in Fig. 7.47. Wong (1995) studied the static load–deformation analysis of the frame with flush end plate connections. The connection type was represented by the Chen–Lui exponential model and appears in the paper by Lui and Chen (1986). In this study, the flush end plate and rigid connection types are considered for comparison. Five elements per beam member and one element per column are employed in modelling of the structure. Gravitational loads are also included and they are considered as additional lumped masses at the nodes on the beam. The frame is assumed to be subjected to the first ten seconds of the 1940 El

Chapter 7. *Cyclic behaviour of flexibly connected elastic steel frames*   255

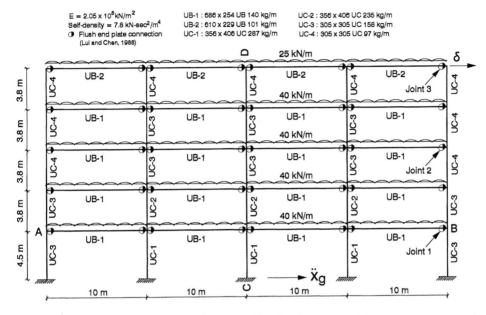

Fig. 7.47. Geometry and property of four-bay five-storey steel frame.

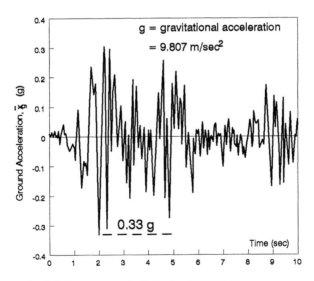

Fig. 7.48. Accelerogram for El Centro (1) (1940) NS component.

Centro N–S earthquake component motion. The earthquake occurred in Imperial Valley, California, USA, on May 18, 1940, and the earthquake ground movements were completely recorded (Dowrick, 1988; and Paz, 1991). The first ten second records are shown in Fig. 7.48 are used in the present study and the peak ground acceleration was

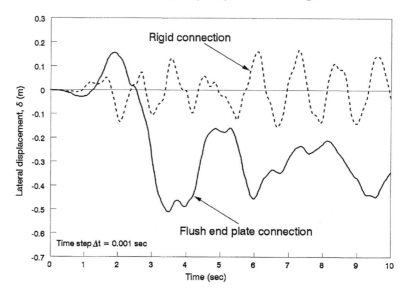

Fig. 7.49. Time history of displacement response of four-bay five-storey frame.

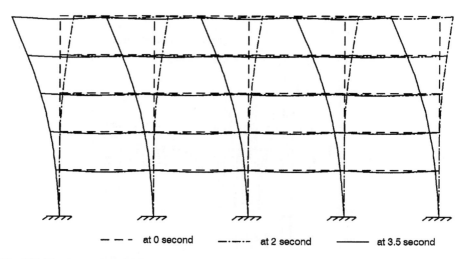

Fig. 7.50. Ten times enlarged deformed shapes of four-bay five-storey frame with non-linear flush end-plate connection type at the specified time.

$0.33\,g$ at about 2 seconds. The transient response of the lateral displacement, δ, at the top of the right-most column is predicted and traced in Fig. 7.49. The deflected shapes of the structure with the non-linear connections at 2 and 3.5 seconds are also depicted in Fig. 7.50. It is recalled that the ground acceleration is absolute (i.e. relative to a stationary point) while the deflection of the frame is relative to the ground. Compared with

Chapter 7. *Cyclic behaviour of flexibly connected elastic steel frames*

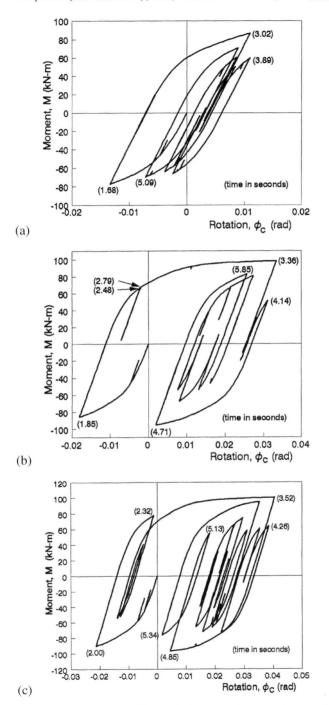

Fig. 7.51. Response of hysteretic M–ϕ_c loops of flush end plate connections up to ten seconds. (a) At joint 1; (b) At joint 2; (c) At joint 3.

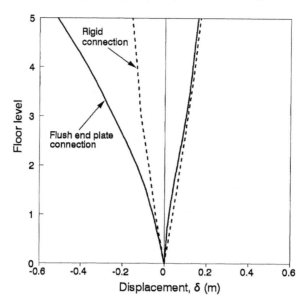

Fig. 7.52. Lateral displacement envelope of four-bay five-storey steel frame up to ten seconds.

the rigid jointed frame, the flexibly connected frame deforms suddenly to a peak value of about -0.5 m at near 3.5 second and then oscillates with this permanent deflection. This indicates that connections possess large rotation after the peak ground acceleration. The hysteretic $M-\phi_c$ loops at joints 1 to 3 are traced in Fig. 7.51 and marked with time records at turning points for clarity. It can be seen that the connections undergo severe rotational deformations during the period from 2 to 3.5 seconds. The lateral displacement envelope of the frame at different floor levels is plotted in Fig. 7.52. The maximum deflection for the flush end plate joint case is approximately three times that for the rigid joint case. It is interesting to note that the backward deflection is much larger than the forward deflection due to permanent rotations formed at the connections of the flexibly connected frame. The bending moment envelopes for the beam A–B and the column C–D are also plotted in Fig. 7.53. Due to the moment being limited by the capacity of the semi-rigid connections, the bending moments at the ends of beam members (i.e. adjacent to columns) are limited to about only one tenth of that of the rigid jointed frame. At mid-span of the beam, the bending moment for the semi-rigid jointed frame is slightly higher because less bending moments are transferred and distributed to the columns through the flexible joints. For the column C–D, it can be seen that less bending moments are induced in the flexibly jointed frame. Based on this example, it can be deduced that flexible connections increase the storey drift but reduce the base shear force significantly under seismic loads. This represents a potentially important role played by connections during earthquakes and also in the design of the structure and its foundation.

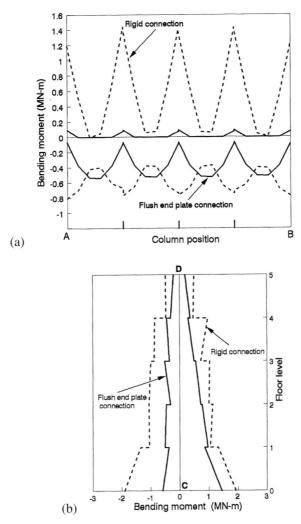

Fig. 7.53. Bending moment envelopes of the four-bay five-storey frame up to ten seconds. (a) The beam along A–B; (b) The column along C–D.

7.9. Conclusions

A robust, stable and reliable incremental-iterative displacement-based finite element method for non-linear large deflection dynamic analysis of steel structures accounting for geometric and connection non-linearities has been presented in this Chapter. A simple algorithm for generating the non-linear connection $M-\phi_c$ looping path under dynamic loads is used. This is applicable to all types of connection models without the need for curve-fitting procedures from one model to another. In cyclic and dynamic analysis, the described algorithm is more simple and powerful than most other methods. The finite element procedure with the algorithm accounting for the cyclic joint behaviour has been coded into the developed computer program GMNAF/D. The present results by the program have been compared with experimental results and several well-known benchmark solutions. It was found that the analysis results are valid and accurate.

Under low harmonic vibrating loads, the displacement response of a structure with non-linear flexible connections will increase. However, displacement is dampened under high harmonic vibrating loads. This is in line with the general practice of avoiding the design of an over-stiff structure subjected to dynamic loads. The non-linear connections of ductile steel frames absorb excitation energy by their hysteretic loops which are known to be an important source of damping in a structure. Apart from the viscous and inelastic damping, the generating process of hysteretic loops of moment–rotation relationship at connections is demonstrated to introduce a major damping to flexibly jointed structures. The connection hysteretic damping will increase the structural energy absorption capacity which is particularly important for a structure to survive under severe dynamic loads.

The dynamic response is shown to be insensitive to the connection model, provided that non-linear connection behaviour is properly fitted. This indicates that simplicity of these models will be the main consideration in their selection. Non-linear hystereic models using the Ramberg–Osgood, the Richard–Abbott and the Chen–Lui exponential models were compared and found to give very close results and without numerical difficulty.

From the results of the numerical examples, it can be concluded that flexible connections will increase the storey drift but reduce the base shear forces under seismic loads. On the other hand, the P-Δ effect due to gravitational loads is noted to be significant for largely deflected structures and should be included. From these observations, geometric change and connection flexibility should be considered as primary sources of non-linearities affecting the dynamic response.

References

AISC (1994): Northridge Steel Update I, AISC, Chicago, IL, USA.

Al-Bermani, F.G.A. and Kitipornchai, S. (1992): Elasto-plastic nonlinear analysis of flexibly jointed space frames. *J. Struct. Div. ASCE* **118**(1), 108–127.

Al-Bermani, F.G.A., Li, B., Zhu, K. and Kitipornchai, S. (1994): Cyclic and seismic response of flexibly jointed frames. *Eng. Struct.* **16**(4), 249–255.

Anderson, D. and Benterkia, Z. (1991): Analysis of semi-rigid frames and criteria for their design. *J. Construct. Steel Res.* **18**, 227–237.

Azizinamini, A. and Radziminski, J.B. (1989): Static and cyclic performance of semirigid steel beam-to-column connections. *J. Struct. Div. ASCE* **115**(12), 2979–2999.

Bathe, K.J. (1996): *Finite Element Procedures*, Prentice Hall, NJ.

Bathe, K.J. and Bolourchi, S. (1979): Large displacement analysis of three-dimensional beam structures. *Int. J. Num. Methods Eng.* **14**, 961–986.

Batho, C. and Lash, S.D. (1936): Further Investigations on Beam and Stanchion Connections Encased in Concrete, Final Report, Together with Lab. Investigation on a Full Scale Steel Frame, Steel Structures Research Committee, Dept. of Scientific and Industrial Research, HMSO, London, pp. 92.

Cardona, A. and Geradin, M. (1988): A beam finite element non-linear theory with finite rotations. *Int. J. Num. Methods Eng.* **26**, 2403–2438.

Carpenter, L.D. and Lu, L.W. (1973): Reversed and Repeated Load Tests of Full-Scale Steel Frames. Bulletin No. 24, American Iron and Steel Institute, New York.

Chan, S.L. (1989): Inelastic post-buckling analysis of tubular beam–columns and frames. *Eng. Struct.* **11**, 23–30.
Chan, S.L. (1992): Large deflection kinematic formulations for three-dimensional framed structures. *Comput. Methods Appl. Mech. Eng.* **95**, 17–36.
Chan, S.L. and Zhou, Z.H. (1994): Pointwise equilibrium polynomial element for nonlinear analysis of frames. *J. Struct. Eng. ASCE* **120**(6), 1703–1717.
Chen, W.F. and Kishi, N. (1989): Semirigid steel beam-to-column connections: Data base and modeling. *J. Struct. Div. ASCE* **115**(1), 105–119.
Chen, W.F. and Zhou, S.P. (1987): Inelastic analysis of steel braced frames with flexible joints. *Int. J. Solids Struct.* **23**(5), 631–649.
Chui, P.P.T. and Chan, S.L. (1996): Transient response of moment-resistant steel frames with flexible and hysteretic joints, *J. Construct. Steel Res.* **39**, 221–243.
Clough, R.W. and Penzien, J. (1975): *Dynamics of Structures*, McGraw-Hill, New York.
Clough, R.W. and Tang, D. (1975): Earthquake Simulator Story of Steel Frame Structure, Vol. I: Experimental Results, Report No. UCB/EERC-75/06, Earthquake Engineering Research Center, Univ. of California, Berkeley.
Colson, A. and Louveau, J.M. (1983): Connections Incidence on the Inelastic Behavior of Steel Structural, *Euromech Colloquium* **174**(October).
Cosenza, E., Luca, D.A. and Faella, C. (1989): Elastic buckling of semi-rigid sway frames. In: R. Narayanan, ed., *Structural Connections, Stability and Strength*, Elsevier Applied Science, pp. 253–295.
COSMOS/M (1990): Finite Element Method Analysis Software (Version 1.61), Structural Research and Analysis Corporation, California, USA.
Cox, M.G. (1972): The Numerical Evaluation of B-Splines. *J. Inst. Math. Applic.* **10**, 134–139.
Dowrick, D.J. (1988): *Earthquake Resistant Design*, 2nd edn., John Wiley and Sons, Singapore.
Frye, M.J. and Morris, G.A. (1975): Analysis of flexibly connected steel frames. *Can. J. Civil Eng.* **2**(3), 280–291.
Gao, L. and Haldar, A. (1995): Nonlinear seismic analysis of space structures with partially restrained connections. *Microcomput. Civil Eng.* **10**, 27–37.
Gerstle, K.H. (1988): Effect of connections on frames. *J. Construct. Steel Res.* **10**, 241–267.
Goto, Y. and Chen, W.F. (1987): On the computer-based design analysis for the flexibly jointed frames. *J. Construct. Steel Res.* **8**, 203–231.
Ho, W.M.G. and Chan, S.L. (1991): Semibifurcation and bifurcation analysis of flexibly connected steel frames. *J. Struct. Eng. ASCE* **117**(8), 2299–2319.
Ho, W.M.G. and Chan, S.L. (1993): An accurate and efficient method for large deflection inelastic analysis of frames with semi-rigid connections. *J. Construct. Steel Res.* **26**, 171–191.
Jaspart, J.P. (1988): Extending of the Merchant–Rankine formula for the assessment of the ultimate load of frames with semi-rigid joints. *J. Construct. Steel Res.* **11**, 283–312.
Jones, S.W., Kirby, P.A. and Nethercot, D.A. (1980): Effect of semi-rigid connections on steel column strength. *J. Construct. Steel Res.* **1**, 38–46.
Kawashima, S. and Fujimoto, T. (1984): Vibration analysis of frames with semi-rigid connections. *Comput. Struct.* **19**, 85–92.
King, W.S. and Chen, W.F. (1993): A LRFD-Based Analysis Method for Semi-Rigid Frame Design, Structural Engineering Report No. CE-STR-93-15, School of Civil Engineering, Purdue Univ., West Lafayette, IN.
Kishi, N. and Chen, W.F. (1986a): Data Base of Steel Beam-to-Column Connections, Structural Engineering Report No. CE-STR-86-26, School of Civil Engineering Purdue Univ., West Lafayette, IN.
Kishi, N. and Chen, W.F. (1986b): Steel Connection Data Bank Program, Structural Engineering Report No. CE-STR-86-18, School of Civil Engineering, Purdue Univ., West Lafayette, IN.
Kishi, N. and Chen, W.F. (1987a): Moment–Rotation Relation of Top- and Seat-Angle Connections, Structural Engineering Report No. CE-STR-87-4, School of Civil Engineering, Purdue Univ., West Lafayette, IN.
Kishi, N. and Chen, W.F. (1987b): Moment–Rotation of Semi-Rigid Connections, Structural Engineering Report No. CE-STR-87-29, School of Civil Engineering, Purdue Univ., West Lafayette, IN.
Korol, R.M., Ghobarah, A. and Osman, A. (1990): Extended end-plate connections under cyclic loading: behaviour and design. *J. Construct. Steel Res.* **16**, 253–280.
Kouhia, R. and Tuomala, M. (1993): Static and dynamic analysis of space frames using simple Timoshenko type elements. *Int. J. Num. Methods Eng.* **36**, 1189–1221.

Krawinkler, H., Bertero, V.V. and Popov, E.P. (1971): Inelastic Behaviour of Steel Beam-to-Column Subassemblages, Report No. UBC/EERC-71/07, Earthquake Engineering Research Center, Univ. of California, Berkeley, CA.

Leger, P. and Dussault, S. (1992): Seismic-energy dissipation in MDOF structures. *J. Struct. Div. ASCE* **118**(5), 1251–1269.

Lui, E.M. and Chen, W.F. (1986): Analysis and behaviour of flexibly-jointed frames. *Eng. Struct.* **8**, 107–118.

Lui, E.M. and Chen, W.F. (1987): Steel frame analysis with flexible joints. *J. Construct. Steel Res.* **8**, 161–202.

Lui, E.M. and Chen, W.F. (1988): Behavior of braced and unbraced semi-rigid frames. *Int. J. Solids Struct.* **24**(9), 893–913.

Mashary, F.A. and Chen, W.F. (1991): Simplified second-order inelastic analysis for steel frames. *Struct. Engineer* **69**(23), 395–399.

Meek, J.L. and Tan, H.S. (1984): Geometrically nonlinear analysis of space frames by an iterative incremental technique. *Comput. Methods Appl. Mech. Eng.* **47**, 261–282.

Mondkar, D.P. and Powell, G.H. (1977): Finite element analysis of non-linear static and dynamic response. *Int. J. Num. Methods in Eng.* **11**, 499–520.

Nader, M.N. and Astaneh, A. (1991): Dynamic behaviour of flexible, semirigid and rigid steel frames. *J. Construct. Steel Res.* **18**, 179–192.

Nethercot, D.A. (1985): Steel Beam-to-Column Connections – A Review of Test Data, Construction Industry Research and Information Association, London, England.

Newmark, N.M. (1959): A method of computation for structural dynamics. *J. Eng. Mech. Div. ASCE* **85**(3), 67–94.

Ostrander, J.R. (1970): An Experimental Investigation of End-Plate Connections, Master's Thesis, Univ. of Saskatchewan, Saskatoon, SK, Canada.

Papadrakakis, M. (1981): Post-buckling analysis of spatial structures by vector iteration methods. *Comput. Struct.* **14**, 393–402.

Paz, M. (1991): *Structural Dynamics: Theory and Computation*, 3rd edn., Van Nostrand Reinhold, New York.

Popov, E.P. (1983): Seismic Moment Connections for Moment-Resisting Steel Frames, Report No. UCB/EERC-83/02, Earthquake Engineering Research Center, Univ. of California, Berkeley, CA.

Popov, E.P. and Bertero, V.V. (1973): Cyclic loading of steel beams and connections. *J. Struct. Div. ASCE* **99**(6), 1189–1204.

Popov, E.P. and Pinkney, R.B. (1969): Cyclic yield reversal in steel building connections. *J. Struct. Div. ASCE* **95**(3), 327–353.

Popov, E.P. and Stephen, R.M. (1970): Cyclic Loading of Full-Size Steel Connections, Report No. UCB/EERC-70/03, Earthquake Engineering Research Centre, Univ. of California, Berkeley, CA (Republished as Bulletin No. 21, American Iron and Steel Institute, 1972.)

Remseth, S.N. (1979): Nonlinear static and dynamic analysis of framed structures. *Comput. Struct.* **10**, 879–897.

Richard, R.M. and Abbott, B.J. (1975): Versatile elastic–plastic stress–strain formula. *J. Eng. Mech. Div. ASCE* **101**(4), 511–515.

Savard, M., Beaulieu, D. and Fafard, M. (1994): Nonlinear finite element analysis of three-dimensional frames. *Can. J. Civil Eng.* **21**, 461–470.

Shi, G. and Atluri, S.N. (1989): Static and dynamic analysis of space frames with nonlinear flexible connections. *Int. J. Num. Methods Eng.* **28**, 2635–2650.

Sivakumaran, K.S. (1988): Seismic response of multi-storey steel buildings with flexible connections. *Eng. Struct.* **10**, 239–248.

So, A.K.W. and Chan, S.L. (1991): Buckling analysis of frames using 1 element/member. *J. Construct. Steel Res.* **20**, 271–289.

Stelmack, T.W., Marley, M.J. and Gerstle, K.H. (1986): Analysis and tests of flexibly connected steel frames. *J. Struct. Eng. ASCE* **112**(7), 1573–1588.

Toma, S. and Chen, W.F. (1992): European calibration frames for second-order inelastic analysis. *Eng. Struct.* **14**(1), 7–14.

Tsai, K.C. and Popov, E.P. (1990): Cyclic behavior of end-plate moment connections. *J. Struct. Div. ASCE* **116**(11), 2917–2930.

Vedero, V.T. and Popov, E.P. (1972): Beam–column subassemblages under repeated loading. *J. Struct. Div. ASCE* **98**(5), 1137–1159.

Vogel, U. (1985): Calibrating frames. *Stahlbau* **54**, 295–301.
Waterloo Maple Software (1992): *Maple V: The Future of Mathematics*, 1st edn., Release 2, Waterloo, ON, Canada.
Williams, F.W. (1964): An approach to the nonlinear behaviour of the members of a rigid jointed place framework with finite deflections. *Quart. J. Mech. Appl. Math.* **17**, 451–469.
Wong, H.P. (1995): A Study on the Behaviour of Realistic Steel Buildings using Design and Rigorous Approaches, Thesis submitted for partial fulfillment for the Degree of Master of Science, Hong Kong Polytechnic Univ., Hong Kong.
Yang, T.Y. and Saigal, S. (1984): A simple element for static and dynamic response of beams with material and geometric nonlinearities. *Int. J. Num. Methods Eng.* **20**, 851–867.
Yau, C.Y. and Chan, S.L. (1994): Inelastic and stability analysis of flexibly connected steel frames by springs-in-series model. *J. Struct. Eng. ASCE* **120**(10), 2803–2819.
Youssef-Agha, W., Aktan, H.M. and Olowokere, O.D. (1989): Seismic response of low-rise steel frames. *J. Struct. Div. ASCE* **115**(3), 594–607.
Zhou, Z.H. and Chan, S.L. (1995): A self equilibrium element for second-order analysis of semi-rigid jointed frames. *J. Eng. Mech. ASCE* **121**(8), 896–902.
Zhu, K., Al-Bermani, F.G.A., Kitipornchai, S. and Li, B. (1995): Dynamic response of flexibly jointed frames. *Eng. Struct.* **17**(8), 575–580.

Chapter 8

TRANSIENT ANALYSIS OF INELASTIC STEEL STRUCTURES WITH FLEXIBLE CONNECTIONS

8.1. Introduction

Although most practical design is based on static or equivalent static response, structures are quite often subjected to dynamic loads. In reality, loadings due to typhoons and earthquakes are dynamic in nature. Unlike the static analysis where monitoring stress and stability can basically insure the safety of a structure, design against dynamic loads is much more complicated. Apart from checking stress and stability at a particular instance, design of a structure against dynamic loads requires mechanism to absorb energy during the attack of dynamic loads. Resonance is another related but undesirable response leading to ever amplifying displacement and stress. Engineers are always attempting to prevent a structural frequency close to the load frequency.

Table 8.1 summarizes the loss of human life in a number of earthquakes. It can be seen that many serious earthquake devastations occurred in the twentieth century. For example, on the 28th July 1976, Tangshan City in China was devastated by a strong earthquake of magnitude 7.8 on the Richter Scale (Gere and Shah, 1980). The whole city was almost completely destroyed and about 240,000 people were killed. This was one of the worst earthquakes in the recent Chinese history. The majority of buildings were not designed to resist heavy lateral forces, so they performed poorly in this earthquake. The lack of adequate seismic resistance in most structures was considered as the main cause for the huge damage. Recently, on the 19th September 1985, Mexico City in Mexico was hit by a major earthquake causing severe local damage. In this disaster, a magnitude of 8.1 on the Richter scale was recorded. In the report by Rosenblueth (1986), it was estimated that 10,000 people were killed, 256 buildings totally collapsed, and the actual financial loss of buildings was between 4 and 6 billion US dollars with an additional 10 billion dollars required for re-construction. The use of the linear elastic approach for solution of critical seismic problems is inadequate, that procedures for capacity analyses should be devised, and the need for ductile construction is clearly evident. More recently, on the 17th January 1995, the Kobe–Osaka region of Japan was struck by a 7.2 magnitude earthquake (Scawthorn and Yanev, 1995; Akiyama and Yamada, 1995). Preliminary estimation was that 5,100 lives were lost and more than 100,000 buildings were badly damaged or destroyed. The economic loss was estimated at over 130 billion US dollars. Significant damage was observed in a wide variety of structures, including buildings, bridges, rail and highways.

Table 8.1
Major earthquakes and associated fatalities from 1905

Date	Location	Magnitude	Number of fatalities
April 4, 1905	Kashmir	8.6	19,000 dead
April 18, 1906	San Francisco, USA	8.3	70,000 dead
December 16, 1920	Ningxia, China	8.5	230,000 dead; one city destroyed
September 1, 1923	Tokyo, Japan	8.3	100,000 ~ 140,000
March 7, 1927	Japan	7.9	3,000
March 2, 1933	Japan	8.9	3,000
May 30, 1935	Quetta, Pakistan	7.5	30,000
January 25, 1939	Chile	8.3	30,000
December 26, 1939	Turkey	7.9	23,000
December 20, 1946	Japan	8.4	2,000 ~ 3,000
July 21, 1952	California, USA	7.7	10
February 29, 1960	Agadir, Morocco	5.6	12,000
March 28, 1964	Alaska, USA	7.9	130
June 16, 1964	Niigata, Japan	7.4	26
March 22, 1966	Hobei, China	7.2	Dead 7,938; injured 8,613; all of buildings
January 5, 1970	Yunnan, China	7.7	Dead 15,621; injured 26,783; 90% buildings
May 31, 1970	Peru	7.7	40,000
October 4, 1972	Iran	7.0	17,000
January 6, 1973	Sichuan, China	7.8	Dead 2,199; injured 2,743; 47,100 houses
July 28, 1976	Tangshan, China	7.8	Dead 240,000; one city destroyed
August 17, 1976	Philippines		8,000
August 4, 1977	Bucharest, Romania		1,541
September 16, 1978	Iran		25,000
November 23, 1980	Italy		2,913
September 19, 1985	Mexico City, Mexico	8.1	10,000 dead; 256 buildings destroyed
November 6, 1988	Yunnan, China	7.6	Dead 748; injured 7,751; 2,242,800 houses
December 7, 1988	Armenia, USSR	7.0	15,000
October 17, 1989	California, USA	7.1	95
March 14, 1992	Turkey	7.3	800 ~ 4,000
October 12, 1992	Egypt	5.8	500
January 17, 1995	Kobe–Osaka, Japan	7.2	5,100 dead; 100,000 buildings
May 10, 1997	Ardakul, Northeastern Iran	7.1	2,400 dead

From the earthquake destroy records, it illustrated clearly the inadequacy of the building design in many cases. Consequently, the use of advanced analysis programs for a more refined and accurate prediction of the structural response of buildings constructed in active seismic zones is needed.

Apart form the requirements in strength and deflection, a structure should also have a ductility property and possess sufficient energy dissipation capacity without a risk of progressive collapse. They should be designed to meet the following requirements.

(1) **Serviceability limit state.** Small loads should be taken by a structure in its elastic range.
(2) **Damage control limit state.** Under a moderate earthquake, some repairable and replaceable elements such as knee bracing and secondary beam members are allowed to yield for absorbing excitation energy, but the main load bearing structure should be protected from any significant damage. In addition, deflection of structures should be controlled within a specified limit for serviceability requirement.
(3) **Survival limit state.** This criterion is particularly important for buildings and structures since it is related to safety. In a severe earthquake which is unlikely to occur within the life of a structure but it is used in the design to examine the ultimate structural safety, some main beams and column bases should be designed to form plastic hinges to absorb excitation energy. To achieve this mechanism, some structural damage is expected. However, the whole structure must be safe and should not collapse. To prevent the total collapse mechanism and to allow the energy dissipation, the location of plastic hinges can be designed to form at all beams and column-bases but not at other locations in the columns.

In spite of some local damage, a structure should survive after a major earthquake without collapse. To achieve this, the structure must be of sufficient ductility with its structural energy absorption capacity greater than the seismic input energy (Housner, 1956; Galambos, 1968; Housner and Jennings, 1977; Kuwamura and Galambos, 1989; Filiatrault et al., 1994). Most structures are therefore designed to develop an inelastic action in critical components that control the amount of seismic energy stored in the systems. Some special components are allowed to yield during an earthquake attack and can be easily replaced by new elements after an earthquake. For example, replaceable knee anchors in kneed braced frames can be designed for this purpose because the flexural yielding will dissipate energy and thus act as "fuse-like" elements to prevent any structural damage to the principal members (Sam et al., 1992; Bourahla and Blakeborough, 1992; Balendra et al., 1994). The seismic input energy is the sum of the kinetic energy, the elastic strain energy, the energy dissipated by the hysteretic yielding action of some structural elements (e.g. knee anchors, beam–column connections, and members at which plastic hinges) and the energy dissipated by other non-yielding mechanisms (e.g. viscous damping). The phenomenon of energy dissipation at plastic hinges and at beam–column connections of ductile steel structures can be observed through the numerical examples in this Chapter.

This Chapter describes non-linear dynamic analysis of inelastic steel structures with semi-rigid joints. Some strain-hardening rules and existing cyclic plasticity models are described in Sections 8.3 and 8.4. Two simple plastic hinge-based models, a refined-plastic hinge and an elastic–plastic hinge, for cyclic analysis are reported and described in Section 8.5. Section 8.6 discusses the modelling of beam-to-column connections under cyclic loads. The developed computer FORTRAN program GMNAF/D described in Chapter 7 is extended by including the dynamic inelastic effects. The algorithm for dynamic analysis of semi-rigid frames allowing for material yielding is incorporated into the computer program. The developed technique is validated through these examples. The advantage of the proposed technique is on its simplicity and fast rate of convergence during iteration for equilibrium. The method is more efficient than the plastic zone approach and can be used in personal computers.

8.2. Review on cyclic plasticity models and methods

Review on the previous work for static inelastic analysis has been given in Chapters 6 and 7. This session gives a summary of the work in dynamic inelastic analysis.

Although extensive research work related to the non-linear static analysis of inelastic frames has been conducted (Johnston and Cheney, 1942; Van Kuren and Galambos, 1964; Alvarez and Birnstiel, 1969; Vogel, 1985; Hsieh et al., 1989; Yau and Chan, 1994; Toma and Chen, 1995; Chen and Chan, 1995; etc.), dynamic studies are relatively few. It may be possible that an inelastic structure under complicated loading, unloading and reloading conditions in cyclic analysis is more complex than the same structure under monotonically increasing loads in static analysis. Further, the validity of the results is more difficult to be checked. In fact, non-linear static analysis can be treated as a particular case for non-linear dynamic analysis without inertia force and hysteretic behaviour.

Numerous cyclic plasticity models have been developed for simulating the constitutive relationship of the material. Commonly used models include the elastic–perfectly plastic model, the bi-linear model, the multi-linear model, the inversed Ramberg–Osgood model, the Ramberg–Osgood model and the strain rate dependent model. In the distributed plasticity approach which adopts the type of stress–strain material model, the analysis can directly employ the basic material law but requires extensive computation on the integration of stresses over the cross-sections for determination of resultant forces. To reduce computational time, the lumped plasticity approach making direct use of the equivalent force–deformation relationship is generally preferred.

The simplest model is the elastic–perfectly-plastic model in which the material constitutive relationship is simply assumed to be either elastic or perfectly plastic without transitional partial yielding between elastic and plastic regimes. The effect of strain-hardening is further ignored. To eliminate this defect, the bi-linear model can be used. However, this model cannot simulate a smooth transition from the elastic to the plastic regime. To obtain a better representation of material yielding, the tri-linear and the multi-linear models were proposed. However, these models destroy the simplicity of the bi-linear model and still contain slope discontinuity within the transition from elastic to plastic regimes, which is undesirable in computer analysis. To provide a smooth curve from elastic to plastic regime, the Ramberg–Osgood and the inversed Ramberg–Osgood models are more commonly adopted.

In the early 1960s, computers were not commonly available and computer time was expensive. To reduce computational effort, simplifications and assumptions were made in inelastic analysis. During this period, the rigid plastic analysis was commonly used

(Lee and Symonds, 1952). In this analysis, an ideally elastic–perfectly-plastic material was assumed and the elastic part of the deformations was neglected with only the plastic deformations considered. The method is simple for initial estimation. However, it is accurate only if the plastic deformations are large compared to the elastic deformations. Based on this theory, Salvadori and Weidlinger (1957) included shear yielding while Rawlings (1964) considered the influence of strain hardening of the material and geometrical change of structure on the motions of the system. For simplicity, many investigators (Berg and DaDeppo, 1960; Heidebrecht et al., 1964; Clough et al., 1965) assumed that the yielding of a member or an element was entirely due to simple flexural bending and the response was obtained on the basis of an assumed curvature relationship of either bilinear or elastic–perfectly-plastic type.

In parallel with the availability of higher speed and larger capacity computers in 1970s, step-by-step numerical procedures such as the finite difference and the finite element methods were developed and became popular. These methods are more systematic and accurate than the rigid plastic method. Generally speaking, the finite difference method is restricted to simple problems with simple boundary conditions. On the other hand, the finite element method is more suitable for complex structures. Arbitrary geometries, material property variations and complex loading functions can be easily allowed for in the finite element method. A complete and a general finite element procedures were documented by well-known texts by Zienkiewicz (1971) and Bathe (1996), among others.

Using the elastic–perfectly-plastic material model is the simplest way to simulate the cyclic inelastic behaviour. Lee and Symonds (1952), Salvadori and Weidlinger (1957), and Rawlings (1964) adopted the rigid plastic method with no elastic deformations considered. Kondoh and Atluri (1986) adopted a complementary energy approach to handle the dynamic, elastic–perfectly-plastic and large deformation analysis of plane frames using the plastic hinge concept. Later, Shi and Atluri (1988) extended the concept of the complementary energy approach to three-dimensional elastic–perfectly-plastic space frames. In their studies, the yield condition was based on an interaction of axial force and bending moments. More recently, Izzuddin and Elnashai (1990) studied the seismic response of plane frames with an elastic–perfectly-plastic material with strain-hardening effect ignored. To employ a more accurate yield surface for a section under axial load and moments, Izzuddin and Elnashai (1993a) employed a three-dimensional yield surface in a force space. When the force point is within the surface, the section is assumed to be ideally elastic. When the force point lies on the surface, fully plastic section will be assumed. However, partial material yielding was not considered in their studies. A lumped plastic hinge method was used for computational efficiency. The technique of automatic mesh refinement of element arrangement was adopted once a plastic hinge was formed. This can reduce the analysis time effectively since the mesh size depends on whether or not a member is yielded. To obtain a more accurate prediction, Izzuddin and Elnashai (1993b) adopted the plastic-zone approach to study the inelastic and cyclic response of space frames. All sections and members were first divided into many fine fibres. The stresses of fibres at different locations were determined by a given uniaxial stress–strain relationship. Spread of yield within structural members was allowed, but a longer computational time was needed.

The bi-linear stress–strain model is a simple function for considering the strain-hardening property of material. Liu and Lin (1979) applied this model to simply supported beams. However, their numerical method requires pre-requisitely the elastic solutions before the dynamic inelastic analysis is carried out. Besides, their method is only suitable for simple structures. Yang and Saigal (1984) studied the static and dynamic response of reinforced concrete beams with material and geometric non-linearities. In their assumptions on the material constitutive law, a non-linear stress–strain material model for compression and a bi-linear stress–strain brittle model for tension were used for concrete, whilst a bi-linear stress–strain relationship for reinforced steel bars under tension and compression was assumed. Most studies assume yielding is concentrated at one location. An attempt to improve this by is the discrete plastic hinge concept where each plastic hinge was subdivided into a series of sub-hinges (Chen, 1981). The force–deformation relationships for each sub-hinge were represented by bi-linear functions. When these functions are combined, a series of sub-hinges is formed. It produces a multi-linear stiffness for each plastic hinge. In other words, this method allows a multi-linear flexibility relationship for the element. However, a numerical divergence problem in the element stiffness formulation may be encountered when the loading/unloading yielding conditions in all sub-hinges are very different.

To produce a smooth transition from the ideally elastic to the fully plastic state, a degradable stress–strain curve can be adopted. The Ramberg–Osgood type (Ramberg and Osgood, 1943) and its inverse are some of these widely used functions. Bockholdt and Weaver (1974) adopted an incremental-load approach in conjunction with the Ramberg–Osgood type of non-linear force–deformation relationships for the static and dynamic analysis of tall buildings loaded beyond the elastic range. Noor and Peters (1980) adopted the Ramberg–Osgood polynomials for the constitutive axial force versus deformation relation in predicting the dynamic response of space trusses allowing for both geometric and material non-linearities. Unlike the traditional displacement-based formulation, they used the mixed formulation in conjunction with the fundamental unknowns consisting of member forces, nodal velocities and nodal displacements. Janatowski and Brinstiel (1970) applied the non-linear force–deformation relationship based on the inverse of the Ramberg–Osgood model to three-dimensional members and a numerical procedure for the static inelastic analysis of space frames. Similarly, Uzgider (1980) employed the inverse of the Ramberg–Osgood function to represent the elasto-plastic force–deformation behaviour at the ends of three-dimensional elements in dynamic analysis of space structures. In this model, the conjugate pair of axial force versus axial deformation and moment versus rotation relations are included.

An alternative way to simulate material yielding is to impose a decaying function from an initial yield surface to a full yield surface in a section. When a force point lies between these two surfaces on the force-space, the section is considered to be partially yielded. The decaying function can control the rate of plasticity of the material considered. Nigam (1970) proposed a pair of yield surfaces in studying the dynamic inelastic response of framed structures. He derived a set of generalized force–displacement relations for typical elements, incorporating the effects of inelastic interaction amongst six forces (i.e. the axial force, shear forces, torsional and bending moments) at a section. However, for simplicity, partial yielding of the material was not considered in his numerical examples.

Similarly, Wen and Farhoomand (1970) proposed two artificially prescribed parabolic and elliptic full-yield functions in the dynamic analysis of inelastic space frames. In order to model the more realistic behaviour of a plastic hinge spreading over a finite length in a member, the size of the plastic hinge was considered and simulated by an elasto-plastic segment with lumped flexibility. The length of plastic hinge is assumed to be one-eighth of the original member. However, partial yielding, strain hardening of the material and the geometric change in structure were ignored in their studies. Kouhia and Tuomala (1993) adopted two simplified mathematical yield functions, which were the hyper-sphere and hyper-cube yield surfaces, for study of the dynamic response of space frames of thin-wall cross-sections using three-dimensional Timoshenko beam elements. Based on the experimental observation of partial degradation of member stiffness at plastic hinges, Soroushian and Alawa (1990) proposed a model of tangent modulus of elasticity to simulate cyclic behaviour of steel struts accounting for stiffness degradation. On the interaction diagram of normalized axial load and bending moment, first-yield and full-yield surfaces were constructed in order to consider the partial yielding when the force-point at a plastic hinge is between these surfaces. In their model, tangent modulus of elasticity changes from a maximum value equal to the elastic modulus, to a minimum value equal to the strain-hardening modulus.

Based on the experiments on the strain rate effect on the cyclic response of steel specimens under reversal uniaxial loads (Manjoine, 1944; Aspden and Campbell, 1966; Mosquera et al., 1985), the stress–strain relations were found to depend on the strain rate. Both dynamic initial yield stress and failure stress rise as the strain rate increases. Cowper and Symonds (1957) developed a formula relating the dynamic yield stress, the quasi-static yield stress and the strain rate for steel material under dynamic loads. The material was assumed to behave linearly elastic before initial yielding, after which material becomes ideally plastic. Mosquera et al. (1985) expressed the dynamic yield stress in terms of plastic strain rate and the corresponding bending moment in terms of curvature rate. For simplicity, they modified the equations to allow for bending moment in terms of the rate of rotation across the plastic hinge and other numerical constraints determined from experiments. The final equation was relatively simple because consideration of strain and curvature rates was avoided.

Recently, Tin-Loi and Vimonsatit (1995) studied the elasto-plastic response of semi-rigid frames under cyclic loads. A lumped plasticity model with a piecewise linearised yield function and a piecewise beam-to-column connection model were used. However, the method was restricted to moderate geometrically non-linear problems. The procedure of their numerical method was based on a mathematical programming approach instead of the commonly used Newton–Raphson type of stiffness matrix method of analysis.

8.3. Strain-hardening rules

Figure 8.1 shows a typical stress–strain curve of a steel specimen under an uniaxial loading. When the material is under an uniaxial compression test, the stress–strain curve is almost identical to a simple tension test. However, when the specimen is pulled under tensile stress to reach the inelastic range and is then reloaded and compressed again, a

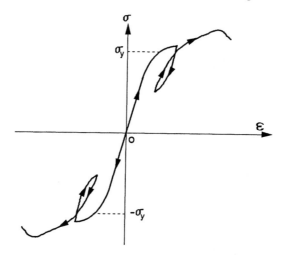

Fig. 8.1. Experimental uniaxial stress–strain curve.

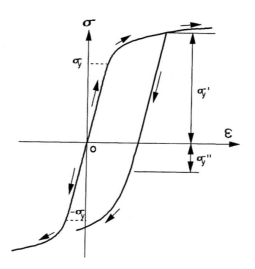

Fig. 8.2. Bauschinger effect.

compressive yield stress will be much smaller than the virgin compressive yield stress obtained in a simple compression test. As illustrated in Fig. 8.2, the specimen with a preloading tensile stress σ'_y, its corresponding compressive yielding stress occurs at a stress σ''_y. The new stress is smaller than the initial yield stress σ_y and much lesser than the subsequent yield point σ'_y. This phenomenon is known as the Bauschinger effect.

As a specimen is under a repeated loading, unloading and reloading process, the yield stress usually increases. This phenomenon is called the strain-hardening or the work-hardening. For a material under reversed loading conditions, the subsequent yield stress is

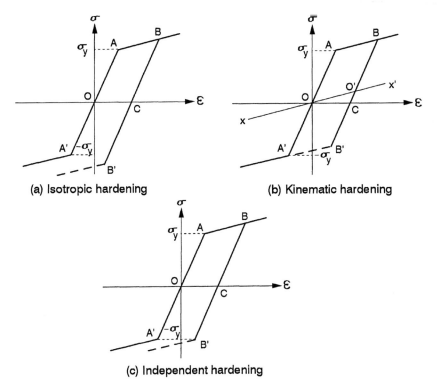

Fig. 8.3. Strain hardening rules.

usually determined by one of the following hardening rules in Sections 8.3.1 to 8.3.3. For simplicity, bi-linear stress–strain models with various strain-hardening rules are discussed as follows.

8.3.1. Isotropic hardening rule

In the isotropic hardening rule, the reversed compressive yield stress is assumed to be equal to the previous tensile yield stress. As shown in Fig. 8.3(a), the reversed compressive yield stress $\sigma_{B'}$ is equal to the tensile yielding stress σ_B before load reversal, where, $|B'C| = |BC|$. Thus, the isotropic hardening rule neglects the Bauschinger effect completely because the reversed compressive yield stress $\sigma_{B'}$ exceeds the initial compressive stress $\sigma_{A'}$.

8.3.2. Kinematic hardening rule

In this rule, the elastic stress range (i.e. $2\sigma_y$) is assumed to be unchanged throughout the strain-hardening process. Since the reversed compressive yield stress $\sigma_{B'}$ is less than the compressive stress $\sigma_{A'}$, the kinematic hardening rule considers the Bauschinger effect to its full extent. From Fig. 8.3(b), $|BB'| = |AA'|$ is obtained. It is noted that there is an inclined line x–x' which has the same slope as the line A–B. The original centre of the elastic region (O in Fig. 8.3(b)) is moved to the interception point between this line and the reversed path.

8.3.3. Independent hardening rule

The reversed compressive stress $\sigma_{B'}$ is assumed to be equal to the virgin compressive stress $\sigma_{A'}$ in subsequent hardening. It implies that the increased tensile stress σ_B has no effect on $\sigma_{A'}$ and the Bauschinger effect cannot be considered. The material is, therefore, hardened independently in tension and in compression. From Fig. 8.3(c), where $|BC| > |OA|$, but $|CB'| = |OA'|$; the material has hardened in tension only, but it behaves like a virgin material under a reversed compressive loading condition.

8.4. Existing cyclic plasticity models

In structural analysis, it is necessary to specify the relationship between stresses and strains. This relationship reflects the physical properties of the material and is usually obtained by appropriate idealization of the experimental stress–strain curves. Some mathematical functions for the stress–strain material relationship have been developed. They are usually used for the plastic zone or distributed plasticity based methods. Although these methods are accurate, they require extensive computer time when compared with the efficient plastic hinge based methods. Some commonly employed models are briefly described as follows.

8.4.1. Elastic–perfectly plastic model

With reference to Fig. 8.4(a), this is the simplest plastic model and requires only two parameters, the Young's modulus E and the yield stress σ_y. Because of its simplicity, it conveniently neglects the effect of strain-hardening in some problems. Plasticity is assumed to occur when the stress has reached the yield stress σ_y. Stress σ along the virgin loading path can, therefore, be expressed in terms of strain ε using the uniaxial stress–strain relationship as

$$\begin{aligned} \varepsilon &= \frac{\sigma}{E} & \text{for } \sigma < \sigma_y, \\ \varepsilon &= \frac{\sigma}{E} + \varepsilon_0 & \text{for } \sigma = \sigma_y, \end{aligned} \quad (8.1)$$

where E is the Young's modulus determined as the slope of the inclined line, and ε_0 is the permanent strain.

8.4.2. Bi-linear strain-hardening model

Referring to Fig. 8.4(b), this model is a modification of the elastic–perfectly plastic model because the second line of non-zero slope E_{s1} is employed instead of a horizon-

Chapter 8. *Transient analysis of inelastic steel structures with flexible connections* 275

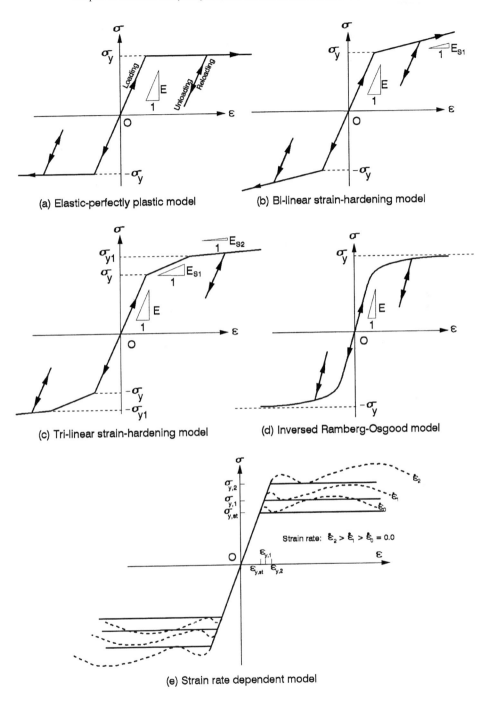

Fig. 8.4. Idealized stress–strain curves.

tal slope assumed in the elastic–perfectly plastic model. This model includes the strain-hardening property of the material. The stress–strain relationship can be expressed as

$$\varepsilon = \frac{\sigma}{E} \qquad \text{for } \sigma \leqslant \sigma_y,$$
$$\varepsilon = \frac{\sigma_y}{E} + \frac{1}{E_{s1}}(\sigma - \sigma_y) \quad \text{for } \sigma > \sigma_y. \tag{8.2}$$

8.4.3. Multi-linear strain-hardening model

In order to improve the bi-linear model, the transition curve from elastic to plastic state can be divided into several linear parts. For the tri-linear model shown in Fig. 8.4(c), an extra line of slope E_{s2} is added to the bi-linear model. Although it is more complicated than the bi-linear model, it is still a simple plasticity model accounting for the strain-hardening effect. The stress–strain relation of the model can be written as

$$\varepsilon = \frac{\sigma}{E} \qquad \qquad \text{for } \sigma \leqslant \sigma_y,$$
$$\varepsilon = \frac{\sigma_y}{E} + \frac{1}{E_{s1}}(\sigma - \sigma_y) \qquad \text{for } \sigma_y < \sigma \leqslant \sigma_{y1}, \tag{8.3}$$
$$\varepsilon = \frac{\sigma_y}{E} + \frac{1}{E_{s1}}(\sigma_{y1} - \sigma_y) + \frac{1}{E_{s2}}(\sigma - \sigma_{y1}) \quad \text{for } \sigma > \sigma_{y1}.$$

Other similar multi-linear models can be formulated by introducing additional linear regions as described as above.

8.4.4. Inversed Ramberg–Osgood model

As shown in Fig. 8.4(d), this model allows a smooth transition from elastic to plastic regime by a mathematical continuous function. Since it only requires three parameters to define the shape of the stress–strain curve, it has been widely employed in material constitutive modeling. In addition, this function guarantees a positive derivative. However, the strain hardening effect cannot be considered in this model because the slope approaches to zero as the strain is large. The mathematical expression of the model for stress in terms of strain is given by

$$\sigma = \frac{E\varepsilon}{(1 + |E\varepsilon/\sigma_y|^n)^{1/n}} \tag{8.4}$$

and its corresponding slope becomes

$$\frac{d\sigma}{d\varepsilon} = \frac{E}{(1 + |E\varepsilon/\sigma_y|^n)^{(n+1)/n}} \tag{8.5}$$

where n is a shape parameter of the curve. The three required parameters are E, σ_y and n which can be interpreted physically.

Equation (8.4) is simple and smooth, and can also be extended to express some force–deformation relationships (e.g. Uzgider, 1980). One of these expressions can be written in terms of moment and rotation as

$$M = \frac{K\theta}{(1 + |K\theta/M_\mathrm{p}|^n)^{1/n}} \quad (8.6)$$

and the slope is given by

$$\frac{\mathrm{d}M}{\mathrm{d}\theta} = \frac{K}{(1 + |K\theta/M_\mathrm{p}|^n)^{(n+1)/n}} \quad (8.7)$$

where M is moment, θ is member end deformation, K is elastic member end stiffness and M_p is plastic moment capacity of the member.

The form of the Ramberg–Osgood model is basically similar to its inverse function as mentioned above. The Ramberg–Osgood model also requires only three parameters to describe the curve. The main difference between these two models is that the strain-hardening effect is included in the Ramberg–Osgood model but not in its inverse. The stress–strain relation of the model can be expressed by

$$\frac{\varepsilon}{\varepsilon_\mathrm{o}} = \frac{\sigma}{\sigma_\mathrm{o}}\left[1 + \left(\frac{\sigma}{\sigma_\mathrm{o}}\right)^{n-1}\right] \quad \text{or} \quad \varepsilon = \frac{\sigma}{E} + \varepsilon_\mathrm{o}\left(\frac{\sigma}{\sigma_\mathrm{o}}\right)^n, \quad (8.8)$$

where ε_o is a reference strain; σ_o is a reference stress; E is the Young's modulus or the initial slope, given by $E = \sigma_\mathrm{o}/\varepsilon_\mathrm{o}$; and n is a parameter defining the sharpness of the curve. The corresponding tangent slope of the curve is given by

$$\frac{\mathrm{d}\sigma}{\mathrm{d}\varepsilon} = \frac{E}{1 + n(\sigma/\sigma_\mathrm{o})^{n-1}}. \quad (8.9)$$

8.4.5. Strain rate dependent model

Under dynamic loads, the yield stress, but not the Young's modulus of elasticity, were found to be dependent on the loading rate (Manjoine, 1944). Both yield stress and failure stress increase monotonously with the strain rate. A formula relating the dynamic yield stress, the quasi-static yield stress and the strain rate was developed by Cowper and Symonds (1957). The equation is given by

$$\begin{aligned}\sigma &= E\varepsilon &&\text{for } \varepsilon < \varepsilon_\mathrm{y} = \varepsilon_{\mathrm{y,st}}\left[1 + \left(\frac{|\dot\varepsilon|}{D}\right)^{1/p}\right], \\ \sigma &= \sigma_\mathrm{y} = \sigma_{\mathrm{y,st}}\left[1 + \left(\frac{|\dot\varepsilon|}{D}\right)^{1/p}\right] &&\text{for } \varepsilon \geqslant \varepsilon_\mathrm{y} = \varepsilon_{\mathrm{y,st}}\left[1 + \left(\frac{|\dot\varepsilon|}{D}\right)^{1/p}\right],\end{aligned} \quad (8.10)$$

where ε_y is the dynamic yield strain; σ_y is the dynamic yield stress corresponding to ε_y; $\dot\varepsilon$ is the strain rate; E is the Young's modulus; $\varepsilon_{\mathrm{y,st}}$ is the quasi-static yield strain; $\sigma_{\mathrm{y,st}}$ is the quasi-static yield stress corresponding to $\varepsilon_{y,st}$; p and D are the material parameters which can be obtained by numerical curve fitting to the experimental results.

The parameter D can be considered as a reciprocal viscosity constant. It can be interpreted that if $\dot{\varepsilon} = D$, the yield stress σ_y will be increased by twice the quasi-static yield stress $\sigma_{y,st}$. In Fig. 8.4(e), the dashed lines represent the actual stress–strain curves under various strain rates, while the solid lines stand for the idealized stress–strain relations given by Eq. (8.10). For steel material, the values of p and D are estimated to be $p = 5$ and $D = 40/\text{sec}$ using curve fitting approximation of experimental results by Cowper and Symonds (1957).

8.5. Proposed cyclic plasticity models used in this study

The stress–strain functions described in Section 8.4 are usually applied to the plastic-zone analysis method. Although these analysis results are accurate, extensive computer time is required for the analysis and these methods are usually only suitable for simple structures. In order to reduce the analysis time effectively, two plasticity models based on the plastic hinge approach are presented in this study. The models are simple and direct because the section yielding of an element node is considered in the force–deformation relationship. In comparison with the plastic zone approach, the greatest merit of the present method is its computational efficiency requiring smaller computer memory storage and analysis time. Basically, the new plasticity models are the natural extension of the plasticity models for the static case in Chapter 6 to the dynamic case. These models are discussed as follows.

8.5.1. Proposed elastic–perfectly plastic hinge model

To compare with the refined-plastic model in Section 8.5.2, this model is aimed to show the structural response without considering partial yielding in members. The material is assumed to be elastic–perfectly plastic (i.e. either perfectly elastic or ideally plastic). The strain-hardening and Bauschinger effects are ignored in this model and the elastic–plastic material property is expressed in the moment–curvature relation. From Fig. 8.5, the dotted line indicates the path traced using the elastic–perfectly plastic model. In the moment–curvature diagram, a section remains elastic along the line OAB' under loading condition. When the plastic moment M_p is reached at point B', a plastic hinge is formed and no further moment can be resisted. For further load, the exceeded moment will be redistributed to adjacent members, and the path will move along a horizontal line B'B. When unloading occurs at point B', the path will follow the line BDC parallel to the virgin slope. When reloading occurs at C, the path will move up along CDBE. Under unloading conditions, the line EFGG'H will be followed where the line GG'H indicates that the negative moment capacity $-M_p$ is reached.

Owing to the presence of axial force, the plastic moment capacity M_p should be changed. The determination of the reduced moment capacity M_{pr} is based on the section assemblage concept proposed in Chapter 6 (see Eqs. (6.41) and (6.42)). The interaction of the axial force and moment for the reduced moment capacity has been plotted by means

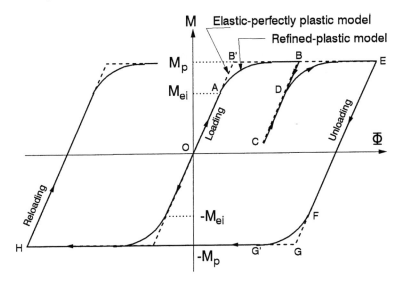

Fig. 8.5. Elastic–perfectly plastic and refined-plastic models employed in the present study.

of a surface shown in Fig. 8.7. This surface is termed as the full-yield surface which is the ultimate strength of a section to resist internal forces. The basic idea of the present section assemblage concept is that moment is taken by the flanges of an I-section and axial and shear loads are taken by the web. The concept was adopted by the Steel Constructional Institute (SCI, 1987) in the published design tables for sectional capacities and it is suitable for uses by the practicing engineers in their design of steel structures. In addition, the advantage of the concept is that the assumptions have been quite widely accepted by the profession which carries a precise and logical physical significance easily visualised by engineers.

8.5.2. Proposed refined-plastic hinge model

A refined-plastic hinge model under cyclic loading is described in this Chapter. Basically, it is similar to the elastic–plastic model except partial yielding is considered. In this model, a normalized plasticity index is used to determine the subsequent plasticity state under reversal loading/unloading conditions. The technique of using the index is simple and powerful in handling the cyclic plasticity problems. The proposed elasto-plastic hinge based model expressed in the force–deformation relationship is simple and efficient when compared with plastic-zone based models.

As illustrated in Fig. 8.5, initial yielding occurs at point A when the first yield moment capacity M_{ei} is attained. On the curve AB, the gradual yielding occurs and the plastic moment capacity M_p is reached at point B. When unloading takes place at point B, gradual yielding characteristics disappears and the path follows the line BDC in which the moment at point C is less than the initial yield moment M_{ei} at point D. On reloading, the path moves along the line CD under the perfectly elastic state and then follows the curve DE

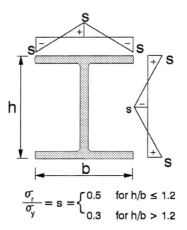

Fig. 8.6. The ECCS residual stress for hot-rolled I-sections.

under the partial yielding state. Similarly, under unloading conditions at point E, the path moves along EFG'H.

In the presence of axial force P, both the initial yield moment M_{ei} and the plastic yield moment M_p would generally be decreased to the reduced initial yield moment M_{er} and the reduced plastic moment M_{pr}, respectively. The determination of M_{pr} is the same as that in the elastic–perfectly plastic hinge model in Section 8.5.1. Since a section is elastic before the initial yielding, a linear function for an initial yielding surface is approximated and suggested as (refer to Fig. 8.7),

$$M_{er} = \left(\sigma_y - \sigma_r - \frac{P}{A}\right) Z_e, \tag{8.11}$$

where σ_y is the yield stress, P is the axial force on the moment, A is the cross-sectional area of the section, Z_e is the elastic modulus of the section and σ_r is the residual stress which can be obtained by measurement or in a recommendation by European Convention for Construction Steelwork (ECCS, 1983) shown in Fig. 8.6.

Under reversal loading/unloading excitation, the plasticity state, defined as the current condition of a plastic hinge in an ideally elastic, partial yielding or fully plastic stage, should be determined so that its loading history data can be saved for predicting the subsequent plasticity state. For example, if the moment at a section is between M_{ei} and M_p, say M_1, under loading condition and then decreases to a value, say M_2, which is above M_{ei} upon unloading, the moment M_1 will be stored. If it is reloaded, the plastic hinge will remain ideally elastic until the moment reaches the past history record M_1. In the following reloading condition, the plastic hinge will partially yield at the moment of M_1. The history of the moment M_1 should, therefore, be saved in a computer analysis for later use. Moment M can be simply normalized to a plasticity index value, μ_p, is given by

$$\mu_p = \frac{|M| - M_{er}}{M_{pr} - M_{er}}, \tag{8.12}$$

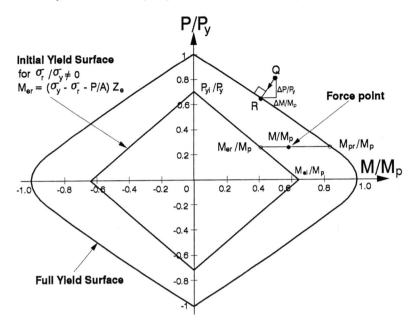

Fig. 8.7. Initial yield and full yield surfaces of a section.

in which M is the absolute of the current updated moment. In the extreme case of an ideally elastic and fully plastic problem, the plasticity index value μ_p is equal to 0 and 1, respectively. The contours for plasticity index equal to 0, 0.25, 0.5, 0.75 and 1 are plotted in Fig. 8.8. In general, the dimensionless plasticity index is stored rather than its corresponding moment M because of simplicity in dealing with various sections. Using the present approach, only the cross-sectional dimensions of members available from section tables by manufacturers or steel research and engineering institutes are required to be stored.

8.6. Beam–column connection modelling

The modelling of beam–column connections has been discussed in Chapter 6 for the static case and in Chapter 7 for the dynamic case. In this Chapter, the formulation will not be re-capitulated. In the present dynamic analysis, a connection is again modelled by a rotational spring called the connection spring. The stiffness of the connection spring element is simply the slope of the connection moment–rotation curve. This spring is then combined with a section spring to give a resultant spring-in-series. The resultant spring is further condensed in the formulation of the element stiffness matrix. The model of a beam–column element attached with section and connection springs is illustrated in Fig. 8.9.

The numerical procedure for modelling hysteretic and flexible connections has been incorporated to the program GMNAF/D (Geometric and Material Non-linear Analysis of

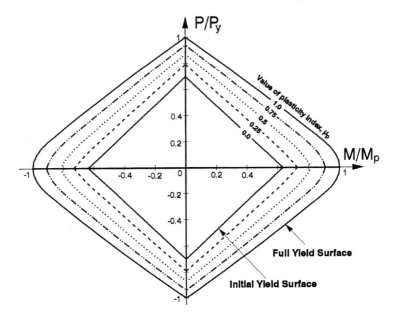

Fig. 8.8. Contour lines of plasticity index values.

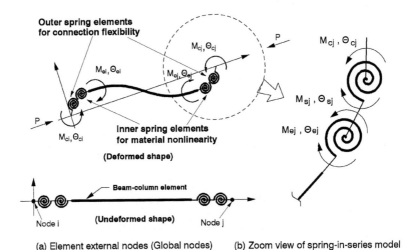

(a) Element external nodes (Global nodes) (b) Zoom view of spring-in-series model

Fig. 8.9. Beam–column element attached with section and connection springs.

Semirigid Frames – Dynamics) described in Chapter 7. The additional numerical procedures for the present inelastic models (i.e. the elastic–plastic and the refined-plastic models) are included in the program. The final program GMNAF/D is, therefore, capable of predicting the dynamic response of steel structures with non-linearities in geometrical change, hysteretic connection flexibility and hysteretic material yielding.

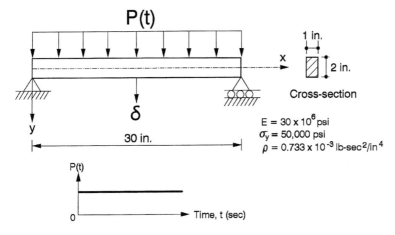

Fig. 8.10. Details of a simply supported beam.

8.7. Numerical examples

In this section, the dynamic inelastic responses using the program GMNAF/D are compared with those obtained by other researchers. Throughout the following examples, viscous damping is not considered (i.e. $[C] = 0$) in order to focus attention on the effect of hysteretic looping actions at beam–column connections and plastic hinges to the overall structural response.

8.7.1. Elastic–plastic dynamic analysis of a simply supported beam

Figure 8.10 shows a simply supported rectangular beam subjected to a step pressure load, $P(t)$. The beam has been investigated by Baron et al. (1961), Nagarajan and Popov (1974), Liu and Lin (1979), and Yang and Saigal (1984). The beam dimensions and properties are indicated in the Figure. The function of step pressure is given by $P(t) = nP_o$, where P_o is the static collapse load and n is the intensity of step pressure. The peak elastic–plastic response of the beam was determined for three values of the intensity, $n = 0.5$, 0.75 and 1.0, respectively. The moment–curvature relation of members is assumed to be elastic–perfectly plastic. Using the present elastic–plastic hinge method, the dynamic elastic–plastic analysis results are compared with nondimensional peak values reported by Baron et al. (1961). In their studies, one quarter of the beam is discreted into five equal 10-noded isoparametric plate elements. In this book, the full length of the beam using 24 beam–column elements is analysed. The mid-span deflection is normalized with respect to the static elastic deflection and its transient response is illustrated in Fig. 8.11. A deviation of paths is noted between the results by the present approach and by Baron et al. (1961). This may be due to the different types of elements and solution methods employed. The peak values by the present method are 1.050, 2.202 and ∞ for $n = 0.5, 0.75$ and 1.0, respectively. The last value of infinity indicates structural failure due to the excessive plastic yielding of material under the large pressure load. The present maximum

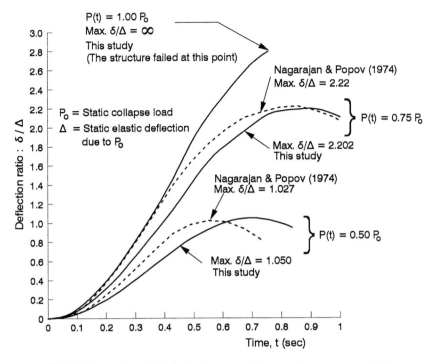

Fig. 8.11. Dynamic analysis of a simply supported beam under step pressure load.

values compare quite closely with the values of 1.027, 2.22 and ∞ given by Baron et al. (1961). Obviously, the present method based on the beam–column element is simpler than the plate element by Baron et al. (1961) and can be used efficiently and effectively in dynamic analysis of skeletal structures.

8.7.2. Dynamic elastic–perfectly plastic response of Lee et al. steel arch

A steel arch fabricated by six pieces of UB 305 × 165 × 40 kg/m Universal Beams is shown in Fig. 8.12(a). The dynamic elastic–perfectly plastic response of the arch has been investigated by Lee et al. (1996). The arch is modelled by six elements and fixed at the supports. The arch is assumed to be subjected to a triangular decaying vertical impact load applied at the top of the arch, as shown in Fig. 8.12(b). It is assumed that a lumped mass of 0.05 kN s^2/m is attached to each node. A time step of 0.1 ms is used in the analysis. In the study by Lee et al. (1996), the material is assumed to be elastic–perfectly plastic and a bi-linear full yield surface on the plane of interaction of axial force and bending moment is assumed. Their yield surface contains sharp interception points where discontinuity of slope occurs. This kink is generally undesirable from computational and physically point of view. Based on the section assemblage concept proposed in Chapter 6, the full yield surface is computed and adopted in the present analysis, as shown in Fig. 8.12(c). From the Figure, the shapes of the full yield surfaces by Lee et al.

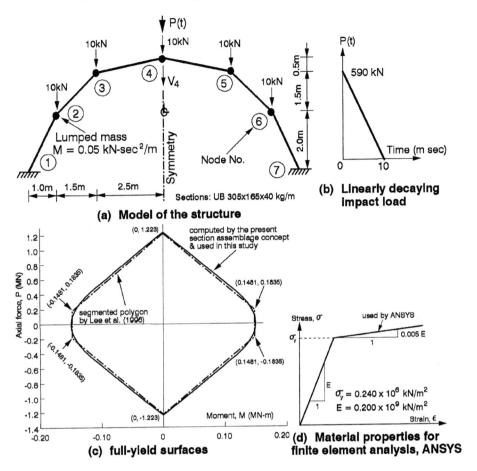

Fig. 8.12. Lee et al. steel arch.

(1996) and by the present studies are noted to be very similar. However, their yield surface (Lee et al., 1996) was obtained by prescribed curve-fitting and therefore their application to a new section type requires another fitting procedure for generation of the a new set of parameters. This procedure carries little physical meaning and inconvenient for a problem with different sections. On the other hand, the present method for determining the yield surface is simple, more general and consistent for I-, H- and rectangular sections. In the paper by Lee et al. (1996), the results by their elastic–perfectly plastic analysis are compared with the finite element solutions obtained from the computer package ANSYS. In the ANSYS analysis, the elasto-plastic 2-dimensional beam element is used to model the stiffness of the member with the material properties shown in Fig. 8.12(d).

The time histories of the vertical deflection and bending moment at node 4 obtained by Lee et al. (1996), the computer package ANSYS and the proposed elastic–perfectly plastic hinge method are illustrated in Figs. 8.13 and 8.14(a). These results are in reasonably

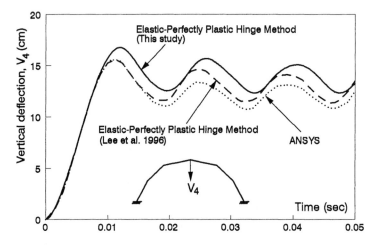

Fig. 8.13. Time history of the vertical deflection at node 4 of the arch.

close agreement. Differences in the results between the proposed method and the Lee et al. (1996) method may be due to the discrepancies in handling of recorrection of force point movement along the full yield surfaces when yielding intensifies. Furthermore, the difference may partly attributed by the ignorance of the axial deformation by Lee et al. (1996) whilst the axial effect is considered in the present study. In this problem with the arching effect, it is expected that the softening $P\text{-}\Delta$ effect in the member stiffness due to sudden impact load and the gravity loads should not be ignored. Consequently, it is not surprising that the maximum deflection predicted by the present analysis is slightly larger than that by Lee et al. (1996). On the other hand, the maximum bending moment obtained by ANSYS is larger than the member plastic moment capacity. It is due to the inclusion of the strain hardening effect in ANSYS, which is ignored by Lee et al. (1996) and in the present study.

The comparison of the variation of bending moments at nodes 1 to 4 by the present approach and the results by Lee et al. (1996) is depicted in Fig. 8.14. The two sets of results are quite close. As expected, all the bending moments are within the plastic moment capacity of the members. However, it is observed that the response period predicted by the present study is longer because the P-delta effect reduces the structural stiffness, resulting a longer natural vibration period. The time histories of bending moments at various nodes are summarized in Fig. 8.15 for ease of comparison.

8.7.3. Dynamic elastic–perfectly plastic response of Toridis–Khozeimeh frame

A steel plane frame subjected to a dynamically applied lateral force of constant magnitude, F, equal to 100 kips acting on the left hand corner of the structure is shown in Fig. 8.16. The mass of steel members is assumed to be multiplied by a factor of 625 and lumped at nodes. Three elements are used to model each beam and column. The dynamic elastic behaviour of the same frame has been studied by Toi and Isobe (1996) and will be re-analysed in the present study. The results are in close agreement (see Fig. 8.17). In the

Chapter 8. *Transient analysis of inelastic steel structures with flexible connections*

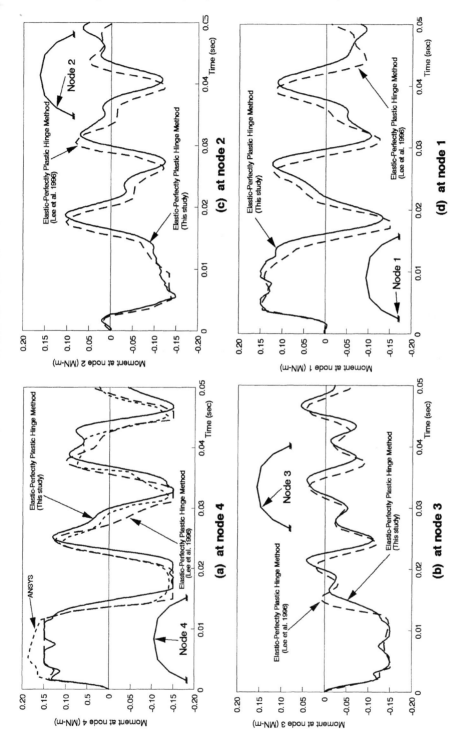

Fig. 8.14. Time history of bending moment at different nodes of the arch.

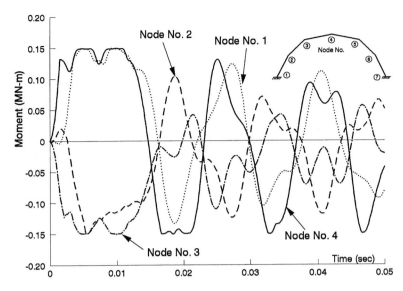

Fig. 8.15. Time history of the bending moment of the arch by the present elastic–perfectly plastic hinge method.

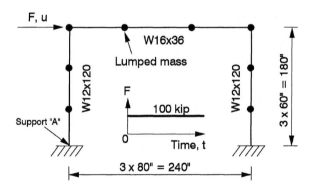

Fig. 8.16. Layout of the Toridis–Khozeimeh frame.

analysis, viscous and hysteretic damping at rigid connections does not exist. In addition, the hysteretic damping at plastic hinges are ignored. The deflection amplitude is, therefore, almost constant without the loss of energy due to damping. As expected, the dynamic elastic deflection amplitude is exactly equal to twice of the static elastic deflection.

The dynamic elastic–perfectly plastic response of the same frame has been investigated by Toridis and Khozeimeh (1971). In their report, a bi-linear stress–strain model with the inelastic branch having a slope of $1/10$ of the elastic branch was used. In the present study, both the elastic–perfectly plastic hinge and the refined-plastic hinge methods are employed. Identical full yield surface is used in both methods. To consider partial yielding in the refined-plastic hinge method, an initial yield surface should be constructed first. Based on the recommendation of ECCS (1983), the maximum magnitude of residual

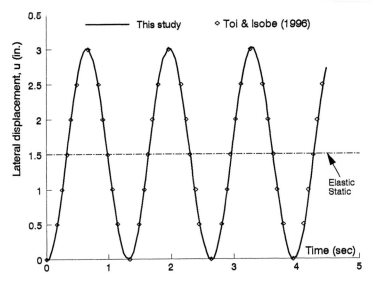

Fig. 8.17. Elastic response of the Toridis–Khozeimeh frame.

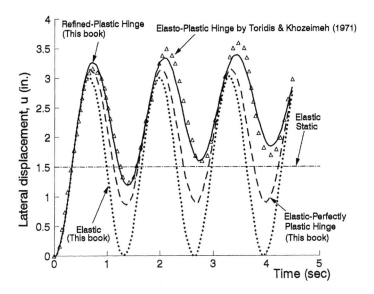

Fig. 8.18. Dynamic response of the Toridis–Khozeimeh frame.

stresses in beam and column members are suggested respectively as 30% and 50% of the yield stress, according to the ratio of depth to breadth of a section shown in Fig. 8.6.

The time histories of lateral deflection at the point of application of the load are plotted in Fig. 8.18. The plastic hinge is firstly formed at a time of about $t = 0.3$ second. However, the plastic deformation is negligible up to a time of about $t = 0.5$ second at which

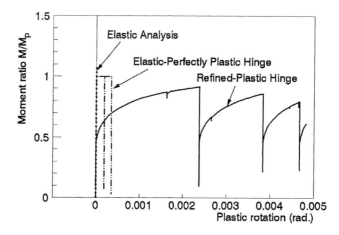

Fig. 8.19. Response of moment vs. plastic rotation at support A of the Toridis–Khozeimeh frame by the present study.

the elastic and the plastic deflections start to deviate. Thereafter, the influence of inelastic action becomes more pronounced, leading to an apparently large deformation. From the Figure, it is observed that the magnitudes of the deformations increase gradually in the subsequent cycles because further plastic deformations occur after considerable permanent deflections form at the plastic hinges.

It can be seen in Fig. 8.18 that the trends for the displacement versus time curves predicted by the bi-linear model by Tordis and Khozeimeh (1971) and the proposed refined-plastic hinge model are similar. In the refined-plastic hinge analysis, the material starts to yield gradually when the yield surface is reached. Therefore, the deformation tends to increase with time until the plastic energy part of the input work done is completely dissipated by the inelastic action while the elastic energy part remains in the system in the form of kinetically elastic oscillations. However, using the elastic–perfectly plastic hinge method, the subsequent oscillation of the deflection has a stabilizing response because the structure is in the elastic range and thus the energy cannot be dissipated through plastic hinges during the oscillations. However, a permanent plastic deformation is formed.

The normalized bending moment versus plastic deformation at the left fixed support A is depicted in Fig. 8.19. It is obvious that there is a part of energy lost in each loop using the refined-plastic hinge method, and this can explain the hysteretic damping effect mentioned above. However, only a little amount of energy dissipated is observed in the elastic–perfectly plastic hinge analysis. This is because the system returns to the elastic state after the plastic deformation occurrence at the first cycle. For elastic analysis, no energy is lost.

8.7.4. Kam–Lin portal frame

The transient response of a simple portal frame loaded with a gravity load of 83.28 kips each at the top of columns has been investigated by Kam and Lin (1988). The frame is

Chapter 8. Transient analysis of inelastic steel structures with flexible connections

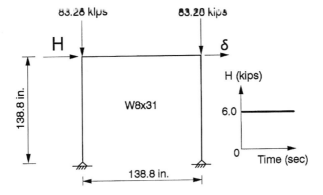

Fig. 8.20. Details of the Kam–Lin frame.

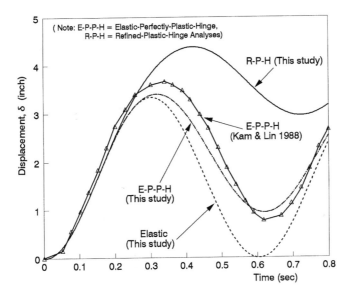

Fig. 8.21. Inelastic response of the Kam–Lin frame.

subjected to an impact lateral load of time function shown in Fig. 8.20. Material is assumed to be elastic–perfectly plastic in the report of Kam and Lin (1988). In the present study, the elastic–plastic hinge and the refined-plastic hinge methods are employed for comparison. The dynamic responses are shown in Fig. 8.21. The elastic–plastic hinge results predicted by the present study and by Kam and Lin (1988) are in close agreement. Their small variation is due to the difference in the numerical approach adopted. As expected, it is observed that the deformation predicted using the refined-plastic hinge method is larger than using the elastic–plastic hinge method. With the exception of elastic analysis, permanent plastic deformations are noted in all types of inelastic analyses.

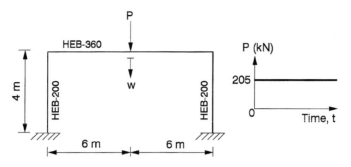

Fig. 8.22. System and loading of the Rothert et al. frame.

8.7.5. Rothert et al. frame

A fixed supported portal frame subjected to a dynamic vertical impact load of 205 kN has been reported by Rothert et al. (1990). Figure 8.22 shows the geometry of the frame and the loading function. Five elements per column member and three elements per beam member are used in the model. A time step of $t = 0.001$ second is employed. In the paper by Rothert el al. (1990), the strain rate dependent material model is used and the stress–strain relationship is expressed as a function of strain rate. Before initial yielding, the material behaves linearly elastic. Once the plastic yielding is reached, the material behaves perfectly plastic. In the present analysis, each member is modelled by six beam–column elements. To account for the effects of the strain rate on the structural response, the yield stress has been increased by 10%, which, in turn, means that the squash load P_y and the plastic moment M_p have been magnified by a factor of 1.1. Viscous damping is not considered in the analysis in order to focus attention on the influence of the hysteretic damping at plastic hinges on the overall structural behaviour. The displacement results at the mid-span of the beam are plotted in Fig. 8.23. It is observed that the elastic–plastic hinge result is close to the solution obtained by Rothert et al. (1990). As expected, the refined-plastic hinge assumption leads to a curve with higher structural flexibility.

8.7.6. Effect of yield stress on dynamic inelastic response

A simple portal frame with lateral sway imperfection of $\psi = 1/200$ at the pinned supports is shown in Fig. 8.24. One element per each member is used in the model and a time step of $t = 0.001$ second is employed. The frame is subjected to a horizontal impact load of 60 kN applied at the top-left of the frame. Two heavy masses of 200 kN each are placed at the top of the columns to magnify the P-Delta effect in the analysis. In other words, an additional P-Δ moment for the swayed frame is induced. Based on the ECCS recommendation (ECCS, 1983), each hot-rolled W8 × 48 steel member is assumed to have residual stresses with the maximum magnitude of 50% of yield stress (i.e. $\sigma_r = 50\% \, \sigma_y$). The dynamic response of lateral deflection of the frame is illustrated in Fig. 8.25. As seen from the graph, when a lower yield stress is employed in the analysis, the magnitude and the plastic permanent deformation of the frame will increase. Similarly, when the yield stress

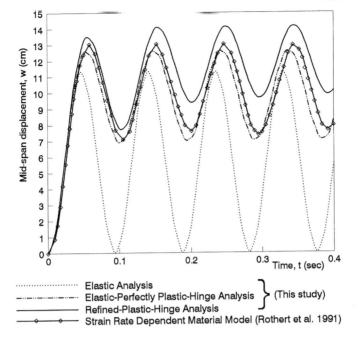

Fig. 8.23. Inelastic responses of the Rothert et al. frame.

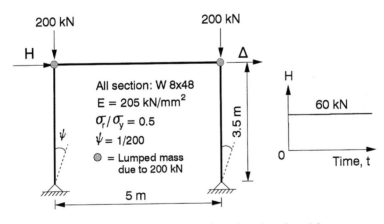

Fig. 8.24. Loading pattern and geometric configuration of portal frame.

is assumed to be infinite (i.e. using the elastic analysis), no plastic deformation exhibits. It is also observed that the refined-plastic hinge analysis gives a larger deflection than the prediction by the elastic–perfectly plastic hinge. This is because, in the refined-plastic hinge analysis, partial yielding occurs prior to the formation of a fully plastic hinge before which the deflection is increased due to yielding in the structure.

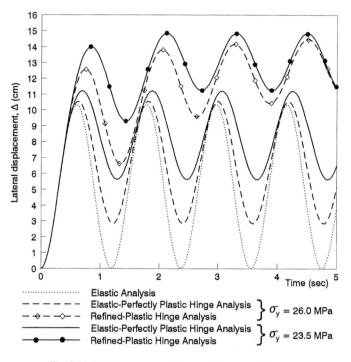

Fig. 8.25. Elastic response of the portal frame by this study.

8.7.7. Dynamic inelastic response of semi-rigid jointed steel frame

In this example, the combined effect of hysteretic member inelastic action and hysteretic semi-rigid joint behaviour on the overall structural response is investigated. A single-bay two-storey steel frame with flexible beam-to-column connections is studied. Details of the geometry and loads are given in Fig. 8.26. Before the dynamic analysis for the suddenly applied horizontal forces at two floors, the frame is initially loaded by the static vertical forces. An imperfection of geometry of $\psi_o = 1/438$ is considered (ECCS, 1983). A time step of $\Delta t = 0.001$ second is chosen in the dynamic analysis.

Three types of beam-to-column joints are considered in this example to investigate the influence of connection type to the structural behaviour. These include the rigid joint, the linear joint and the non-linear joint. Since very close results are obtained if the non-linear moment–rotation curve of connections is well represented by the Richard–Abbott model (Richard and Abbott, 1975), the Ramberg–Osgood model (Ramberg and Osgood, 1943) and the Lui–Chen Exponential model (Lui and Chen, 1988) as discussed in Chapter 7, the choice among these connection models is not important. The moment–rotation curve relationship of the non-linear joint case is thus simulated by the Richard–Abbott model and the required coefficients of the model are listed in Fig. 8.27(a). The structural response is not sensitive to the non-linear and relatively stiff joint. A relatively soft connection is assumed in this example and the connection moment capacity is less than the moment

Chapter 8. *Transient analysis of inelastic steel structures with flexible connections*

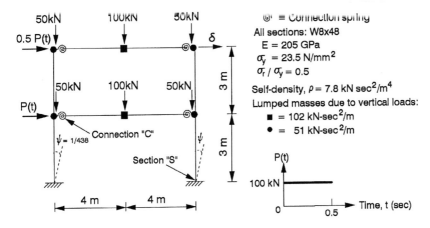

Fig. 8.26. Two-storey steel frame.

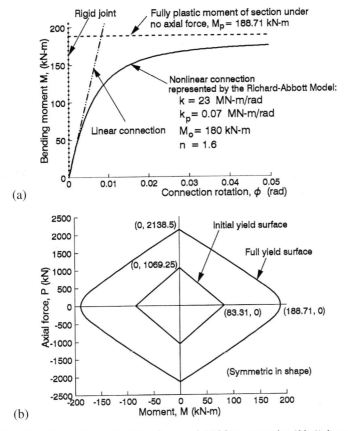

Fig. 8.27. Information of connections and yield surfaces used. (a) Moment–rotation (M–ϕ) data of connections; (b) Initial- and full-yield surfaces of members.

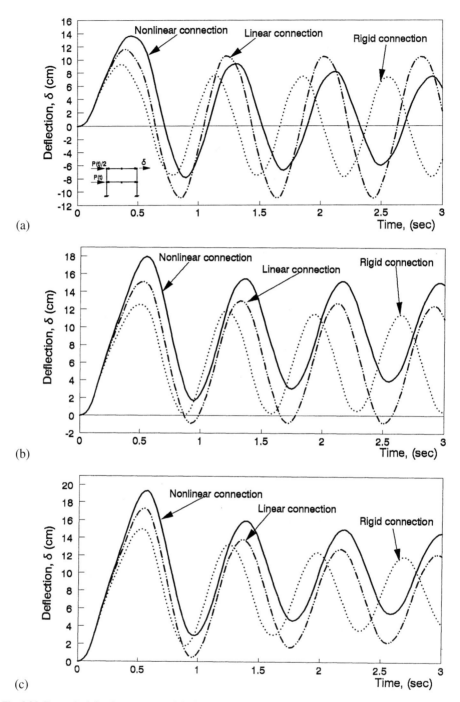

Fig. 8.28. Dynamic deflection response of the 2-storey frame. (a) Elastic analysis; (b) Elastic–perfectly plastic hinge analysis; (c) Refined-plastic hinge analysis.

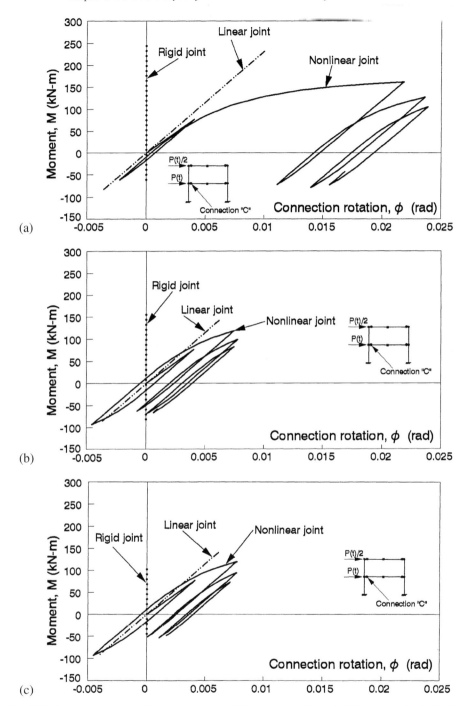

Fig. 8.29. Connection hysteretic loops at connection C for various analyses. (a) Elastic analysis; (b) Elastic–perfectly plastic hinge analysis; (c) Refined-plastic hinge analysis.

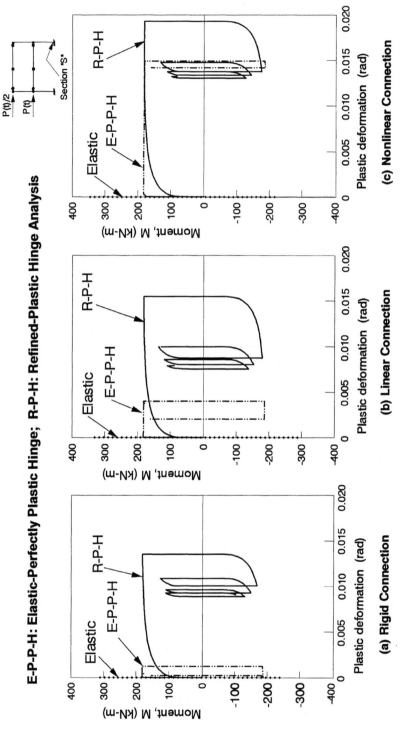

Fig. 8.30. Hysteretic force–deformation loops at plastic hinge S for various connection types.

capacity of frame member of 188.71 kN m. Material non-linearity of members is also considered. For the section of W8 × 48, the maximum magnitude of residual stress is assumed to be 50% of the yield stress. Based on the developed section assemblage concept, the corresponding initial-yield and full-yield surfaces are determined and plotted in Fig. 8.27(b).

The dynamic deflections are traced in Fig. 8.28. From the results using the elastic analysis shown in Fig. 8.28(a), a permanent displacement and damping effects are observed in the non-linear connection which is not noted in either the rigid or the linear connection type cases. This is because of the hysteretic moment–rotation loops and the shifted rotations at the non-linear connections. These observations have been explained in Chapter 7. On the other hand, from Fig. 8.28(a) to (c), it can be easily seen that the effect of permanent deflection and energy dissipation is more pronounced in the refined-plastic hinge analysis. This is because an early yielding in a plastic hinge leads to a larger plastic rotation. The maximum deflection is noted to be larger and the response period is longer for the frame with non-linear joints and using the refined-plastic hinge method. This is the results from the structural stiffness being reduced by the degrading flexible connections and the softening plastic hinges.

The hysteretic moment–rotation loops at connection C are traced in Fig. 8.29. In the inelastic analyses, bending moments at all beam-to-column connections have been limited within the plastic capacity of members. For the non-linear joint cases, it can be seen that less hysteretic energy is dissipated at the non-linear connections in the elastic analysis than in the refined-plastic hinge analysis. This observation is the result of the energy dissipation by the inelastic action of plastic hinges.

The hysteretic paths of moment versus plastic deformation at support S is depicted in Fig. 8.30. It can be seen that the input energy has been gradually dissipated by the amount of area enclosed in hysteretic loops in the refined-plastic hinge analysis. The energy dissipating rate reduces as the loop size decreases. Damping continues until the overall system has transformed to an elastic state with the constant elastic kinematic and potential energy stored in the structure. On the other hand, the energy has been dissipated only in the first few cycles when the elastic–perfectly plastic hinge analysis is employed. It is because damping due to inelastic action of plastic hinges occurs only when the bending moment is equal to the plastic moment capacity. However, no such damping can be seen in the elastic analysis.

8.8. Conclusions

Non-linear dynamic analysis of inelastic steel structures with semi-rigid connections has been presented in this chapter. The transient response of the structures has been investigated and discussed. A new refined-plastic hinge method based on the concept of plasticity index and section assemblage is proposed. Partial yielding in the transition from the initial yield moment to the fully plastic moment has been included in the proposed method. The plasticity state under reversal loading/unloading conditions is expressed in terms of a normalized plasticity index. The concept of the index is found to be robust in handling path-dependent plasticity problems. The present elasto-plastic hinge model is, generally, simpler and more efficient than plastic-zone-based plasticity models. The full yield surface of a cross-section is determined conveniently by the section assemblage concept. Unlike the prescribed full yield surfaces, the present method is a general and consistent approach for I-, H- and rectangular sections. For the sake of comparison with the refined-plastic hinge model, a simple elastic–perfectly plastic hinge method ignoring partial yielding is also considered. The flexibility of beam-to-column connections is also taken into account in the present analysis. The resultant springs-in-series model was found to be applicable to both static and dynamic analyses accounting for material yielding and connection flexibility. This spring-in-series model has been successfully incorporated into the developed computer program GMNAF/D. The present numerical results have been compared with the results by other researchers and it was found that they are generally in close agreement. The newly developed refined-plastic hinge model was proved to be computational stable without divergent problem and can be used as a reliable tool for large deflection and inelastic analysis of steel frames.

Demonstrated from the numerical examples, the effects of connection flexibility and member yielding change the structural response significantly. These effects will induce the hysteretic damping which is particularly important in assessing the energy absorption capacity of the structure against dynamic loads due to machine, typhoons and seismic forces. To this, non-linearities due to flexibility in connections, inelastic action of plastic hinges, and geometrical change of structure can be considered and included conveniently in sophisticated and rigorous non-linear dynamic analysis and design of steel frames.

References

Akiyama, H. and Yamada, S. (1995): Damage to steel buildings in the Hyogoken–Nanbu earthquake. In: *EASEC-5, Proceedings of the Fifth East Asia–Pacific Conference on Structural Engineering and Construction*, Queensland, Australia, 25–27 July, Vol. 3, pp. 2463–2474.

Alvarez, R.J. and Birnstiel, C. (1969): Inelastic analysis of multistory multibay frames. *J. Struct. Div. ASCE* **95**(11), 2477–2503.

Aspden, R.J. and Campbell, J.D. (1966): The effect of loading rate on the elasto-plastic flexure of steel beams. *Proc. Roy. Soc. London* **A290**, 266–285.

Balendra, T., Lim, E.L. and Lee, S.L. (1994): Ductile knee braced frames with shear yielding knee for seismic resistant structures. *Eng. Struct.* **16**(4), 263–271.

Baron, M.L., Bleich, H.H. and Weidlinger, P. (1961): Dynamic elastic–plastic analysis of structures. *J. Eng. Mech. Div. ASCE* **87**(1), Proc. Paper No. 2731, 23–42.

Bathe, K.J. (1996): *Finite Element Procedures*, Prentice Hall, NJ.

Berg, G.V. and DaDeppo, D.A. (1960): Dynamic analysis of elasto-plastic structures. *J. Eng. Mech. Div. ASCE* **86**(2), Proc. Paper No. 2440, 35–58.

Bockholdt, J.L. and Weaver, J.W. (1974): Inelastic dynamic analysis of tier buildings. *Comput. Struct.* **4**, 627–645.

Bourahla, N. and Blakeborough, A. (1992): Seismic tests on model knee braced frames. In: J.L. Clarke, F.K. Garas and G.S.T. Armer, eds., *Structural Design for Hazardous Loads*, Inst. of Structural Engineers, pp. 71–79.

Chen, F.P. (1981): Generalized Plastic Hinge Concepts for 3-D Beam-Column Elements, Ph.D. Dissertation, Univ. of California, Berkeley, CA.

Chen, W.F. and Chan, S.L. (1995): Second-order inelastic analysis of steel frames using element with midspan and end springs. *J. Struct. Eng. ASCE* **121**(3), 530–541.

Clough, R.W., Benuska, K.L. and Wilson, E.L. (1965): Inelastic earthquake response of tall buildings. In: *Proc. Third World Conference on Earthquake Engineering*, Vol. II, Auckland and Wellington, New Zealand, pp. 68–69.

Cowper, G.R. and Symonds, P.S. (1957): Strain Hardening and Strain Rate Effects in the Impact Loading of Cantilever Beams, Technical Report 28, Brown University.

European Convention for Constructional Steelwork (ECCS) (1983): Ultimate Limit State Calculation of Sway Frames with Rigid Joints. ECCS, Technical Working Group B.2, Systems Publication No. 33.

Filiatrault, A., Leger, P. and Tinawi, R. (1994): On the computation of seismic energy in inelastic structures. *Eng. Struct.* **16**(6), 425–436.

Galambos, T.V. (1968): Deformation and energy absorption capacity of steel structures in the inelastic range. *Am. Iron Steel Inst. Bull.* **8**(March).

Gere, J.M. and Shah, H.C. (1980): *The 1976 Tangshan China Earthquake*, E.E.R.I., March.

Heidebrecht, A.C., Lee, S.L. and Fleming, J.F. (1964): Dynamic analysis of elastic–plastic frames. *J. Struct. Div. ASCE* **90**(2), 315–343.

Housner, G.W. (1956): Limit design of structures to resist earthquakes. In: *Proc. World Conference on Earthquake Engineering*, pp. 5.1–5.13.

Housner, G.W. and Jennings, P.C. (1977): The capacity of extreme earthquake motions to damage structures. In: W.J. Hall, ed., *Structural and Geotechnical Mechanic*, Prentice-Hill, Englewood Cliffs, NJ, pp. 102–116.

Hsieh, S.H., Deierlein, G.G., McGuire, W. and Abel, J.F. (1989): Technical Manual for CU-STAND, Structural Engineering Report No. 89-12, School of Civil and Environmental Engineering, Cornell University, Ithaca, NY.

Izzuddin, B.A. and Elnashai, A.S. (1990): Refined large displacement nonlinear dynamic analysis of steel structures. In: *Ninth European Conference on Earthquake Engineering*, Moscow, pp. 199–206.

Izzuddin, B.A. and Elnashai, A.S. (1993a): Adaptive space frame analysis, Part I: A plastic hinge approach. *Proc. Inst. Civil Eng. Struct. Bldgs.* **99**, 303–316.

Izzuddin, B.A. and Elnashai, A.S. (1993b): Adaptive space frame analysis, Part II: A distributed plasticity approach. *Proc. Inst. Civil Eng. Struct. Bldgs.* **99**, 317–326.

Johnston, B.G. and Cheney, L. (1942): Steel Columns of Rolled Wide Flange Section, Progress Report No. 2, American Institute of Steel Constructions (AISC), Chicago, IL.

Jonatowski, J.J. and Brinstiel, C. (1970): Inelastic stiffened suspension space structures. *J. Struct. Div. ASCE* **96**(ST6), 1143–1166.

Kam, T.Y. and Lin, S.C. (1988): Nonlinear dynamic analysis of inelastic steel plane frames. *Comput. Struct.* **28**(4), 535–542.

Kondoh, K. and Atluri, S.N. (1986): Large-deformation, elasto-plastic analysis of frames under nonconservative loading, using explicitly derived tangent stiffness based on assumed stresses. *Comp. Mech.* **1**, in press.

Kouhia, R. and Tuomala, M. (1993): Static and dynamic analysis of space frames using simple Timoshenko type elements. *Int. J. Numer. Methods Eng.* **36**, 1189–1221.

Kuwamura, H. and Galambos, T.V. (1989): Earthquake loads for structural reliability. *J. Struct. Eng. ASCE* **115**, 2166–2183.

Lee, E.H. and Symonds, P.S. (1952): Large plastic deformations of beams under transverse impact. *J. Appl. Mech. Trans. ASME* **74**, 308.

Lee, S.L., Swaddiwudhipong, S. and Alwis, W.A.M. (1996): Dynamic response of elasto-plastic planar arches. *Struct. Eng. Mech.* **4**(1), 9–23.

Liu, S.C. and Lin, T.H. (1979): Elastic–plastic dynamic analysis of structures using known elastic solutions. *Earthquake Eng. Struct. Dynamics* **7**, 147–159.

Lui, E.M. and Chen, W.F. (1988): Behavior of braced and unbraced semi-rigid frames. *Int. J. Solids Struct.* **24**(9), 893–913.

Manjoine, M.J. (1944): Influence of rate of strain and temperature on yield stress of mild steel. *J. Appl. Mech.* **11**, 211–218.

Mosquera, J.M., Kolsky, H. and Symonds, P.S. (1985): Impact tests on frames and elastic–plastic solutions. *J. Eng. Mech. ASCE* **111**(11), 1380–1401.

Nagarajan, S. and Popov, E.P. (1974): Elastic–plastic dynamic analysis of axisymmetric solids. *Comput. Struct.* **4**, 1117–1134.

Nigam, N.C. (1970): Yielding in framed structures under dynamic loads. *J. Eng. Mech. Div. ASCE* **96**(5), 687–709.

Noor, A.K. and Peters, J.M. (1980): Nonlinear dynamic analysis of space trusses. *Comput. Methods Appl. Mech. Eng.* **21**, 131–151.

Ramberg, W. and Osgood, W.R. (1943): Description of Stress–Strain Curves by Three Parameters, Tech. Report No. 902, Nat. Advisory Committee for Aeronautics, Washington, DC.

Rawlings, B. (1964): Dynamic plastic analysis of steel frames. *J. Struct. Div. ASCE* **90**(3), 265–283.

Richard, R.M. and Abbott, B.J. (1975): Versatile elastic–plastic stress–strain formula. *J. Eng. Mech. Div. ASCE* **101**(4), 511–515.

Rosenblueth, E. (1986): The Mexican earthquake: A first-hand report. *Civil Eng.* **56**(1), 38–40.

Rothert, H., Gebbeken, N. and Brandt, V. (1990): Dynamic inelastic analysis of plane steel structures taking into account the effect of rate-dependent behaviour. In: *European Conference on Structural Dynamics EURODYN '90*, pp. 1023–1029.

Salvadori, M.G. and Weidlinger, P. (1957): On the dynamic strength of rigid-plastic beams under blast loads. *Proc. ASCE* **83**(4).

Sam, M.T., Balendra, T. and Liew, C.Y. (1992): Knee brace frame: An economical structural system for aseismic steel buildings. In: J.L. Clarke, F.K. Garas and G.S.T. Armer, eds., *Structural Design for Hazardous Loads*, Inst. of Structural Engineers, pp. 107–113.

Scawthorn, C. and Yanev, P.I. (1995): Preliminary Report 17 January 1995, Hyogo-ken Nambu, Japanese, Earthquake. *Eng. Struct.* **17**(3), 146–157.

SCI (1987): *Steelwork Design Guide to BS5950*, Part 1: 1985, Vol. 1, Section Properties and Member Capacities, 2nd edn., The Steel Construction Institute.

Shi, G. and Atluri, S.N. (1988): Elasto-plastic large deformation analysis of space-frames: A plastic-hinge and stress-based explicit derivation of tangent stiffnesses. *Int. J. Num. Methods Eng.* **26**, 589–615.

Soroushian, P. and Alawa, M.S. (1990): Hysteretic modeling of steel struts: Refined physical theory approach. *J. Struct. Eng. ASCE* **116**(11), 2903–2916.

Tin-Loi, F. and Vimonsatit, V. (1995): Elastoplastic analysis of semi-rigid frames under cyclic loading. *Mech. Struct. Mach.* **23**(1), 17–33.

Toi, Y. and Isobe, D. (1996): Finite element analysis of quasi-static and dynamic collapse behaviors of framed structures by the adaptively shifted integration technique. *Comput. Struct.* **58**(5), 947–955.

Toma, S. and Chen, W.F. (1995): A selection of calibration frames in North America for second-order inelastic analysis. *Eng. Struct.* **17**(2), 104–112.

Toridis, T.G. and Khozeimeh, K. (1971): Inelastic response of frames to dynamic loads. *J. Eng. Mech. Div. ASCE* **97**(3), 847–863.

Uzgider, E.A. (1980): Inelastic response of space frames to dynamic loads. *Comput. Struct.* **11**, 97–112.

van Kuren, R.C. and Galambos, T.V. (1964): Beam–column experiments. *J. Struct. Div. ASCE* **90**(2), 223–255.

Vogel, U. (1985): Calibrating frames. *Stahlbau* **54** (October) 295–311.

Wen, R.K. and Farhoomand, F. (1970): Dynamic analysis of inelastic space frames. *J. Eng. Mech. Div. ASCE* **96**(5), 667–686.

Yang, T.Y. and Saigal, S. (1984): A simple element for static and dynamic response of beams with material and geometric nonlinearities. *Int. J. Num. Methods Eng.* **20**, 851–867.

Yau, C.Y. and Chan, S.L. (1994): Inelastic and stability analysis of flexibly connected steel frames by spring-in-series model. *J. Struct. Eng. ASCE* **120**(10), 2803–2819.

Zienkiewicz, O.C. (1971): *The Finite Element Method in Engineering Science*, McGraw-Hill, London.

Appendix A

A.1. The 12 × 12 linear stiffness matrix, $[K_L]$

$KL_{1,1} = EA/L,$

$KL_{1,7} = -EA/L,$

$KL_{2,2} = KL_{11}^z = EI_z \Big\{ 3(C_{kl}^z)^{-1}$
$\qquad \times \big\{ [384(S_i^z)^2 + 192k_{11}^z S_i^z + 96(k_{11}^z)^2](S_j^z)^2$
$\qquad + [192k_{11}^z(S_i^z)^2 - 96(k_{11}^z)^2 S_j^z]S_i^z + 96(k_{11}^z)^2(S_i^z)^2 \big\} \Big\},$

$KL_{2,6} = KL_{12}^z = EI_z \Big\{ 2(C_{kl}^z)^{-1} \big\{ 192(S_i^z)^2(S_j^z)^2$
$\qquad + [192k_{11}^z(S_i^z)^2 - 48(k_{11}^z)^2 S_i^z]S_j^z + 96(k_{11}^z)^2(S_i^z)^2 \big\} \Big\},$

$KL_{2,8} = KL_{13}^z = EI_z \Big\{ 3(C_{kl}^z)^{-1} \big\{ [384(S_i^z)^2 + 192k_{11}^z S_i^z + 96(k_{11}^z)^2](S_j^z)^2$
$\qquad + [192k_{11}^z(S_i^z)^2 - 96(k_{11}^z)^2 S_i^z]S_j^z + 96(k_{11}^z)^2(S_i^z)^2 \big\} \Big\},$

$KL_{2,12} = KL_{14}^z = EI_z \Big\{ 2(C_{kl}^z)^{-1} \big\{ [192(S_i^z)^2 + 192k_{11}^z S_i^z + 96(k_{11}^z)^2](S_j^z)^2$
$\qquad - 48(k_{11}^z)^2 S_i^z S_j^z \big\} \Big\},$

$KL_{3,3} = KL_{11}^y = EI_y \Big\{ 3(C_{kl}^y)^{-1} \big\{ [384(S_i^y)^2 + 192k_{11}^y S_i^y + 96(k_{11}^y)^2](S_j^y)^2$
$\qquad + [192k_{11}^y(S_i^y)^2 - 96(k_{11}^y)^2 S_i^y]S_j^y + 96(k_{11}^y)^2(S_i^y)^2 \big\} \Big\},$

$KL_{3,5} = KL_{12}^y = -EI_y \Big\{ 2(C_{kl}^y)^{-1} \big\{ 192(S_i^y)^2(S_j^y)^2$
$\qquad + [192k_{11}^y(S_i^y)^2 - 48(k_{11}^y)^2 S_i^y]S_j^y + 96(k_{11}^y)^2(S_i^y)^2 \big\} \Big\},$

$KL_{3,9} = KL_{13}^y = EI_y \Big\{ 3(C_{kl}^y)^{-1} - \big\{ [384(S_i^y)^2 + 192k_{11}^y S_i^y + 96(k_{11}^y)^2](S_j^y)^2$
$\qquad + [192k_{11}^y(S_i^y)^2 - 96(k_{11}^y)^2 S_i^y]S_j^y + 96(k_{11}^y)^2(S_i^y)^2 \big\} \Big\},$

$KL_{3,11} = KL_{14}^y = -EI_y \Big\{ 2(C_{kl}^y)^{-1} \big\{ [192(S_i^y)^2 + 192k_{11}^y S_i^y + 96(k_{11}^y)^2](S_j^y)^2$
$\qquad - 48(k_{11}^y)^2 S_i^y S_j^y \big\} \Big\},$

$KL_{4,4} = GJ/L,$

$KL_{4,7} = -GJ/L,$

$KL_{5,5} = KL_{22}^y = EI_y \Big\{ 1(C_{kl}^y)^{-1}$
$\qquad \times [128(S_i^y)^2(S_j^y)^2 + 192k_{11}^y(S_i^y)^2 S_j^y + 96(k_{11}^y)^2(S_i^y)^2] \Big\},$

$KL_{5,9} = KL_{23}^y = -EI_y \Big\{ 2(C_{kl}^y)^{-1} - \big\{ 192(S_i^y)^2(S_j^y)^2$
$\qquad + [192k_{11}^y(S_i^y)^2 - 48(k_{11}^y)^2 S_i^y]S_j^y + 96(k_{11}^y)^2(S_i^y)^2 \big\} \Big\},$

$$KL_{5,11} = KL_{24}^y = EI_y\{{_1(C_{kl}^y)}^{-1}[64(S_i^y)^2(S_j^y)^2 - 48(k_{11}^y)^2 S_i^y S_j^y]\},$$

$$KL_{6,6} = KL_{22}^z = EI_z\{{_1(C_{kl}^z)}^{-1} \\ \times [128(S_i^z)^2(S_j^z)^2 + 192k_{11}^z(S_i^z)^2 S_j^z + 96(k_{11}^z)^2(S_i^z)^2]\},$$

$$KL_{6,8} = KL_{23}^z = EI_z\{{_2(C_{kl}^z)}^{-1} - \{192(S_i^z)^2(S_j^z)^2 \\ + [192k_{11}^z(S_i^z)^2 - 48(k_{11}^z)^2 S_i^z]S_j^z + 96(k_{11}^z)^2(S_i^z)^2\}\},$$

$$KL_{6,12} = KL_{24}^z = EI_z\{{_1(C_{kl}^z)}^{-1}[64(S_i^z)^2(S_j^z)^2 - 48(k_{11}^z)^2 S_i^z S_j^z]\},$$

$$KL_{7,7} = EA/L,$$

$$KL_{8,8} = KL_{33}^z = EI_z\{{_3(C_{kl}^z)}^{-1}\{[384(S_i^z)^2 + 192k_{11}^z S_i^z + 96(k_{11}^z)^2](S_j^z)^2 \\ + [192k_{11}^z(S_i^z)^2 - 96(k_{11}^z)^2 S_i^z]S_j^z + 96(k_{11}^z)^2(S_i^z)^2\}\},$$

$$KL_{8,12} = KL_{34}^z = EI_z\{{_2(C_{kl}^z)}^{-1} - \{[192(S_i^z)^2 + 192k_{11}^z S_i^z + 96(k_{11}^z)^2](S_j^z)^2 \\ - 48(k_{11}^z)^2 S_i^z S_j^z\}\},$$

$$KL_{9,9} = KL_{33}^y = EI_y\{{_3(C_{kl}^y)}^{-1}\{[384(S_i^y)^2 + 192k_{11}^y S_i^y + 96(k_{11}^y)^2](S_j^y)^2 \\ + [192k_{11}^y(S_i^y)^2 - 96(k_{11}^y)^2 S_i^y]S_j^y + 96(k_{11}^y)^2(S_i^y)^2\}\},$$

$$KL_{9,11} = KL_{34}^y = -EI_y\{{_2(C_{kl}^y)}^{-1} - \{[192(S_i^y)^2 + 192k_{11}^y S_i^y \\ + 96(k_{11}^y)^2](S_i^y)^2 - 48(k_{11}^y)^2 S_i^y S_j^y\}\},$$

$$KL_{10,10} = GJ/L,$$

$$KL_{11,11} = KL_{44}^y = EI_y\{{_1(C_{kl}^y)}^{-1}[128(S_i^y)^2 + 192k_{11}^y S_i^y + 96(k_{11}^y)^2](S_j^y)^2\},$$

$$KL_{12,12} = KL_{44}^z = EI_z\{{_1(C_{kl}^z)}^{-1}[128(S_i^z)^2 + 192k_{11}^z S_i^z + 96(k_{11}^z)^2](S_j^z)^2\},$$

where

$$k_{11}^z = 4EI_z/L,$$
$$k_{11}^y = 4EI_y/L,$$

$$_1(C_{kl}^y) = 2\{[16L(S_i^y)^2 + 32k_{11}^y L S_i^y + 16(k_{11}^y)^2 L](S_j^y)^2 \\ + [32k_{11}^y L(S_i^y)^2 + 56(k_{11}^y)^2 L S_i^y + 24(k_{11}^y)^3 L]S_j^y \\ + 16(k_{11}^y)^2 L(S_i^y)^2 + 24(k_{11}^y)^3 L S_i^y + 9(k_{11}^y)^4 L\},$$

$$_2(C_{kl}^y) = 2\{[16L^2(S_i^y)^2 + 32k_{11}^y L^2 S_i^y + 16(k_{11}^y)^2 L^2](S_j^y)^2 \\ + [32k_{11}^y L^2(S_i^y)^2 + 56(k_{11}^y)^2 L^2 S_i^y + 24(k_{11}^y)^3 L^2]S_j^y \\ + 16(k_{11}^y)^2 L^2(S_i^y)^2 + 24(k_{11}^y)^3 L^2 S_i^y + 9(k_{11}^y)^4 L^2\},$$

$$_3(C_{kl}^y) = 2\{[16L^3(S_i^y)^2 + 32k_{11}^y L^3 S_i^y + 16(k_{11}^y)^2 L^3](S_j^y)^2 \\ + [32k_{11}^y L^3(S_i^y)^2 + 56(k_{11}^y)^2 L^3 S_i^y + 24(k_{11}^y)^3 L^3]S_j^y \\ + 16(k_{11}^y)^2 L^3(S_i^y)^2 + 24(k_{11}^y)^3 L^3 S_i^y + 9(k_{11}^y)^4 L^3\},$$

Appendix A

$$\begin{aligned}
{}_1(C_{kl}^z) = {}&2\Big\{\big[16L(S_i^z)^2 + 32k_{11}^z L S_i^z + 16(k_{11}^z)^2 L\big](S_j^z)^2\\
&+ \big[32k_{11}^z L(S_i^z)^2 + 56(k_{11}^z)^2 L S_i^z + 24(k_{11}^z)^3 L\big]S_j^z\\
&+ 16(k_{11}^z)^2 L(S_i^z)^2 + 24(k_{11}^z)^3 L S_i^z + 9(k_{11}^z)^4 L\Big\},\\
{}_2(C_{kl}^z) = {}&2\Big\{\big[16L^2(S_i^z)^2 + 32k_{11}^z L^2 S_i^z + 16(k_{11}^z)^2 L^2\big](S_j^z)^2\\
&+ \big[32k_{11}^z L^2(S_i^z)^2 + 56(k_{11}^z)^2 L^2 S_i^z + 24(k_{11}^z)^3 L^2\big]S_j^z\\
&+ 16(k_{11}^z)^2 L^2(S_i^z)^2 + 24(k_{11}^z)^3 L^2 S_i^z + 9(k_{11}^z)^4 L^2\Big\},\\
{}_3(C_{kl}^z) = {}&2\Big\{\big[16L^3(S_i^z)^2 + 32k_{11}^z L^3 S_i^z + 16(k_{11}^z)^2 L^3\big](S_j^z)^2\\
&+ \big[32k_{11}^z L^3(S_i^z)^2 + 56(k_{11}^z)^2 L^3 S_i^z + 24(k_{11}^z)^3 L^3\big]S_j^z\\
&+ 16(k_{11}^z)^2 L^3(S_i^z)^2 + 24(k_{11}^z)^3 L^3 S_i^z + 9(k_{11}^z)^4 L^3\Big\}.
\end{aligned}$$

A.2. The 12 × 12 geometric stiffness matrix, $[K_G]$

The 12 × 12 geometric stiffness matrix, $[KG]$ is symmetric. The non-zero terms in the upper triangle (in local co-ordinates) are given below.

$$KG_{1,1} = 0$$

$$\begin{aligned}
KG_{2,2} = KG_{11}^z = {}&P\Big\{{}_1(C_{kg}^z)^{-1}\big\{\big[192(S_i^z)^2 + 336k_{11}^z S_i^z + 192(k_{11}^z)^2\big](S_j^z)^2\\
&+ \big[336k_{11}^z(S_i^z)^2 + 504(k_{11}^z)^2 S_i^z + 240(k_{11}^z)^3\big]S_j^z\\
&+ 192(k_{11}^z)^2(S_i^z)^2 + 240(k_{11}^z)^3 S_i^z + 90(k_{11}^z)^4\big\}\Big\},
\end{aligned}$$

$$\begin{aligned}
KG_{2,6} = KG_{12}^z = {}&P\Big\{{}_2(C_{kg}^z)^{-1}\big\{\big[16(S_i^z)^2 - 16k_{11}^z S_i^z\big](S_j^z)^2\\
&+ \big[32k_{11}^z(S_i^z)^2 - 28(k_{11}^z)^2 S_i^z\big]S_j^z + 32(k_{11}^z)^2(S_i^z)^2\big\}\Big\},
\end{aligned}$$

$$\begin{aligned}
KG_{2,8} = KG_{13}^z = {}&P\Big\{{}_1(C_{kg}^z)^{-1} - \big\{\big[192(S_i^z)^2 + 336k_{11}^z S_i^z + 192(k_{11}^z)^2\big](S_j^z)^2\\
&+ \big[336k_{11}^z(S_i^z)^2 + 504(k_{11}^z)^2 S_i^z + 240(k_{11}^z)^3\big]S_j^z\\
&+ 192(k_{11}^z)^2(S_i^z)^2 + 240(k_{11}^z)^3 S_i^z + 90(k_{11}^z)^4\big\}\Big\},
\end{aligned}$$

$$\begin{aligned}
KG_{2,12} = KG_{14}^z = {}&P\Big\{{}_2(C_{kg}^z)^{-1}\big\{\big[16(S_i^z)^2 + 32k_{11}^z S_i^z + 32(k_{11}^z)^2\big](S_j^z)^2\\
&+ \big[-16k_{11}^z(S_i^z)^2 - 28(k_{11}^z)^2 S_i^z\big]S_j^z\big\}\Big\},
\end{aligned}$$

$$\begin{aligned}
KG_{3,3} = KG_{11}^y = {}&P\Big\{{}_1(C_{kg}^y)^{-1}\big\{\big[192(S_i^y)^2 + 336k_{11}^y S_i^y + 192(k_{11}^y)^2\big](S_j^y)^2\\
&+ \big[336k_{11}^y(S_i^y)^2 + 504(k_{11}^y)^2 S_i^y + 240(k_{11}^y)^3\big]S_j^y\\
&+ 192(k_{11}^y)^2(S_i^y)^2 + 240(k_{11}^y)^3 S_i^y + 90(k_{11}^y)^4\big\}\Big\},
\end{aligned}$$

$$\begin{aligned}
KG_{3,5} = KG_{12}^y = {}&-P\Big\{{}_2(C_{kg}^y)^{-1}\big\{\big[16(S_i^y)^2 - 16k_{11}^y S_i^y\big](S_j^y)^2\\
&+ \big[32k_{11}^y(S_i^y)^2 - 28(k_{11}^y)^2 S_i^y\big]S_j^y + 32(k_{11}^y)^2(S_i^y)^2\big\}\Big\},
\end{aligned}$$

$$KG_{3,9} = KG_{13}^y = P\left\{_1(C_{kg}^y)^{-1} - \left\{[192(S_i^y)^2 + 336k_{11}^y S_i^y + 192(k_{11}^y)^2](S_j^y)^2\right.\right.$$
$$\left.\left. + [336k_{11}^y(S_i^y)^2 + 504(k_{11}^y)^2 S_i^y + 240(k_{11}^y)^3]S_j^y\right.\right.$$
$$\left.\left. + 192(k_{11}^y)^2(S_i^y)^2 + 240(k_{11}^y)^3 S_i^y + 90(k_{11}^y)^4\right\}\right\},$$

$$KG_{3,11} = KG_{14}^y = -P\left\{_2(C_{kg}^y)^{-1}\left\{[16(S_i^y)^2 + 32k_{11}^y S_i^y + 32(k_{11}^y)^2](S_j^y)^2\right.\right.$$
$$\left.\left. + [-16k_{11}^y(S_i^y)^2 - 28(k_{11}^y)^2 S_i^y]S_j^y\right\}\right\},$$

$$KG_{4,4} = Pr^2/L,$$
$$KG_{4,8} = P/L,$$
$$KG_{4,9} = P/L,$$

$$KG_{5,5} = KG_{12}^y = P\left\{_3(C_{kg}^y)^{-1}[64L(S_i^y)^2(S_j^y)^2 + 144k_{11}^y L(S_i^y)^2 S_j^y\right.$$
$$\left. + 96(k_{11}^y)^2 L(S_i^y)^2]\right\},$$

$$KG_{5,9} = KG_{12}^y = -P\left\{_2(C_{kg}^y)^{-1} - \left\{[16(S_i^y)^2 - 16k_{11}^y S_i^y](S_j^y)^2\right.\right.$$
$$\left.\left. + [32k_{11}^y(S_i^y)^2 - 28(k_{11}^y)^2 S_i^y]S_j^y + 32(k_{11}^y)^2(S_i^y)^2\right\}\right\},$$

$$KG_{5,11} = KG_{14}^y = P\left\{_3(C_{kg}^y)^{-1} - \left\{[16L(S_i^y)^2 + 48k_{11}^y LS_i^y](S_j^y)^2\right.\right.$$
$$\left.\left. + [48k_{11}^y L(S_i^y)^2 + 84(k_{11}^y)^2 LS_i^y]S_j^y\right\}\right\},$$

$$KG_{6,6} = KG_{12}^z = P\left\{_3(C_{kg}^z)^{-1}[64L(S_i^z)^2(S_j^z)^2 + 144k_{11}^z L(S_i^z)^2 S_j^z\right.$$
$$\left. + 96(k_{11}^z)^2 L(S_i^z)^2]\right\},$$

$$KG_{6,8} = KG_{12}^z = P\left\{_2(C_{kg}^z)^{-1} - \left\{[16(S_i^z)^2 - 16k_{11}^z S_i^z](S_j^z)^2\right.\right.$$
$$\left.\left. + [32k_{11}^z(S_i^z)^2 - 28(k_{11}^z)^2 S_i^z]S_j^z + 32(k_{11}^z)^2(S_i^z)^2\right\}\right\},$$

$$KG_{6,12} = KG_{14}^z = P\left\{_3(C_{kg}^z)^{-1}\left\{[16L(S_i^z)^2 + 48k_{11}^z LS_i^z](S_j^z)^2\right.\right.$$
$$\left.\left. + [48k_{11}^z L(S_i^z)^2 + 84(k_{11}^z)^2 LS_i^z]S_j^z\right\}\right\},$$

$$KG_{7,7} = 0,$$

$$KG_{8,8} = KG_{13}^z = P\left\{_1(C_{kg}^z)^{-1}\left\{[192(S_i^z)^2 + 336k_{11}^z S_i^z + 192(k_{11}^z)^2](S_j^z)^2\right.\right.$$
$$\left.\left. + [336k_{11}^z(S_i^z)^2 + 504(k_{11}^z)^2 S_i^z + 240(k_{11}^z)^3]S_j^z\right.\right.$$
$$\left.\left. + 192(k_{11}^z)^2(S_i^z)^2 + 240(k_{11}^z)^3 S_i^z + 90(k_{11}^z)^4\right\}\right\},$$

$$KG_{8,12} = KG_{14}^z = P\left\{_1(C_{kg}^z)^{-1} - \left\{[16(S_i^z)^2 + 32k_{11}^z S_i^z + 32(k_{11}^z)^2](S_j^z)^2\right.\right.$$
$$\left.\left. + [-16k_{11}^z(S_i^z)^2 - 28(k_{11}^z)^2 S_i^z]S_j^z\right\}\right\},$$

$$KG_{9,9} = KG_{13}^y = P\left\{_1(C_{kg}^y)^{-1}\left\{[192(S_i^y)^2 + 336k_{11}^y S_i^y + 192(k_{11}^y)^2](S_j^y)^2\right.\right.$$
$$\left.\left. + [336k_{11}^y(S_i^y)^2 + 504(k_{11}^y)^2 S_i^y + 240(k_{11}^y)^3]S_j^y\right.\right.$$
$$\left.\left. + 192(k_{11}^y)^2(S_i^y)^2 + 240(k_{11}^y)^3 S_i^y + 90(k_{11}^y)^4\right\}\right\},$$

$$KG_{9,11} = KG_{14}^y = P\left\{_1(C_{kg}^z)^{-1} - \left\{[16(S_i^y)^2 + 32k_{11}^y S_i^y + 32(k_{11}^y)^2](S_j^y)^2\right.\right.$$
$$\left.\left. + [-16k_{11}^y(S_i^y)^2 - 28(k_{11}^y)^2 S_i^y]S_j^y\right\}\right\},$$

$$KG_{10,10} = Pr^2/L,$$
$$KG_{11,11} = KG_{14}^y = P\left\{3(C_{kg}^y)^{-1}[64L(S_i^y)^2 + 144k_{11}^y LS_i^y + 96(k_{11}^y)^2 L](S_j^y)^2\right\},$$
$$KG_{12,12} = KG_{14}^z = P\left\{3(C_{kg}^z)^{-1}[64L(S_i^z)^2 + 144k_{11}^z LS_i^z + 96(k_{11}^z)^2 L](S_j^z)^2\right\},$$

where

$$k_{11}^z = 4EI_z/L,$$
$$k_{11}^y = 4EI_y/L,$$
$$_1(C_{kg}^z) = 2\Big\{[80L(S_i^z)^2 + 160k_{11}^z LS_i^z + 80(k_{11}^z)^2 L](S_j^z)^2 \\ + [160k_{11}^z L(S_i^z)^2 + 280(k_{11}^z)^2 LS_i^z + 120(k_{11}^z)^3 L]S_j^z \\ + 80(k_{11}^z)^2 L(S_i^z)^2 + 120(k_{11}^z)^3 LS_i^z + 45(k_{11}^z)^4 L\Big\},$$
$$_2(C_{kg}^z) = 2\Big\{[80(S_i^z)^2 + 160k_{11}^z S_i^z + 80(k_{11}^z)^2](S_j^z)^2 \\ + [160k_{11}^z (S_i^z)^2 + 280(k_{11}^z)^2 S_i^z + 120(k_{11}^z)^3]S_j^z \\ + 80(k_{11}^z)^2 (S_i^z)^2 + 120(k_{11}^z)^3 S_i^z + 45(k_{11}^z)^4\Big\},$$
$$_3(C_{kg}^z) = 2\Big\{[240(S_i^z)^2 + 480k_{11}^z S_i^z + 240(k_{11}^z)^2](S_j^z)^2 \\ + [480k_{11}^z (S_i^z)^2 + 840(k_{11}^z)^2 S_i^z + 360(k_{11}^z)^3]S_j^z \\ + 240(k_{11}^z)^2 (S_i^z)^2 + 360(k_{11}^z)^3 S_i^z + 135(k_{11}^z)^4\Big\},$$
$$_1(C_{kg}^y) = 2\Big\{[80L(S_i^y)^2 + 160k_{11}^y LS_i^y + 80(k_{11}^y)^2 L](S_j^y)^2 \\ + [160k_{11}^y L(S_i^y)^2 + 280(k_{11}^y)^2 LS_i^y + 120(k_{11}^y)^3 L]S_j^y \\ + 80(k_{11}^y)^2 L(S_i^y)^2 + 120(k_{11}^y)^3 LS_i^y + 45(k_{11}^y)^4 L\Big\},$$
$$_2(C_{kg}^y) = 2\Big\{[80(S_i^y)^2 + 160k_{11}^y S_i^y + 80(k_{11}^y)^2](S_j^y)^2 \\ + [160k_{11}^y (S_i^y)^2 + 280(k_{11}^y)^2 S_i^y + 120(k_{11}^y)^3]S_j^y \\ + 80(k_{11}^y)^2 (S_i^y)^2 + 120(k_{11}^y)^3 S_i^y + 45(k_{11}^y)^4\Big\},$$
$$_3(C_{kg}^y) = 2\Big\{[240(S_i^y)^2 + 480k_{11}^y S_i^y + 240(k_{11}^y)^2](S_j^y)^2 \\ + [480k_{11}^y (S_i^y)^2 + 840(k_{11}^y)^2 S_i^y + 360(k_{11}^y)^3]S_j^y \\ + 240(k_{11}^y)^2 (S_i^y)^2 + 360(k_{11}^y)^3 S_i^y + 135(k_{11}^y)^4\Big\}.$$

A.3. The 12 × 12 consistent mass matrix, [M]

The 12 × 12 consistent mass matrix, $[M]$, is symmetric. The non-zero terms in the upper triangle (in local co-ordinates) are given below.

$$M_{1,1} = 280\overline{m}L/420$$
$$M_{1,7} = 140\overline{m}L/420$$

$$M_{2,2} = M_{11}^z = \overline{m}\Big\{{}_1(C_m^z)^{-1}\{[416L(S_i^z)^2 + 656k_{11}^z LS_i^z + 264(k_{11}^z)^2 L](S_j^z)^2$$
$$+ [936k_{11}^z L(S_i^z)^2 + 1312(k_{11}^z)^2 LS_i^z + 462(k_{11}^z)^3 L]S_j^z$$
$$+ 544(k_{11}^z)^2 L(S_i^z)^2 + 672(k_{11}^z)^3 LS_i^z + 210(k_{11}^z)^4 L\}\Big\},$$

$$M_{2,6} = M_{12}^z = \overline{m}\Big\{{}_1(C_m^z)^{-1}\{[176L^2(S_i^z)^2 + 128k_{11}^z L^2 S_i^z](S)^2$$
$$+ [440k_{11}^z L^2(S_i^z)^2 + 288(k_{11}^z)^2 L^2 S_i^z]S_j^z$$
$$+ 288(k_{11}^z)^2 L^2(S_i^z)^2 + 168(k_{11}^z)^3 L^2 S_i^z\}\Big\},$$

$$M_{2,8} = M_{13}^z = \overline{m}\Big\{{}_1(C_m^z)^{-1}\{[144L(S_i^z)^2 + 324k_{11}^z LS_i^z + 156(k_{11}^z)^2 L](S_j^z)^2$$
$$+ [324k_{11}^z L(S_i^z)^2 + 648(k_{11}^z)^2 LS_i^z + 273(k_{11}^z)^3 L]S_j^z$$
$$+ 156(k_{11}^z)^2 L(S_i^z)^2 + 273(k_{11}^z)^3 LS_i^z + 105(k_{11}^z)^4 L\}\Big\},$$

$$M_{2,12} = M_{14}^z = \overline{m}\Big\{{}_2(C_m^z)^{-1}$$
$$\times \{-[104L^2(S_i^z)^2 + 260k_{11}^z L^2 S_i^z + 132(k_{11}^z)^2 L^2](S_j^z)^2$$
$$+ [152k_{11}^z L^2(S_i^z)^2 + 342(k_{11}^z)^2 L^2 S_i^z + 147(k_{11}^z)^3 L^2]S_j^z\}\Big\},$$

$$M_{3,3} = M_{11}^y = \overline{m}\Big\{{}_1(C_m^y)^{-1}\{[416L(S_i^y)^2 + 656k_{11}^y LS_i^y + 264(k_{11}^y)^2 L](S_j^y)^2$$
$$+ [936k_{11}^y L(S_i^y)^2 + 1312(k_{11}^y)^2 LS_i^y + 462(k_{11}^y)^3 L]S_j^y$$
$$+ 544(k_{11}^y)^2 L(S_i^y)^2 + 672(k_{11}^y)^3 LS_i^y + 210(k_{11}^y)^4 L\}\Big\},$$

$$M_{3,5} = M_{12}^y = -\overline{m}\Big\{{}_2(C_m^y)^{-1}\{[176L^2(S_i^y)^2 + 128k_{11}^y L^2 S_i^y](S)^2$$
$$+ [440k_{11}^y L^2(S_i^y)^2 + 288(k_{11}^y)^2 L^2 S_i^y]S_j^y$$
$$+ 288(k_{11}^y)^2 L^2(S_i^y)^2 + 168(k_{11}^y)^3 L^2 S_i^y\}\Big\},$$

$$M_{3,9} = M_{13}^y = \overline{m}\Big\{{}_1(C_m^y)^{-1}\{[144L(S_i^y)^2 + 324k_{11}^y LS_i^y + 156(k_{11}^y)^2 L](S_j^y)^2$$
$$+ [324k_{11}^y L(S_i^y)^2 + 648(k_{11}^y)^2 LS_i^y + 273(k_{11}^y)^3 L]S_j^y$$
$$+ 156(k_{11}^y)^2 L(S_i^y)^2 + 273(k_{11}^y)^3 LS_i^y + 105(k_{11}^y)^4 L\}\Big\},$$

$$M_{3,11} = M_{14}^y = -\overline{m}\Big\{{}_2(C_m^y)^{-1}$$
$$\times \{-[104L^2(S_i^y)^2 + 260k_{11}^y L^2 S_i^y + 132(k_{11}^y)^2 L^2](S_j^y)^2$$
$$+ [152k_{11}^y L^2(S_i^y)^2 + 342(k_{11}^y)^2 L^2 S_i^y + 147(k_{11}^y)^3 L^2]S_j^y\}\Big\},$$

$$M_{4,4} = 280\overline{m}LJ/(420A),$$
$$M_{4,10} = 140\overline{m}LJ/(420A),$$

$$M_{5,5} = M_{22}^y = \overline{m}\Big\{{}_2(C_m^y)^{-1}$$
$$\times [32L^3(S_i^y)^2(S_j^y)^2 + 88k_{11}^y L^3(S_i^y)^2 S_j^y + 64(k_{11}^y)^2 L^3(S_i^y)^2]\Big\},$$

$$M_{5,9} = M_{23}^y = -\overline{m}\Big\{{}_2(C_m^y)^{-1}\{[104L^2(S_i^y)^2 + 152k_{11}^y L^2 S_i^y](S_j^y)^2$$
$$+ [260k_{11}^y L^2(S_i^y)^2 + 342(k_{11}^y)^2 L^2 S_i^y]S_j^y$$
$$+ 132(k_{11}^y)^2 L^2(S_i^y)^2 + 147(k_{11}^y)^3 L^2 S_i^y\}\Big\},$$

Appendix A

$$M_{5,11} = M_{24}^y = \overline{m}\Big\{2(C_m^y)^{-1}\big\{-[24L^3(S_i^y)^2 + 40k_{11}^y L^3 S_i^y](S_j^y)^2$$
$$+ [40k_{11}^y L^3 (S_i^y)^2 + 62(k_{11}^y)^2 L^3 S_i^y]S_j^y\big\}\Big\},$$

$$M_{6,6} = M_{22}^z = \overline{m}\Big\{2(C_m^z)^{-1}$$
$$\times [32L^3(S_i^z)^2(S_j^z)^2 + 88k_{11}^z L^3(S_i^z)^2 S_j^z + 64(k_{11}^z)^2 L^3(S_i^z)^2]\Big\},$$

$$M_{6,8} = M_{22}^z = \overline{m}\Big\{2(C_m^z)^{-1}\big\{[104L^2(S_i^z)^2 + 152k_{11}^z L^2 S_i^z](S_j^z)^2$$
$$+ [260k_{11}^z L^2(S_i^z)^2 + 342(k_{11}^z)^2 L^2 S_i^z]S_j^z$$
$$+ 132(k_{11}^z)^2(S_i^z)^2 + 147(k_{11}^z)^3 L^2 S_i^z\big\}\Big\},$$

$$M_{6,12} = M_{22}^z = \overline{m}\Big\{2(C_m^z)^{-1}\big\{-[24L^3(S_i^z)^2 + 40k_{11}^z L^3 S_i^z](S_j^z)^2$$
$$+ [40k_{11}^z L^3(S_i^z)^2 + 62(k_{11}^z)^2 L^3 S_i^z]S_j^z\big\}\Big\},$$

$$M_{7,7} = 280\overline{m}L/420,$$

$$M_{8,8} = M_{33}^z = \overline{m}\Big\{{}_1(C_m^z)^{-1}\big\{[416L(S_i^z)^2 + 936k_{11}^z L S_i^z + 544(k_{11}^z)^2 L](S_j^z)^2$$
$$+ [656k_{11}^z L(S_i^z)^2 + 1312(k_{11}^z)^2 L S_i^z + 672(k_{11}^z)^3 L]S_j^z$$
$$+ 264(k_{11}^z)^2 L(S_i^z)^2 + 462(k_{11}^z)^3 L S_i^z + 210(k_{11}^z)^4 L\big\}\Big\},$$

$$M_{8,12} = M_{34}^z = \overline{m}\Big\{2(C_m^z)^{-1}$$
$$\times \big\{-[176L^2(S_i^z)^2 + 440k_{11}^z L^2 S_i^z + 288(k_{11}^z)^2 L^2](S_j^z)^2$$
$$+ [128k_{11}^z L^2(S_i^z)^2 + 288(k_{11}^z)^2 L^2 S_i^z + 168(k_{11}^z)^3 L^2]S_j^z\big\}\Big\},$$

$$M_{9,9} = M_{33}^y = \overline{m}\Big\{{}_1(C_m^y)^{-1}\big\{[416L(S_i^y)^2 + 936k_{11}^y L S_i^y + 544(k_{11}^y)^2 L](S_j^y)^2$$
$$+ [656k_{11}^y L(S_i^y)^2 + 1312(k_{11}^y)^2 L S_i^y + 672(k_{11}^y)^3 L]S_j^y$$
$$+ 264(k_{11}^y)^2 L(S_i^y)^2 + 462(k_{11}^y)^3 L S_i^y + 210(k_{11}^y)^4 L\big\}\Big\},$$

$$M_{9,11} = M_{34}^y = -\overline{m}\Big\{2(C_m^y)^{-1}$$
$$\times \big\{-[176L^2(S_i^y)^2 + 440k_{11}^y L^2 S_i^y + 288(k_{11}^y)^2 L^2](S_j^y)^2$$
$$+ [128k_{11}^y L^2(S_i^y)^2 + 288(k_{11}^y)^2 L^2 S_i^y + 168(k_{11}^y)^3 L^2]S_j^y\big\}\Big\},$$

$$M_{10,10} = 280\overline{m}LJ/(420A),$$

$$M_{11,11} = M_{44}^y = \overline{m}\Big\{2(C_m^y)^{-1}\big\{[32L^3(S_i^y)^2 + 88k_{11}^y L^3 S_i^y + 64(k_{11}^y)^2 L^3](S_j^y)^2\big\}\Big\},$$

$$M_{12,12} = M_{44}^z = \overline{m}\Big\{2(C_m^z)^{-1}\big\{[32L^3(S_i^z)^2 + 88k_{11}^z L^3 S_i^z + 64(k_{11}^z)^2 L^3](S_j^z)^2\big\}\Big\},$$

where

$$k_{11}^z = 4EI_z/L,$$
$$k_{11}^y = 4EI_y/L,$$
$${}_1(C_m^z) = 2\big[560(S_i^z)^2 + 1120k_{11}^z S_i^z + 560(k_{11}^z)^2\big](S_j^z)^2$$
$$+ [1120k_{11}^z(S_i^z)^2 + 1960(k_{11}^z)^2 S_i^z + 840(k_{11}^z)^3]S_j^z$$
$$+ 560(k_{11}^z)^2(S_i^z)^2 + 840(k_{11}^z)^3 S_i^z + 315(k_{11}^z)^4,$$

$$_2(C_m^z) = 2\big[1680(S_i^z)^2 + 3360k_{11}^z S_i^z + 1680(k_{11}^z)^2\big](S_j^z)^2$$
$$+ \big[3360k_{11}^z(S_i^z)^2 + 5880(k_{11}^z)^2 S_i^z + 2520(k_{11}^z)^3\big]S_j^z$$
$$+ 1680(k_{11}^z)^2(S_i^z)^2 + 2520(k_{11}^z)^3 S_i^z + 945(k_{11}^z)^4,$$

$$_1(C_m^y) = 2\big[560(S_i^y)^2 + 1120k_{11}^y S_i^y + 560(k_{11}^y)^2\big](S_j^y)^2$$
$$+ \big[1120k_{11}^y(S_i^y)^2 + 1960(k_{11}^y)^2 S_i^y + 840(k_{11}^y)^3\big]S_j^y$$
$$+ 560(k_{11}^y)^2(S_i^y)^2 + 840(k_{11}^y)^3 S_i^y + 315(k_{11}^y)^4,$$

$$_2(C_m^y) = 2\big[1680(S_i^y)^2 + 3360k_{11}^y S_i^y + 1680(k_{11}^y)^2\big](S_j^y)^2$$
$$+ \big[3360k_{11}^y(S_i^y)^2 + 5880(k_{11}^y)^2 S_i^y + 2520(k_{11}^y)^3\big]S_j^y$$
$$+ 1680(k_{11}^y)^2(S_i^y)^2 + 2520(k_{11}^y)^3 S_i^y + 945(k_{11}^y)^4.$$

Appendix B

USER'S MANUAL OF GMNAF

B.1. Introduction

This Chapter aims to give readers an instruction menu for the understanding and the use of the developed computer program GMNAF-ED (Geometric and Material Nonlinear Analysis of Frames for EDucational use), which can be downloaded from website http://www.cse.polyu.edu.hk/SLCHAN/chan.htm. This web address is valid at time of publication. However, no guarantees are provided regarding access in perpetuity.

This computer program uses an incremental-iterative Newton–Raphson procedure to trace the load versus deflection path of a framed structure. In summary, the tangent stiffness is used to predict a displacement increment due to an applied load as

$$[\Delta u] = [K_T]^{-1}[\Delta F], \tag{B.1}$$

in which $[\Delta u]$ is the displacement increment due to the applied load $[\Delta F]$ and $[K_T]$ is the tangent stiffness matrix.

The displacement increment is then added to the total accumulated displacement to obtained the updated displacement and also to the geometry of the structure to obtain the updated geometry as

$$[u]_{i+1} = [u]_i + [\Delta u]_i, \tag{B.2}$$

$$[x]^j_{i+1} = [x]^j_i + [\Delta u]_i, \tag{B.3}$$

in which the subscript refers to a particular iteration or increment and the superscript is the coordinate axes ($j = 1, 2$ and 3 for X, Y and Z coordinate).

The internal force increment, $[\Delta R_e]$, is then computed as

$$[\Delta R_e] = [k_e][\Delta u], \tag{B.4}$$

in which $[k_e]$ is the element stiffness matrix.

Hence the updated resistance for the element can be determined as

$$[R_e]_{i+1} = [R_e]_i + [\Delta R_e]. \tag{B.5}$$

The total resistance, $[R]$, can be computed by a standard assembling process and compared against the applied load as

$$[R] = \sum_{}^{N_{\text{ele}}} [R_e], \tag{B.6}$$

$$[\Delta F] = [F] - [R], \tag{B.7}$$

in which $[F]$ is the applied load, N_{ele} is the number of elements in the system and $[R]$ is the resistance of the complete structure. $[\Delta F]$ is the unbalanced, out-of-balanced or residual forces.

The unbalanced force will then be substituted into Eq. (B.1) and the process is repeated until the norms as follows are less than a certain tolerance.

$$\begin{aligned} [\Delta F]^{\text{T}}[\Delta F] &< Toler * [F]^{\text{T}}[F] \\ [\Delta u]^{\text{T}}[\Delta u] &< Toler * [u]^{\text{T}}[u] \end{aligned} \tag{B.8}$$

Notes

(1) In the Newton–Raphson method, the tangent stiffness matrix $[K_{\text{T}}]$ in equation (B.1) is reformed every iteration and in the Modified Newton–Raphson method, it is only reformed in the first iteration. In the program, the number of iteration is assigned as a parameter and when it is infinite and 1, the method is automatically converted to the Newton–Raphson and the modified Newton–Raphson methods, respectively.
(2) As the Newton–Raphson method or its modified form iterates at a constant load, it diverges when a limit point is encountered. To this end, the minimum residual displacement method is used to traverse the limit point.

B.2. Data preparation

The data file must be fully compatible with the read statement in the program. It must begin with a title which can be several lines and terminated by an asterisk *, which indicates that it is the end of the title. After this *, the input data follows. The following line number ignores those lines occupied by the data lines.

1st line (2 data)
(1) NMEM   No. of elements
(2) NNODE   No. of nodes

2nd line (7 data)
(1) NCYC   No. of load cycles
(2) NITER   No. of iterations allowed for each cycles. When the iteration number exceeds this value, it will go to the next load cycle irrespective of whether or not the convergence has been satisfied
(3) NTAN   No. of iterations for the tangent stiffness to reform
(4) TRILAM   Load factor
(5) YSTES   Material yield stress
(6) IFLAG8   Activation of line search technique (put 0 in this program)
(7) IFLAG2   Activation of the BFGS method (put 0 in this program)

Appendix B

3rd line (5 data)

(1) IFLAG7   A number for control of nonlinear solution method. Only 11 and 43 is allowed of which the first one means the Newton–Raphson method and the second one means the Minimum Residual Displacement method. If this number is 111 or 143, it means a non-proportional load case is applied. The Dead Load vector in the following will first be imposed in the first load cycle and the carrying load factor will then be imposed and incremented in subsequent cycles

(2) LOADPA   The desired number of iterations in each iteration. This is used to control the load size

(3) CONTPA   The maximum distance allowed to be travelled by each load step

(4) MRECYC   A parameter to control re-solution (put 0 in this program)

(5) MRETAN   Another parameter to control re-solution (put 0 in this program)

4th line to (Number of elements + 3)-th line, 12 data each line

(1) ELAST   Young's modulus of elasticity
(2) GRIG   Shear modulus of elasticity
(3) AREA   Cross sectional area
(4) YIY   Second moment of area about y-axis
(5) ZIZ   Second moment of area about z-axis
(6) PIP   Polar moment of area
(7) JOT1   Node number for joint 1
(8) JOT2   Node number for joint 2
(9) IIY   Connection type of joint 1 about y-axis
(10) IIZ   Connection type of joint 1 about z-axis
(11) JJY   Connection type of joint 2 about y-axis
(12) JJZ   Connection type of joint 2 about z-axis

Note: For connection types from (9) to (12), each number is composed of 3 digit integer (e.g. -170, 260). The first number (-1 for -170 and 2 for 260) refers to the connection type and the second two digits (70 for -170 and 60 for 260) refer to the percentage (%) of the applied moment to the plastic moment causing the moment vs. rotation curve to bend.

Line (Number of elements + 4) to line ($2 \times$ Number of elements + 3), 5 data each line

(1) MPY   Plastic modulus about y-axis
(2) MPZ   Plastic modulus about z-axis
(3) PY   Cross sectional area
(4) K_L_R   Slenderness ratio (kL/r)
(5) UPE   Twisted angle of the principle axis to the local element axis (in degree)

Line ($2 \times$ No. of elements + 4) to line ($2 \times$ No. of elements + No. of nodes + 3), 3 data each line

(1) XNODE X coordinate
(2) YNODE Y coordinate
(3) ZNODE Z coordinate

Line ($2 \times$ No. of elements + No. of nodes + 4) to line ($2 \times$ No. of elements + $2 \times$ No. of nodes + 3), 6 data each line

(1) BORI $((I-1)*\text{NFREE}+1)$ Force in X-axis
(2) BORI $((I-1)*\text{NFREE}+2)$ Force in Y-axis
(3) BORI $((I-1)*\text{NFREE}+3)$ Force in Z-axis
(4) BORI $((I-1)*\text{NFREE}+4)$ Moment in X-axis
(5) BORI $((I-1)*\text{NFREE}+5)$ Moment in Y-axis
(6) BORI $((I-1)*\text{NFREE}+6)$ Moment in Z-axis

Note: This set of force is the increasing set of load, in contrast to the below. *NFREE* is the number of degrees of freedom at a node and is set to be 6 for three dimensional problem in the program. I is the node number.

Line ($2 \times$ No. of elements + $2 \times$ No. of nodes + 4) to line ($2 \times$ No. of elements + $3 \times$ No. of nodes + 3), 6 data each line

(1)	FBORI $((I-1)*\text{NFREE}+1)$	Force in X-axis
(2)	FBORI $((I-1)*\text{NFREE}+2)$	Force in Y-axis
(3)	FBORI $((I-1)*\text{NFREE}+3)$	Force in Z-axis
(4)	FBORI $((I-1)*\text{NFREE}+4)$	Moment in X-axis
(5)	FBORI $((I-1)*\text{NFREE}+5)$	Moment in Y-axis
(6)	FBORI $((I-1)*\text{NFREE}+6)$	Moment in Z-axis

Note: When the first number in line 3 exceeds 100, the present set of load vector will be applied only in the first load cycle using the constant load method. After the first load cycle, the previous set of load will be incremented.

Line (2 × No. of elements + 3 × No. of nodes + 4) to line (2×No. of elements + 4×No. of nodes + 3), 6 data each line

(1)	KBOUND $((I-1)*\text{NFREE}+1)$	Translational boundary condition in X-axis
(2)	KBOUND $((I-1)*\text{NFREE}+2)$	Translational boundary condition in Y-axis
(3)	KBOUND $((I-1)*\text{NFREE}+3)$	Translational boundary condition in Z-axis
(4)	KBOUND $((I-1)*\text{NFREE}+4)$	Rotational boundary condition about X-axis
(5)	KBOUND $((I-1)*\text{NFREE}+5)$	Rotational boundary condition about Y-axis
(6)	KBOUND $((I-1)*\text{NFREE}+6)$	Rotational boundary condition about Z-axis

Note: Set 0 for allowing movement in the direction considered while 1 for not allowing movement.

B.3. Activation of the program

In the data file, you should have a 3L.EXE, DRAW.EXE and the program GMNAF-ED.EXE. Copy them all in a directory.

After preparation of the data file of file extension .DAT type GMNAF-ED to activate the program. It will then ask you for the data file name. Type in your prepared data file name and the program should start running (try to type in 6story for trial run).

After the analysis is completed, type 3L for plotting of the load vs. deflection curve. Use the cursor to move to the position of your input data file name and then press return, you will then see a screen asking for node and x, y and z rotations and translational displacement. Use TAB to move to the desired position and then press the SPACEBAR for a X to appear inside the bracket. Press CRTL-D to draw the curve. You can also print the curve by pressing H. Type ESC key repeatedly to quit the plotting program.

You can also view the simulation process of the analysis results. Type DRAW in DOS environment and the program will ask you for the file name. Type in your file name (without extension) and the structure will then appear on screen. Use the menu driven command behind to rotate etc. the structure until the desired position is obtained. Press A for animation and, if you want to see once more, press n to get the structure back to original position and then press A again.

B.4. Example

An example is selected to demonstrate the application of the developed computer program. The problem is the well-known Vogel's six-story frame which is an European calibrating frame for inelastic analysis models. The result has been plotted in Fig. 6.43 and

Appendix B

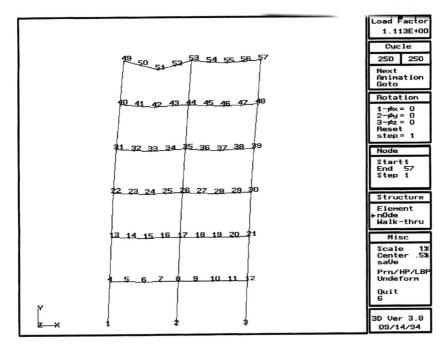

Fig. B.1. Graphical animation of Vogel frame.

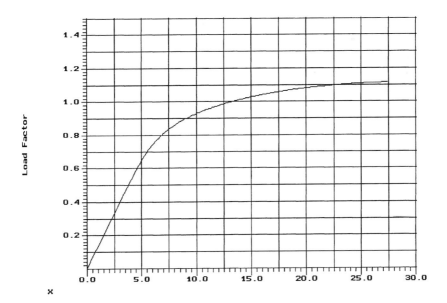

Fig. B.2. Load–deflection curve of Vogel frame.

it can be observed that the present result is compared well with those obtained by Vogel (1985).

In executing the program GMNAF-ED.EXE, a data input file name is asked and just type "6story" and press RETURN. The program is then running until the prescribed number of cycles has been performed and it stops. The output files named "6story.out" and "6story.pgd" are automatically generated. These files can be used for checking whether the input data is typed correctly using DRAW and other text editors. The second file can be further used to animate the deformation of the structure step by step using DRAW as illustrated in Fig. B.1 and to plot the load-deflection curve at node 57 in X-direction using 3L as indicated in Fig. B.2. The content of the data file is shown as follows for reference.

Data input filename: 6story.dat
Vogel's six-story frame (2 bays)
*

66 57
250 50 3.1 23.5 0 0
43 3 0.5 0 0

2.05e4	1.e1	9.10e+1	8.091e+3	8.091e+3	76.6e+8	1	4	−150	−150	−150	−150
2.05e4	1.e1	8.45e+1	2.313e+4	2.313e+4	51.1e+8	4	5	−170	−170	−170	−170
2.05e4	1.e1	8.45e+1	2.313e+4	2.313e+4	51.1e+8	5	6	−170	−170	−170	−170
2.05e4	1.e1	8.45e+1	2.313e+4	2.313e+4	51.1e+8	6	7	−170	−170	−170	−170
2.05e4	1.e1	8.45e+1	2.313e+4	2.313e+4	51.1e+8	7	8	−170	−170	−170	−170
2.05e4	1.e1	1.18e+2	1.492e+4	1.492e+4	124.0e+8	2	8	−150	−150	−150	−150
2.05e4	1.e1	8.45e+1	2.313e+4	2.313e+4	51.1e+8	8	9	−170	−170	−170	−170
2.05e4	1.e1	8.45e+1	2.313e+4	2.313e+4	51.1e+8	9	10	−170	−170	−170	−170
2.05e4	1.e1	8.45e+1	2.313e+4	2.313e+4	51.1e+8	10	11	−170	−170	−170	−170
2.05e4	1.e1	8.45e+1	2.313e+4	2.313e+4	51.1e+8	11	12	−170	−170	−170	−170
2.05e4	1.e1	9.10e+1	8.091e+3	8.091e+3	76.6e+8	3	12	−150	−150	−150	−150
2.05e4	1.e1	9.10e+1	8.091e+3	8.091e+3	76.6e+8	4	13	−150	−150	−150	−150
2.05e4	1.e1	7.27e+1	1.627e+4	1.627e+4	37.3e+8	13	14	−170	−170	−170	−170
2.05e4	1.e1	7.27e+1	1.627e+4	1.627e+4	37.3e+8	14	15	−170	−170	−170	−170
2.05e4	1.e1	7.27e+1	1.627e+4	1.627e+4	37.3e+8	15	16	−170	−170	−170	−170
2.05e4	1.e1	7.27e+1	1.627e+4	1.627e+4	37.3e+8	16	17	−170	−170	−170	−170
2.05e4	1.e1	1.18e+2	1.492e+4	1.492e+4	124.0e+8	8	17	−150	−150	−150	−150
2.05e4	1.e1	7.27e+1	1.627e+4	1.627e+4	37.3e+8	17	18	−170	−170	−170	−170
2.05e4	1.e1	7.27e+1	1.627e+4	1.627e+4	37.3e+8	18	19	−170	−170	−170	−170
2.05e4	1.e1	7.27e+1	1.627e+4	1.627e+4	37.3e+8	19	20	−170	−170	−170	−170
2.05e4	1.e1	7.27e+1	1.627e+4	1.627e+4	37.3e+8	20	21	−170	−170	−170	−170
2.05e4	1.e1	9.10e+1	8.091e+3	8.091e+3	76.6e+8	12	21	−150	−150	−150	−150
2.05e4	1.e1	9.10e+1	8.091e+3	8.091e+3	76.6e+8	13	22	−150	−150	−150	−150
2.05e4	1.e1	6.26e+1	1.177e+4	1.177e+4	28.1e+8	22	23	−170	−170	−170	−170
2.05e4	1.e1	6.26e+1	1.177e+4	1.177e+4	28.1e+8	23	24	−170	−170	−170	−170
2.05e4	1.e1	6.26e+1	1.177e+4	1.177e+4	28.1e+8	24	25	−170	−170	−170	−170
2.05e4	1.e1	6.26e+1	1.177e+4	1.177e+4	28.1e+8	25	26	−170	−170	−170	−170
2.05e4	1.e1	1.06e+2	1.126e+4	1.126e+4	103.0e+8	17	26	−150	−150	−150	−150
2.05e4	1.e1	6.26e+1	1.177e+4	1.177e+4	28.1e+8	26	27	−170	−170	−170	−170
2.05e4	1.e1	6.26e+1	1.177e+4	1.177e+4	28.1e+8	27	28	−170	−170	−170	−170
2.05e4	1.e1	6.26e+1	1.177e+4	1.177e+4	28.1e+8	28	29	−170	−170	−170	−170
2.05e4	1.e1	6.26e+1	1.177e+4	1.177e+4	28.1e+8	29	30	−170	−170	−170	−170
2.05e4	1.e1	9.10e+1	8.091e+3	8.091e+3	76.6e+8	21	30	−150	−150	−150	−150

Appendix B

2.05e4	1.e1	9.10e+1	8.091e+3	8.091e+3	76.6e+8	22	31	−150	−150	−150	−150
2.05e4	1.e1	5.38e+1	8.356e+3	8.356e+3	20.1e+8	31	32	−170	−170	−170	−170
2.05e4	1.e1	5.38e+1	8.356e+3	8.356e+3	20.1e+8	32	33	−170	−170	−170	−170
2.05e4	1.e1	5.38e+1	8.356e+3	8.356e+3	20.1e+8	33	34	−170	−170	−170	−170
2.05e4	1.e1	5.38e+1	8.356e+3	8.356e+3	20.1e+8	34	35	−170	−170	−170	−170
2.05e4	1.e1	1.06e+2	1.126e+4	1.126e+4	103.0e+8	26	35	−150	−150	−150	−150
2.05e4	1.e1	5.38e+1	8.356e+3	8.356e+3	20.1e+8	35	36	−170	−170	−170	−170
2.05e4	1.e1	5.38e+1	8.356e+3	8.356e+3	20.1e+8	36	37	−170	−170	−170	−170
2.05e4	1.e1	5.38e+1	8.356e+3	8.356e+3	20.1e+8	37	38	−170	−170	−170	−170
2.05e4	1.e1	5.38e+1	8.356e+3	8.356e+3	20.1e+8	38	39	−170	−170	−170	−170
2.05e4	1.e1	9.10e+1	8.091e+3	8.091e+3	76.6e+8	30	39	−150	−150	−150	−150
2.05e4	1.e1	5.43e+1	2.492e+3	2.492e+3	31.2e+8	31	40	−150	−150	−150	−150
2.05e4	1.e1	5.38e+1	8.356e+3	8.356e+3	20.1e+8	40	41	−170	−170	−170	−170
2.05e4	1.e1	5.38e+1	8.356e+3	8.356e+3	20.1e+8	41	42	−170	−170	−170	−170
2.05e4	1.e1	5.38e+1	8.356e+3	8.356e+3	20.1e+8	42	43	−170	−170	−170	−170
2.05e4	1.e1	5.38e+1	8.356e+3	8.356e+3	20.1e+8	43	44	−170	−170	−170	−170
2.05e4	1.e1	7.81e+1	5.696e+3	5.696e+3	59.3e+8	35	44	−150	−150	−150	−150
2.05e4	1.e1	5.38e+1	8.356e+3	8.356e+3	20.1e+8	44	45	−170	−170	−170	−170
2.05e4	1.e1	5.38e+1	8.356e+3	8.356e+3	20.1e+8	45	46	−170	−170	−170	−170
2.05e4	1.e1	5.38e+1	8.356e+3	8.356e+3	20.1e+8	46	47	−170	−170	−170	−170
2.05e4	1.e1	5.38e+1	8.356e+3	8.356e+3	20.1e+8	47	48	−170	−170	−170	−170
2.05e4	1.e1	5.43e+1	2.492e+3	2.492e+3	31.2e+8	39	48	−150	−150	−150	−150
2.05e4	1.e1	5.43e+1	2.492e+3	2.492e+3	31.2e+8	40	49	−150	−150	−150	−150
2.05e4	1.e1	3.91e+1	3.982e+3	3.982e+3	20.1e+8	49	50	−170	−170	−170	−170
2.05e4	1.e1	3.91e+1	3.982e+3	3.982e+3	20.1e+8	50	51	−170	−170	−170	−170
2.05e4	1.e1	3.91e+1	3.982e+3	3.982e+3	20.1e+8	51	52	−170	−170	−170	−170
2.05e4	1.e1	3.91e+1	3.982e+3	3.982e+3	20.1e+8	52	53	−170	−170	−170	−170
2.05e4	1.e1	7.81e+1	5.696e+3	5.696e+3	59.3e+8	44	53	−150	−150	−150	−150
2.05e4	1.e1	3.91e+1	3.982e+3	3.982e+3	12.9e+8	53	54	−170	−170	−170	−170
2.05e4	1.e1	3.91e+1	3.982e+3	3.982e+3	12.9e+8	54	55	−170	−170	−170	−170
2.05e4	1.e1	3.91e+1	3.982e+3	3.982e+3	12.9e+8	55	56	−170	−170	−170	−170
2.05e4	1.e1	3.91e+1	3.982e+3	3.982e+3	20.1e+8	56	57	−170	−170	−170	−170
2.05e4	1.e1	5.43e+1	2.492e+3	2.492e+3	31.2e+8	48	57	−150	−150	−150	−150

827.	827.	91.	15.	0.
1307.	1307.	84.5	15.	0.
1307.	1307.	84.5	15.	0.
1307.	1307.	84.5	15.	0.
1307.	1307.	84.5	15.	0.
1283.	1283.	118.	15.	0.
1307.	1307.	84.5	15.	0.
1307.	1307.	84.5	15.	0.
1307.	1307.	84.5	15.	0.
1307.	1307.	84.5	15.	0.
827.	827.	91.	15.	0.
827.	827.	91.	15.	0.
1019.	1019.	72.7	15.	0.
1019.	1019.	72.7	15.	0.
1019.	1019.	72.7	15.	0.
1019.	1019.	72.7	15.	0.
1283.	1283.	118.	15.	0.
1019.	1019.	72.7	15.	0.
1019.	1019.	72.7	15.	0.
1019.	1019.	72.7	15.	0.

1019.	1019.	72.7	15.	0.
827.	827.	91.	15.	0.
827.	827.	91.	15.	0.
804.	804.	62.6	15.	0.
804.	804.	62.6	15.	0.
804.	804.	62.6	15.	0.
804.	804.	62.6	15.	0.
1053.	1053.	106.	15.	0.
804.	804.	62.6	15.	0.
804.	804.	62.6	15.	0.
804.	804.	62.6	15.	0.
804.	804.	62.6	15.	0.
827.	827.	91.	15.	0.
827.	827.	91.	15.	0.
628.	628.	53.8	15.	0.
628.	628.	53.8	15.	0.
628.	628.	53.8	15.	0.
628.	628.	53.8	15.	0.
1053.	1053.	106.	15.	0.
628.	628.	53.8	15.	0.
628.	628.	53.8	15.	0.
628.	628.	53.8	15.	0.
628.	628.	53.8	15.	0.
827.	827.	91.	15.	0.
354.	354.	54.3	15.	0.
628.	628.	53.8	15.	0.
628.	628.	53.8	15.	0.
628.	628.	53.8	15.	0.
628.	628.	53.8	15.	0.
643.	643.	78.1	15.	0.
628.	628.	53.8	15.	0.
628.	628.	53.8	15.	0.
628.	628.	53.8	15.	0.
628.	628.	53.8	15.	0.
354.	354.	54.3	15.	0.
354.	354.	54.3	15.	0.
367.	367.	39.1	15.	0.
367.	367.	39.1	15.	0.
367.	367.	39.1	15.	0.
367.	367.	39.1	15.	0.
643.	643.	78.1	15.	0.
367.	367.	39.1	15.	0.
367.	367.	39.1	15.	0.
367.	367.	39.1	15.	0.
367.	367.	39.1	15.	0.
354.	354.	54.3	15.	0.

0.	0.	0.
6.000000e + 2	0.	0.
12.000000e + 2	0.	0.
0.008333e + 2	3.750000e + 2	0.
1.508333e + 2	3.750000e + 2	0.
3.008333e + 2	3.750000e + 2	0.
4.508333e + 2	3.750000e + 2	0.

6.008333e + 2	3.750000e + 2	0.
7.508333e + 2	3.750000e + 2	0.
9.008333e + 2	3.750000e + 2	0.
10.508333e + 2	3.750000e + 2	0.
12.008333e + 2	3.750000e + 2	0.
.016667e + 2	7.500000e + 2	0.
1.516667e + 2	7.500000e + 2	0.
3.016667e + 2	7.500000e + 2	0.
4.516667e + 2	7.500000e + 2	0.
6.016667e + 2	7.500000e + 2	0.
7.516667e + 2	7.500000e + 2	0.
9.016667e + 2	7.500000e + 2	0.
10.516667e + 2	7.500000e + 2	0.
12.016667e + 2	7.500000e + 2	0.
.025000e + 2	11.250000e + 2	0.
1.525000e + 2	11.250000e + 2	0.
3.025000e + 2	11.250000e + 2	0.
4.525000e + 2	11.250000e + 2	0.
6.025000e + 2	11.250000e + 2	0.
7.525000e + 2	11.250000e + 2	0.
9.025000e + 2	11.250000e + 2	0.
10.525000e + 2	11.250000e + 2	0.
12.025000e + 2	11.250000e + 2	0.
.033333e + 2	15.000000e + 2	0.
1.533333e + 2	15.000000e + 2	0.
3.033333e + 2	15.000000e + 2	0.
4.533333e + 2	15.000000e + 2	0.
6.033333e + 2	15.000000e + 2	0.
7.533333e + 2	15.000000e + 2	0.
9.033333e + 2	15.000000e + 2	0.
10.533333e + 2	15.000000e + 2	0.
12.033333e + 2	15.000000e + 2	0.
.041667e + 2	18.750000e + 2	0.
1.541667e + 2	18.750000e + 2	0.
3.041667e + 2	18.750000e + 2	0.
4.541667e + 2	18.750000e + 2	0.
6.041667e + 2	18.750000e + 2	0.
7.541667e + 2	18.750000e + 2	0.
9.041667e + 2	18.750000e + 2	0.
10.541667e + 2	18.750000e + 2	0.
12.041667e + 2	18.750000e + 2	0.
.050000e + 2	22.500000e + 2	0.
1.550000e + 2	22.500000e + 2	0.
3.050000e + 2	22.500000e + 2	0.
4.550000e + 2	22.500000e + 2	0.
6.050000e + 2	22.500000e + 2	0.
7.550000e + 2	22.500000e + 2	0.
9.050000e + 2	22.500000e + 2	0.
10.550000e + 2	22.500000e + 2	0.
12.050000e + 2	22.500000e + 2	0.

0.	0.	0.	0.	0.	0.
0.	0.	0.	0.	0.	0.
0.	0.	0.	0.	0.	0.

20.44	−36.825	0.	0.	0.	0.
0.	−73.65	0.	0.	0.	0.
0.	−73.65	0.	0.	0.	0.
0.	−73.65	0.	0.	0.	0.
0.	−73.65	0.	0.	0.	0.
0.	−73.65	0.	0.	0.	0.
0.	−73.65	0.	0.	0.	0.
0.	−73.65	0.	0.	0.	0.
0.	−36.825	0.	0.	0.	0.
20.44	−36.825	0.	0.	0.	0.
0.	−73.65	0.	0.	0.	0.
0.	−73.65	0.	0.	0.	0.
0.	−73.65	0.	0.	0.	0.
0.	−73.65	0.	0.	0.	0.
0.	−73.65	0.	0.	0.	0.
0.	−73.65	0.	0.	0.	0.
0.	−73.65	0.	0.	0.	0.
0.	−36.825	0.	0.	0.	0.
20.44	−36.825	0.	0.	0.	0.
0.	−73.65	0.	0.	0.	0.
0.	−73.65	0.	0.	0.	0.
0.	−73.65	0.	0.	0.	0.
0.	−73.65	0.	0.	0.	0.
0.	−73.65	0.	0.	0.	0.
0.	−73.65	0.	0.	0.	0.
0.	−73.65	0.	0.	0.	0.
0.	−36.825	0.	0.	0.	0.
20.44	−36.825	0.	0.	0.	0.
0.	−73.65	0.	0.	0.	0.
0.	−73.65	0.	0.	0.	0.
0.	−73.65	0.	0.	0.	0.
0.	−73.65	0.	0.	0.	0.
0.	−73.65	0.	0.	0.	0.
0.	−73.65	0.	0.	0.	0.
0.	−73.65	0.	0.	0.	0.
0.	−36.825	0.	0.	0.	0.
20.44	−36.825	0.	0.	0.	0.
0.	−73.65	0.	0.	0.	0.
0.	−73.65	0.	0.	0.	0.
0.	−73.65	0.	0.	0.	0.
0.	−73.65	0.	0.	0.	0.
0.	−73.65	0.	0.	0.	0.
0.	−73.65	0.	0.	0.	0.
0.	−73.65	0.	0.	0.	0.
0.	−36.825	0.	0.	0.	0.
10.23	−23.775	0.	0.	0.	0.
0.	−47.55	0.	0.	0.	0.
0.	−47.55	0.	0.	0.	0.
0.	−47.55	0.	0.	0.	0.
0.	−47.55	0.	0.	0.	0.
0.	−47.55	0.	0.	0.	0.
0.	−73.65	0.	0.	0.	0.
0.	−73.65	0.	0.	0.	0.
0.	−73.65	0.	0.	0.	0.

0.	−73.65	0.	0.	0.	0.
0.	−73.65	0.	0.	0.	0.
0.	−36.825	0.	0.	0.	0.
10.23	−23.775	0.	0.	0.	0.
0.	−47.55	0.	0.	0.	0.
0.	−47.55	0.	0.	0.	0.
0.	−47.55	0.	0.	0.	0.
0.	−47.55	0.	0.	0.	0.
0.	−47.55	0.	0.	0.	0.
0.	−47.55	0.	0.	0.	0.
0.	−47.55	0.	0.	0.	0.
0.	−23.775	0.	0.	0.	0.
0.	0.	0.	0.	0.	0.
0.	0.	0.	0.	0.	0.
0.	0.	0.	0.	0.	0.
0.	0.	0.	0.	0.	0.
0.	0.	0.	0.	0.	0.
0.	0.	0.	0.	0.	0.
0.	0.	0.	0.	0.	0.
0.	0.	0.	0.	0.	0.
0.	0.	0.	0.	0.	0.
0.	0.	0.	0.	0.	0.
0.	0.	0.	0.	0.	0.
0.	0.	0.	0.	0.	0.
0.	0.	0.	0.	0.	0.
0.	0.	0.	0.	0.	0.
0.	0.	0.	0.	0.	0.
0.	0.	0.	0.	0.	0.
0.	0.	0.	0.	0.	0.
0.	0.	0.	0.	0.	0.
0.	0.	0.	0.	0.	0.
0.	0.	0.	0.	0.	0.
0.	0.	0.	0.	0.	0.
0.	0.	0.	0.	0.	0.
0.	0.	0.	0.	0.	0.
0.	0.	0.	0.	0.	0.
0.	0.	0.	0.	0.	0.
0.	0.	0.	0.	0.	0.
0.	0.	0.	0.	0.	0.
0.	0.	0.	0.	0.	0.
0.	0.	0.	0.	0.	0.
0.	0.	0.	0.	0.	0.
0.	0.	0.	0.	0.	0.
0.	0.	0.	0.	0.	0.
0.	0.	0.	0.	0.	0.
0.	0.	0.	0.	0.	0.
0.	0.	0.	0.	0.	0.
0.	0.	0.	0.	0.	0.

0.	0.	0.	0.	0.	0.
0.	0.	0.	0.	0.	0.
0.	0.	0.	0.	0.	0.
0.	0.	0.	0.	0.	0.
0.	0.	0.	0.	0.	0.
0.	0.	0.	0.	0.	0.
0.	0.	0.	0.	0.	0.
0.	0.	0.	0.	0.	0.
0.	0.	0.	0.	0.	0.
0.	0.	0.	0.	0.	0.
0.	0.	0.	0.	0.	0.
0.	0.	0.	0.	0.	0.
0.	0.	0.	0.	0.	0.
0.	0.	0.	0.	0.	0.
0.	0.	0.	0.	0.	0.

1	1	1	1	1	1
1	1	1	1	1	1
1	1	1	1	1	1
0	0	1	1	1	0
0	0	1	1	1	0
0	0	1	1	1	0
0	0	1	1	1	0
0	0	1	1	1	0
0	0	1	1	1	0
0	0	1	1	1	0
0	0	1	1	1	0
0	0	1	1	1	0
0	0	1	1	1	0
0	0	1	1	1	0
0	0	1	1	1	0
0	0	1	1	1	0
0	0	1	1	1	0
0	0	1	1	1	0
0	0	1	1	1	0
0	0	1	1	1	0
0	0	1	1	1	0
0	0	1	1	1	0
0	0	1	1	1	0
0	0	1	1	1	0
0	0	1	1	1	0
0	0	1	1	1	0
0	0	1	1	1	0
0	0	1	1	1	0
0	0	1	1	1	0
0	0	1	1	1	0
0	0	1	1	1	0
0	0	1	1	1	0
0	0	1	1	1	0
0	0	1	1	1	0
0	0	1	1	1	0
0	0	1	1	1	0

0	0	1	1	1	0
0	0	1	1	1	0
0	0	1	1	1	0
0	0	1	1	1	0
0	0	1	1	1	0
0	0	1	1	1	0
0	0	1	1	1	0
0	0	1	1	1	0
0	0	1	1	1	0
0	0	1	1	1	0
0	0	1	1	1	0
0	0	1	1	1	0
0	0	1	1	1	0
0	0	1	1	1	0
0	0	1	1	1	0
0	0	1	1	1	0
0	0	1	1	1	0
0	0	1	1	1	0
0	0	1	1	1	0

References

Vogel, U. (1985): Calibrating frames. *Stahlbau* **54**(October), 295–311.

LIST OF SYMBOLS

A	Cross-sectional area
c_1, c_2	Stability functions
$[C]$	Damping matrix
d, t, T, B, D	Dimensions of a section: depth of web, web thickness, flange thickness, breadth of flange, depth of section
$[D]$	Matrix of Hookean's constant for the material properties
E	Young's modulus of elasticity
E_t	Tangent modulus
$F, \{\Delta F\}$	External force and Unbalanced forces or incremental applied forces
G	Shear modulus of elasticity
I_w	Warping section constant
I_y, I_z	Second moments of area about the two principal axes
J	St. Venant torsion constant
$[K_T]$	Tangent stiffness matrix
$[K_G]$	Geometric stiffness matrix
$[K_L]$	Linear stiffness matrix
$[K_o]$	Large displacement matrix
$K_{11}, K_{12}, K_{21}, K_{22}$ or $K_{ii}, K_{ij}, K_{ji}, K_{jj}$	Elements of element flexural stiffness
L	Length of an element or member
L_o	Original length of an element or member
L_e	Effective length
$[L]$	The 12×12 local to global transformation matrix
ΔL	Axial lengthening/shortening
$\overline{m}$	Mass per unit length
M	Moment
ΔM	Moment increment
$\Delta M_c, \Delta M_e, \Delta M_s$	Incremental applied moments at the connection, element and section nodes, respectively
M_{pr}	Reduced plastic moment capacity of section in the presence of axial force
M_{er}	Reduced first yield moment in the presence of axial force
M_x, M_y, M_z	Applied moments about the x, y and z
m_x, m_y, m_z	Element moments about the x, y and z
M_p	Plastic moment capacity of cross-section
$[M]$	Mass matrix
$[N]$	12×12 translational matrix for nodal displacements
P	Axial force
P_{cr}	Euler's buckling load

List of symbols

P_s	Shear buckling load
P_y	Squash load of cross-section
r_p	Polar radius of gyration
$\{R\}, R$	Resistance of structure
S	Section spring stiffness
S_c^o	Initial connection stiffness
S_c	Tangent connection stiffness
S_s	Tangent stiffness of section spring
$[T]$	12×6 transformation matrix from axes joining the two end nodes to local co-ordinate system
Δt	Time step
t_1, t_2	Resultant spring stiffness of connection spring and section spring at end nodes 1 and 2
U	Strain energy stored within the volume of a continuum
$[\Delta u], [\Delta \dot{u}], [\Delta \ddot{u}]$	Incremental displacement, velocity and acceleration vectors, respectively
$\{u\}, \{\dot{u}\}, \{\ddot{u}\}$	Displacement, velocity and acceleration vectors
v	Lateral displacement function
V	External work done
vol	Volume of a continuum
$\Delta \ddot{x}_g$	Increment of absolute ground acceleration
x, y, z	Element centroidal coordinate system
Z_e, Z_p	Elastic and plastic moduli
δ_a	Axial lengthening due to relative nodal displacements
Δ	Increment operator, or axial displacement, or deflection
Δ_b, δ_b	Axial lengthening due to bowing
λ	Load factor
Π	Total potential energy
ψ_o	Initial out-of-plumb
$\{\sigma\}$	Stress tensor
$\{\varepsilon\}$	Strain tensor
$\theta_x, \theta_y, \theta_z$	Rotations about x-, y- and z-axes, respectively
$\Delta \theta_c, \Delta \theta_b, \Delta \theta_s$	Incremental nodal rotations at the connection, section and beam nodes, respectively
τ_{xy}, τ_{xz}	Shear stress in y and z directions
σ_y, ε_y	Yield stress and corresponding yield strain
σ_r	Maximum residual stress
μ	Plasticity index
σ, ε	Stress and strain
μ_p	Plasticity index value
$\dot{\varepsilon}$	Strain rate
ϕ_c	Connection rotation
ϕ_p	Permanent rotation
ν	Poisson ratio
λ_{cr}	Buckling load factor

AUTHOR INDEX

[Plain numbers refer to text pages on which the author (or his/her work) is cited. Boldface numbers refer to the pages where bibliographic references are listed.]

Abbott, B.J.   99, 100, 112 **120** 202 **262** 294 **302**
Abdel-Rohman, M.   147 **192**
Abel, J.F.   137, 175, 186, 187 **192 193** 268 **301**
Akiyama, H.   265 **300**
Aktan, H.M.   99, 100 **121** 198, 204, 228 **263**
Al-Bermani, F.G.A.   99, 100, 107 **118 121** 125 **193** 197, 204, 226, 228 **260 263**
Al-Mashary, F.   127, 137 **191**
Alawa, M.S.   271 **302**
Almroth, B.O.   20 **27**
Alvarez, R.J.   125, 147 **191** 268 **300**
Alwis, W.A.M.   284–286 **301**
Anderson, D.   197 **260**
Ang, K.M.   95, 100–112 **118**
Antonio, M.   94 **119**
Arbabi, F.   100 **118**
Argyris, J.H.   77 **91**
Aspden, R.J.   271 **300**
Astaneh, A.   93 **120** 197 **262**
Atluri, S.N.   66 **76** 90 **92** 100 **120** 162 **193** 198, 204, 234 **262** 269 **301 302**
Atsuta   61 **61**
Avery, L.K.   100, 106 **119**
Azizinamini, A.   94, 109 **118** 179 **191** 196–198, 204 **260**

Baigent, A.H.   29 **52** 316 **323**
Bailey, J.R.   94 **118**
Baker, J.F.   100 **118**
Balendra, T.   267 **300 302**
Baron, M.L.   283, 284 **300**
Basu, P.K.   126 **191**
Bathe, K.J.   17, 20 **27** 47 **52** 59, 70 **75** 78, 83 **91** 206, 223 **260** 269 **300**
Batho, C.   100, 106, 112 **118** 204 **260**
Batoz, J.L.   78, 81 **91**
Beaulieu, D.   112 **118** 197 **262**

Beedle, L.S.   125 **193**
Bell, W.G.   94, 105 **118 120**
Benterkia, Z.   197 **260**
Benuska, K.L.   269 **300**
Berg, G.V.   269 **300**
Bergan, P.G.   78, 80 **91**
Bertero, V.V.   197 **262**
Birnstiel, C.   125, 147 **191 192** 268 **300**
Bjorhovde, R.   94 **118** 125 **192**
Blakeborough, A.   267 **300**
Bleich, H.H.   283, 284 **300**
Bockholdt, J.L.   270 **300**
Bolourchi, S.   70 **75** 223 **260**
Bourahla, N.   267 **300**
Bradburn, J.H.   94, 109 **118** 179 **191**
Brandes, J.L.   105 **118**
Brandt, V.   292 **302**
Bridge, R.Q.   126 **192**
Brinstiel, C.   270 **301**
Brun, P.   112 **118**
Brush, D.O.   20 **27**

Campbell, J.D.   271 **300**
Campus, F.   125 **193**
Cannon, J.C.   105 **119**
Cardinal, J.W.   101 **120**
Cardona, A.   223, 226 **260**
Carpenter, L.D.   197 **260**
Chajes, A.   20 **27**
Chan, S.L.   11, 12, 24 **27 28** 57, 66, 68, 70 **76** 77–79, 86, 90 **91** 99, 100 **118 120** 124, 125, 127, 142, 145 **192–194** 197, 207, 228, 243, 244, 247 **261–263** 268 **300 302**
Chandler, D.B.   133 **193**
Chen,   61 **76**
Chen, F.P.   270 **300**
Chen, P.F.S.   126 **192 193**
Chen, W.F.   9 **27** 41, 43 **52 53** 66, 68 **76** 95, 99–101, 106, 108–110 **118–121** 125–127,

	129–137, 142, 166, 169–171, 173, 178–180, 185, 187 **191 192–194** 197, 201, 204, 220, 226, 247, 249, 254 **261 262** 268, 294 **300–302**
Cheney, L.	125 **193** 268 **301**
Chesson, E.J.	94, 105, 112 **118–120**
Chu, K.H.	125 **192**
Chui, P.P.T.	142, 145 **192** 228 **261**
Clarke, M.J.	68 **76** 79 **91** 126 **192**
Clough, R.W.	25 **27** 210 **261** 269 **300**
Cohn, M.Z.	147 **192**
Colson, A.	99, 100, 106 **118** 204 **261**
Cosenza, E.	197 **261**
Cowper, G.R.	271, 277, 278 **301**
Cox, M.G.	100, 103 **118** 204 **261**
Crisfield, M.A.	78 **91**
DaDeppo, D.A.	269 **300**
Davison, J.B.	95 **118**
Deierlein, G.G.	175, 186, 187 **192** 268 **301**
Dhatt, G.	78, 81 **91**
Douty, R.T.	105 **118**
Dowling, P.J.	90 **91**
Dowrick, D.J.	255 **261**
Duan, L.	135, 137, 171, 173 **192**
Dussault, S.	211 **262**
Dvorkin, E.N.	78, 83 **91**
El-Zanaty, M.H.	125 **192**
Elnashai, A.S.	66 **76** 90 **91** 269 **301**
Faella, C.	197 **261**
Fafard, M.	197 **262**
Farhoomand, F.	271 **302**
Felippa, C.A.	46, 47 **52**
Filiatrault, A.	267 **301**
Fisher, G.P.	125 **193**
Fleming, J.F.	269 **301**
Freitas, J.A.T.	147 **192**
Frye, M.J.	100, 101, 103–105 **119** 204 **261**
Fujii, F.	68 **76**
Fujimoto, T.	100 **119** 198, 232, 233 **261**
Galambos, T.V.	125, 126 **192** 267, 268 **301 302**
Gao, L.	100 **119** 198, 202, 204 **261**
Gaylord, E.H.	78 **92**
Gebbeken, N.	292 **302**
George, A.	47 **52**
Geradin, M.	223, 226 **260**
Gere, J.M.	20 **28** 265 **301**
Gerstle, K.H.	100, 101 **119 120** 197, 204, 228, 230 **261 262**
Ghobarah, A.	197, 198 **261**
Giroux, Y.M.	112 **118**
Goldberg, J.E.	106 **119**
Goto, Y.	135 **192** 197 **261**
Goverdhan, A.V.	95 **119**
Guralnick, S.A.	127, 147 **192**
Hafez, M.A.	94 **119**
Haldar, A.	100 **119** 198, 202, 204 **261**
Hancock, G.J.	29 **52** 61, 68 **76** 79 **91** 126 **192** 316 **323**
Harstead, G.A.	147 **192**
He, J.	127, 147 **192**
Hechtman, R.A.	94, 105, 112 **119**
Heidebrecht, A.C.	269 **301**
Ho, G.W.M.	79 **91** 197 **261**
Hormby, D.E.	109 **120** 179 **193**
Horne, M.R.	20 **27**
Hou, F.Y.	125 **192**
Housner, G.W.	267 **301**
Howlett, J.N.	101 **119**
Hsiao, K.M.	125 **192**
Hsieh, S.H.	175, 186, 187 **192** 268 **301**
Huang, H.T.	100, 106 **119**
Iron, B.M.	46 **52**
Isobe, D.	286 **302**
Izzuddin, B.A.	66 **76** 90 **91** 125 **192** 269 **301**
Jackson, K.B.	94 **121**
Jaspart, J.P.	197 **261**
Jeffrey, P.K.	100, 106 **119**
Jenkins, W.M.	101 **119**
Jennings, A.	66 **76**
Jennings, P.C.	267 **301**
Johnson, L.G.	105 **119**
Johnson, N.D.	95, 109 **119** 179 **192**
Johnston, B.G.	94, 105, 112 **119** 125 **193** 268 **301**
Jonatowski, J.J.	270 **301**
Jones, S.W.	100, 103 **119** 204 **261**

Kam, T.Y.	290, 291 **301**	Lionberger, S.R.	101 **119**
Kaminsky, E.L.	125 **193**	Lipson, S.L.	94, 105, 112 **119**
Karamanlidis, D.	125 **193**	Liu, J.W.	47 **52**
Kassimali, A.	147 **193**	Liu, S.C.	283 **301**
Kato, S.	127 **193**	Livesley, R.K.	66 **76** 133 **193**
Kaur, D.	101 **119**	Lock, A.C.	78 **92**
Kawashima, S.	100 **119** 198, 232, 233 **261**	Loganathan, S.	125 **193**
Kennedy, D.J.L.	94 **119**	Louveau, J.M.	99, 100, 106 **118** 204 **261**
Ketter, R.L.	125 **192 193**	Lu, L.W.	**193** 197 **260**
Khozeimeh, K.	288, 290 **302**	Luca, D.A.	197 **261**
King, W.S.	100, 106 **119** 126, 134–136 **193** 204 **261**	Lui, E.M.	43 **53** 66, 68 **76** 99, 100, 101, 108–110 **119** 130, 139, 141, 142, 166, 178–180, 185, 187 **192 193** 197, 201, 204, 220, 226, 249, 254 **262** 294 **301**
Kirby, P.A.	95, 100, 103 **118–120** 204 **261**		
Kishi, N.	41 **53** 95, 100, 106, 108 **118 119** 201, 204 **261**		
Kitipornchai, S.	11, 12 **27** 66 **76** 99, 100, 107 **118 121** 125 **193** 197, 204, 226, 228 **260 263**	Mains, R.M.	105 **118**
		Majid	66 **76**
		Mallet, R.H.	68 **76**
		Manjoine, M.J.	271, 277 **301**
Koiter, W.T.	89 **91**	Marcal, P.V.	68 **76**
Kolsky, H.	271 **301**	Marley, M.J.	100, 101 **120** 197, 228, 230 **262**
Kolstein, M.H.	94 **121**		
Kondoh, K.	66 **76** 90 **92** 162 **193** 269 **301**	Mashary, F.A.	247 **262**
Koon, C.M.	57 **76**	Mason, R.E.	125 **193**
Korol, R.M.	197, 198 **261**	Massonnet, C.	125 **193**
Kouhia, R.	223, 225, 226 **261** 271 **301**	Maxwell, S.M.	101 **119**
Krawinkler, H.	197 **262**	McGuire, W.	78, 83 **92** 137, 175, 186, 187 **192 193** 268 **301**
Kreigh, J.D.	109 **120** 179 **193**		
Krishnamurthy, N.	100, 106 **119**	McKee, R.J.	94 **120**
Kuwamura, H.	267 **301**	Meek, J.L.	36 **53** 68 **76** 125 **193** 244 **262**
		Melchers, R.E.	101, 109 **119 121**
Lash, S.D.	100, 106 **118** 204 **260**	Merchant, W.	22 **28**
Lee, E.H.	269 **301**	Moncarz, P.D.	101 **120**
Lee, H.W.	12 **27**	Mondkar, D.P.	221, 223 **262**
Lee, S.L.	267, 269, 284–286 **300 301**	Monforton, A.R.	100 **120**
Leger, P.	211 **262** 267 **301**	Moore, D.B.	95 **120**
LeMessurier, A.P.	101 **119**	Moore, H.F.	94 **121**
Leu, K.C.	147 **192**	Morris, G.A.	95, 100, 101, 103–105, 111, 112 **118 119** 204 **261**
Lewitt, C.W.	112 **119**		
Li, B.	99, 100, 107 **118 121** 204, 226, 228 **260 263**	Mosquera, J.M.	271 **301**
		Munse, W.H.	94, 105, 112 **118–120**
Li, Y.	139, 141 **193**	Murray, D.M.	125 **192**
Liapunov, S.	139 **193**	Mutoh, I.	127 **193**
Liew, C.Y.	267 **302**		
Liew, J.Y.R.	126, 129, 131–134, 169–171, 173 **193**	Nader, M.N.	93 **120** 197 **262**
		Nagarajan, S.	283 **301**
		Narayanan, R.	14 **28**
Lightfoot, F.	101 **119**	Nedergaard, H.	126 **193**
Lim, E.L.	267 **300**	Nethercot, D.A.	95, 100, 103 **118–120** 201, 204 **261 262**
Lin, S.C.	290, 291 **301**		
Lin, T.H.	283 **301**	Newmark, N.M.	206, 211 **262**

Nigam, N.C.	270 **301**	Saka, T.	161 **194**
Noor, A.K.	270 **301**	Salvadori, M.G.	269 **302**
		Sam, M.T.	267 **302**
Okazawa, S.	68 **76**	Savard, M.	197 **262**
Olowokere, O.D.	99, 100 **121** 198, 204, 228 **263**	Scawthorn, C.	265 **302**
		Shah, H.C.	265 **301**
Oran, C.	66 **76** 79 **92**	Shanmugam, N.E.	101 **121**
Orbison, J.G.	137 **193**	Sharifi, P.	78 **92**
Osgood, W.R.	95, 100, 111 **120** 270, 294 **302**	Sharman, P.W.	29 **53**
		Sherbourne, A.N.	94, 105 **120**
Osman, A.	197, 198 **261**	Shi, G.	100 **120** 198, 204, 234 **262** 269 **302**
Ostrander, J.R.	94, 95, 105, 109 **120** 179 **193** 249 **262**	Shi, Y.J.	99 **120**
		Shomura, M.	127 **193**
		Shuku, Y.	161 **194**
Pabarcius, A.	125 **192**	Sivakumaran, K.S.	100 **120** 198, 204 **262**
Pan, A.D.	9 **27**	Smith, D.L.	125 **192**
Papadrakakis, M.	61 **76** 244 **262**	So, A.K.W.	207 **262**
Paz, M.	217, 218, 255 **262**	Sommer, W.H.	94, 95, 101, 105, 112 **120**
Pedersen, P.T.	126 **193**	Soroushian, P.	271 **302**
Penzien, J.	25 **27** 210 **261**	Spiliopoulos, K.V.	125 **192**
Peters, J.M.	270 **301**	Spooner, L.A.	105 **119**
Picard, A.	112 **118**	Stelmack, T.W.	100, 101 **120** 197, 228, 230 **262**
Pinkney, R.B.	196–198, 203 **262**		
Popov, E.P.	78 **92** 93 **120** 196–198, 203, 204 **262** 283 **301**	Stephen, R.M.	93 **120** 197 **262**
		Subramanian, C.V.	101 **120**
Powell, G.H.	126 **192 193** 221, 223 **262**	Sugimoto, H.	101 **120**
		Sun, S.	57 **76**
Radziminski, J.B.	94, 109 **118** 179 **191** 196, 197, 198, 204 **260**	Swaddiwudhipong, S.	284–286 **301**
		Symonds, P.S.	269, 271, 277, 278 **301**
Ramberg, W.	95, 100, 111 **120** 270, 294 **302**		
Ramm, E.	78 **92**	Tan, H.S.	68 **76** 244 **262**
Rankine, W.J.M.	22 **28**	Tang, D.	**261**
Rathbun, J.C.	94, 100, 105 **120**	Taniguchi, Y.	161 **194**
Rawlings, B.	269 **302**	Tarpy, T.S.	101 **120**
Razzaq, Z.	101 **120**	Thompson, L.E.	94 **120**
Remseth, S.N.	246, 247 **262**	Timoshenko, S.P.	20 **28**
Ribeiro, A.C.B.S.	147 **192**	Tin-Loi, F.	127, 147 **194** 271 **302**
Richard, R.M.	99, 100, 106, 109, 112 **119 120** 179 **193** 202 **262** 294 **302**	Tinawi, R.	267 **301**
		Toi, Y.	286 **302**
Riks, E.	78 **92**	Toma, S.	41 **52** 125, 126 **194** 247 **262** 268 **302**
Romstad, K.M.	101 **120**		
Rosenblueth, E.	265 **302**	Toridis, T.G.	288, 290 **302**
Rothert, H.	292 **302**	Trahair, N.S.	11 **28** 126 **192**
Rowan, H.C.	100, 112 **118**	Tsai, K.C.	196–198, 204 **262**
		Tuomala, M.	223–226 **261** 271 **301**
Sabir, A.B.	78 **92**		
Saigal, S.	125 **194** 221, 223 **263** 270, 283 **302**	Usami, T.	12 **28**
		Uzgider, E.A.	270, 277 **302**

van Kuren, R.C. 268 **302**
Vedero, V.T. 196, 198 **262**
Vimonsatit, V. 271 **302**
Vinnakota, S. 101 **120**
Visintainer, D.A. 94 **120**
Vogel, U. 29, 50 **52 53** 174, 175, 186 **194** 247 **263** 268 **302** 316 **323**

Walpole, W.R. 95, 109 **119** 179 **192**
Washizu, K. 37 **53**
Weaver, J.W. 101 **119** 270 **300**
Weidlinger, P. 269, 283, 284 **300 302**
Wempner, G.A. 78 **92**
Wen, R.K. 271 **302**
White, D.W. 125, 126, 129, 131–136, 169–171, 173, 175, 186 **193 194**
Williams, F.W. 87, 88 **92** 243 **263**
Wilson, E.L. 269 **300**
Wilson, W.M. 94 **121**
Winter, G. 125 **193**
Wong, H.P. 254 **263**
Wong, M.B. 127, 147 **194**
Wong, Y.L. 99 **120**
Wright, E.W. 78 **92**
Wu, F.H. 110 **121**

Wu, T.S. 100 **120**

Yamada, S. 265 **300**
Yanev, P.I. 265 **302**
Yang, T.Y. 125 **194** 221, 223 **263** 270, 283 **302**
Yang, Y.B. 78, 83 **92**
Yau, C.Y. 24 **28** 127 **194** 197, 247 **263** 268 **302**
Yee, Y.L. 109 **121**
Yen, T. 9 **27**
Young, C.R. 94 **121**
Youssef-Agha, W. 99, 100 **121** 198, 204, 228 **263**
Yu, C.H. 101 **121**

Zhou, S.P. 171, 173 **192** 197, 226 **261**
Zhou, Z.H. 66, 70 **76** 90 **91** 207, 243 **261 263**
Zhu, K. 99, 100, 107 **118 121** 204, 226, 228 **260 263**
Ziemian, R.D. 126, 138 **194**
Zienkiewicz, O.C. 17 **28** 68 **76** 269 **302**
Zoetemeijer, P. 94 **121**

SUBJECT INDEX

advanced analysis   24–26, 124, 191
allowable stress design method   1
analytical models   98–100
arc-length method   78, 83–85, 91

band-width technique   45
base shear forces   260
basic sectional parameters   143, 170
Bauschinger effect   272–274
beam–column connections   267, 281, 283
beam–column element   128, 129, 137, 148, 150, 153, 154, 156, 159, 162
beam–column joints   93
beam–column stiffness   130, 133, 134, 147
benchmark solutions   125, 138, 191
bi-linear strain-hardening model   274
bifurcation analysis   58, 59
bounding surface method   201
bowing effect   215
brittle materials   196
buckling load   59, 71, 73, 75, 162, 165–168, 188
buckling mode   162

central cored web   123, 143
Chen–Lui exponential model   204, 226, 228, 229, 237, 249, 254
Cholesky's method   46
collapse analysis   123, 127
collapse mechanism   169, 173, 175, 267
column softening   238
column tangent modulus   130
compatibility   30–32, 37, 40
computer memory   138, 147
connection data bank   41
connection flexibility   101, 117, 123, 125, 153, 155, 167, 169, 176, 180, 182, 188, 191, 195, 204, 208, 217, 252, 253, 260
connection property   199
connection spring   43, 113–115, 117, 123, 191, 281

connection stiffness   31, 44, 49–51, 93, 94, 99–101, 103, 105–108, 110, 111, 114
consistent mass matrix   210
constant work method   83, 84, 91
constant-average-acceleration assumption   206, 212
convergence rate   213
critical damping   246
cross-section plastic strength   128, 129
cross-section plastification   124, 127, 128, 133, 139, 148
cubic B-spline model   100, 103, 105, 108
cubic Hermitian function   207
current stiffness parameter method   81
curve-fitting   99, 100, 103, 105, 106, 108, 109, 112
curve-fitting procedure   144, 147, 170
cyclic loading   279
cyclic plasticity models   267, 268, 274, 278

damage control limit state   267
damping   267, 283, 288, 290, 292, 299, 300
damping ratio   217
degradable stiffness   123, 190
direct iterative method   60, 62
displacement control method   78, 81–83, 91
displacement-based finite element method   198, 206, 209, 234, 259
dome frame   246
drift-off error   62
ductility   2, 4, 7, 17, 24, 25, 42, 43, 52, 94, 196, 197, 267
dynamic and cyclic analysis   24
dynamic behaviour   223, 225, 226, 239, 247
dynamic response   195, 196, 223, 224, 232, 234, 236, 240, 242, 243, 246, 248–253, 260
dynamic yield stress   271, 277

earthquakes   195, 216, 253, 258, 265, 266

effective length 9–11, 17, 19, 20, 24, 124, 160–162, 166, 168
eigen-vector 216
El Centro N–S earthquake component motion 255
elastic bifurcation analysis 19
elastic–perfectly plastic model 268
elasto-plastic buckling 8, 9, 12
element stiffness matrix 32, 36, 41
end-spring method 137
energy absorption capacity 195, 260
energy dissipation 197, 238, 250, 252, 267, 299
energy dissipation capacity 267
equations of motions 196, 206, 211
equilibrium 30–32, 36–38, 40, 43, 44, 46, 48
equilibrium check 145
equilibrium condition 212
equilibrium error 48, 66
equilibrium path 59, 62, 63, 77, 78, 80–83, 87, 89, 90
Euler buckling 9
European calibration frame 186
extended end-plate connections 95

fibre 128, 138
fibre stress distribution 124, 147
finite element method 155, 156
force distribution 186
force–displacement relationship 129, 132, 133, 154
force point 126, 146, 148, 152, 170, 171
frame action 180
full yield surface 135, 144, 146, 148, 152, 171

gable frame 125, 126, 174, 175
geometric imperfections 97
geometric stiffness matrix 57, 59, 68, 209, 210, 217, 218
geometrical second-order effect 123
global coordinate system 32, 157
global stiffness matrix 41, 46, 47
global structural stiffness matrix 155
ground acceleration 211, 255, 256, 258

harmonic vibrating loads 260
header plate connections 94, 95

hexagonal frame 244, 245
hybrid element 43, 115–117, 153, 204, 205, 207, 208
hysteretic connection behaviour 201
hysteretic damping 195, 211, 234, 238, 250, 252, 260

incremental secant stiffness method 244
independent hardening method 199, 201
independent hardening rule 274
initial connection stiffness 201, 203
initial geometric imperfections 145, 176
initial stress 209, 216, 217
initial yield surface 134–136
instability 124, 155, 166, 168, 169, 176, 180, 181
instability effect 209, 210, 241
integration procedure 124, 147
inversed Ramberg–Osgood model 268, 276
isotropic hardening rule 273
iterative scheme 212

kinematic hardening method 201
kinematic hardening rule 273

Lagrangian formulation 17
large deflection problem 223
large displacement matrix 68
lateral buckling 10, 11
lateral deflection 101
linear analysis 2, 5, 16, 18, 19, 21, 23, 26, 30, 32, 45, 46, 48, 51
linear stiffness matrix 57, 59, 68, 209, 210, 216
load-carrying capacity 123–125, 178, 191
load cycle 78, 80–83
load–deformation curve 123, 127, 145, 147, 175, 178
load versus deflection analysis 57, 59
local buckling 13
local coordinate systems 32
lumped mass 210, 249

material constitutive relationship 30, 268
material hardening 201
material yielding 123–128, 145, 152, 155, 156, 166, 168, 177, 182, 186, 191, 268–270, 282, 300

Subject index

mathematical models   98–100
mesh   138
minimum residual displacement method   78, 85–87, 89, 91
mixed models   98–100, 117
modified element stiffness   130
modified element stiffness matrix   115, 117
moment–curvature relation   127, 128, 278
moment equilibrium condition   114
moment–rotation curve   101, 103, 111, 117
multi-linear model   101
multi-linear strain-hardening model   276

natural frequency   217–220, 232, 239, 240, 242
Newmark numerical integration method   206
Newton–Raphson method   55, 62–65, 67, 72–75, 77–80, 82
non-linear analysis   55, 57, 59, 60, 62, 66, 70, 71, 74, 75
non-linearity   124, 126, 177, 188
normalized plasticity index   279, 299

partial yielding   129, 130, 133–135, 137, 145, 147, 148, 188, 190, 268, 270, 271, 278–280, 288, 293, 299, 300
passive damping   195
permanent rotation   200
Perry–Robertson formula   57
plastic hinge   123, 124, 126–130, 133, 137, 138, 142, 147, 149, 150, 152, 155, 169–171, 175, 186, 188, 190, 191, 195, 267, 271, 283, 288, 290, 292, 299, 300
plastic hinge approach   124, 128, 139
plastic moment capacity   129, 130, 135, 137, 148, 173, 177, 180, 187
plastic permanent deformation   292
plastic zone approach   124–126, 137–139, 147
plate assemblage concept   190
polynomial model   100, 101, 103–105
post-buckling   244
pre-buckling   244
principal axes   32, 35, 36, 52, 213
principle of stationary potential energy   36
principle of superposition   19

proportional damping effect   216
pure incremental method   60–62

Ramberg–Osgood model   100, 111, 112, 204, 229, 234, 235, 268, 270, 277, 294
reduced plastic moment   140, 141, 143, 152
refined-plastic hinge method   123, 124, 141, 145, 147, 171, 175, 178, 180, 185–187, 189–191
residual displacement   212, 213
residual stress   97, 125, 126, 131, 136, 140, 142, 143, 145, 146, 167–169, 171, 175, 186, 190
resonance   265
Richard–Abbott model   99, 100, 112, 113
rigid plastic analysis   268
riveted connections   94
rotational deformation   97, 98, 108, 113

secant stiffness   45, 61, 62, 66, 70, 75
second-order effects   124, 191
second-order elastic analysis   22–25
second-order inelastic analysis   24–26
section assemblage concept   278, 279, 284, 299, 300
section assemblage method   142, 147
section spring   123, 144–148, 190
section spring stiffness   144, 146, 148, 153
seismic input energy   267
seismic zones   195, 267
semi-rigid connections   196–198, 204, 210, 220, 221, 223, 225, 226, 229, 243, 244, 249, 258
semi-rigid joint   51, 52
serviceability limit state   1, 267
shape function   38
shear buckling   14
single web angle connection   95
skyline storage scheme   46
slenderness ratio   17, 19
slope discontinuity   101
smooth transition   123, 145, 147, 191
snap-through buckling   166
snap-through buckling mode   243
snap-through problems   78
solution point   77, 86
springs-in-series   123, 125, 152, 191, 300
squash load   129, 135, 140

stability function	66	tolerance	212, 217
stability functions	130, 133, 147	torsional buckling	9, 11, 12, 17, 26
static analysis	196, 223–225, 238	total potential energy functional	36
static condensation procedure	123, 154, 191	transformation matrix	157, 213
static monotonic loading tests	94	transient analysis	265
stiffness	29–32, 36, 40–47, 49–52	transverse displacement function	205
stiffness matrix assembly	45	truss action	180, 181
strain energy	36, 37, 39	ultimate limit state	1, 2, 4, 5, 9
strain rate dependent model	268, 277	ultimate load	124, 127, 147, 170, 175–178, 182, 186
strain-hardening	108, 109, 113	unbalanced force	64, 66
strain-hardening effect	127, 138	unbalanced residual force	212
strain-hardening rules	267, 271, 273		
strength	30, 42, 43, 50, 52	vibration	195, 198, 216–218, 220, 233, 238–242, 244, 250
stress–strain relationship	124, 138, 139, 269, 270, 274, 276, 292	vibration analysis	196, 216, 218, 219, 221, 238
structural collapse	187	viscous damping	195, 205, 209, 211, 221, 248
structural safety	195	viscous damping matrix	210
survival limit state	267	Vogel six-storey frame	186, 187
tangent connection stiffness	103, 108, 111, 114, 117, 199, 200, 206, 234	William's toggle frame	87, 243
tangent stiffness	46	yield strain	138
tangent stiffness matrix	61, 62, 64, 66–68, 77, 78, 80, 81, 90	yield stress	128, 129, 138, 140, 143, 145, 167, 169, 171, 178, 271–274, 277, 278, 280, 289, 292, 299
time history	206, 213, 225, 227, 256, 286–288	yield surface	126, 145, 148, 152
time step	206, 212, 221, 223, 238, 244, 284, 292, 294		
Timoshenko beam–column theory	16		